# TEMPERATURE AND TOXICOLOGY

## An Integrative, Comparative, and Environmental Approach

# TEMPERATURE AND TOXICOLOGY

## An Integrative, Comparative, and Environmental Approach

Christopher J. Gordon

Taylor & Francis
Taylor & Francis Group

Boca Raton London New York Singapore

A CRC title, part of the Taylor & Francis imprint, a member of the Taylor & Francis Group, the academic division of T&F Informa plc.

**Library of Congress Cataloging-in-Publication Data**

Gordon, Christopher J.
Temperature and toxicology : an integrative, comparative, and environmental approach / Christopher J. Gordon.
p. cm.
Includes bibliographical references.
ISBN 0-8493-3024-6
1. Toxicity, Effect of temperature on. 2. Body temperature. I. Title.

RA1216.G67 2005
571.9'519--dc22 2004054475



Direct all inquiries to CRC Press, 2000 N.W. Corporate Blvd., Boca Raton, Florida 33431.



**Visit the CRC Press Web site at www.crcpress.com**


International Standard Book Number 0-8493-3024-6
Library of Congress Card Number 2004054475
Printed in the United States of America 1 2 3 4 5 6 7 8 9 0
Printed on acid-free paper

*To my parents,*

*Brian G. Gordon*

*and*

*Nancy L. Gordon*

# Preface

In the summer of 2002, Dr. Sam Kacew, the editor of the *Journal of Toxicology and Environmental Health*, invited me to prepare a book on the subject of toxicology and temperature regulation. A large portion of my research at the U.S. Environmental Protection Agency has focused on studying the integrative thermoregulatory responses of laboratory mammals when exposed to pesticides and other toxicants. Hence, I was very excited at the prospect of preparing a book in my primary area of study but wondered if there would be enough material in the field to justify an entire book. However, as I began the background and literature searches, I quickly realized that a book on toxicology and temperature regulation should include coverage on how temperature affects the toxicity of chemicals and drugs. Indeed, this area of study is vast and abounds with *in vitro* and *in vivo* toxicological studies in mammals, amphibians, fish, and invertebrates. All together, the temperature and toxicology literature could fill several texts.

The book's title evolved to its final version as I discovered that mammalian thermoregulatory responses would be best presented with a strong comparative and environmental physiological perspective. Some of the earliest work in temperature and toxicology was performed with non-mammalian species because their sensitivity to toxicants and drugs can be easily manipulated by altering the water temperature. There is a wealth of comparative physiological data on temperature and toxicology that merits presentation in a book of this nature. Environmental physiology is also crucial in the toxicological responses of mammals and other species. The reader will discover how human health and susceptibility to toxicants can be strongly linked to environmental temperature. Heat stress in the work place, the greenhouse effect, and the Gulf War syndrome are examples of current major health issues in toxicology that call for a strong environmental physiological approach. All together, it can be seen that a three-pronged physiological approach — integrative, comparative, and environmental — would be an ideal framework for a treatise on temperature and toxicology.

Other chapters in the book are linked to these major themes. Studies in wildlife, including endocrine disruptor research, is inexorably linked to environmental temperature. Natural toxins, such as fungal and algal toxins, have marked effects on body temperature at extremely low doses. Moreover, these toxins present a significant health and economic impact on humans and agricultural species. Studies in heat shock and related stress proteins are intimately related to temperature and toxicology. The ongoing revolution of techniques to study the mechanisms of stress proteins and other molecular markers of toxicology call for a better understanding of the relationships between thermal stress and toxicology.

Temperature, toxics, and life are inseparable is the message I want to convey to the readers of this book. I have incorporated the data from hundreds of researchers into tables and figures with the sincere desire that their work be presented and discussed as they intended.

**Christopher J. Gordon**

# Acknowledgments

I would like to thank the U.S. Environmental Protection Agency (EPA) for its support in the preparation of the book. I am appreciative of the following colleagues for providing reviews and comments of selected chapters: Drs. Jeff McKee, Amir Rezvani, Ying Yang, David Dubose, Richard Peterson, Jennifer Sorenson, David Herr, James Allen, Donald Spiers, Glen Tattersall, Diane Miller, Ginger Moser, Robert Carroll, Justin Brown, Luiz Branco, Larry Crawshaw, Christopher Barney, Vernon Benignus, Ken Bowler, and Lawrence Katz. I am very appreciative for the advice and encouragement from Dr. Lisa Leon that was given throughout in the preparation of the book. The collaborative research and advice from Drs. Pam Rowsey, Cina Mack, and Edward Smith are greatly appreciated. I thank Renee Bosman and Hannah Rogers of the U.S. EPA library services for performing literature searches. I am indebted to Peggy Becker for the tremendous effort she put forth in the preparation of the figures, literature searches, and proofing of the manuscript. I thank my mentors, Drs. J. Homer Ferguson and James E. Heath, for their guidance. Finally, I am thankful for the enduring support from my loving wife, Susie, and children, Kevin and Karen.

# Author

Christopher Gordon received his Ph.D. degree in physiology from the University of Illinois—Urbana–Champaign; and an M.S. degree in zoology and a B.S. degree in biology, both from the University of Idaho. He currently holds the position of Research Physiologist in the Neurotoxicology Division in the National Health Effects and Environmental Research Laboratory of the U.S. Environmental Protection Agency located in Research Triangle Park, North Carolina. He has published over 160 research papers, review articles, and book chapters in the fields of temperature regulation and toxicology. He also published a book in 1993 entitled *Temperature Regulation in Laboratory Rodents*. He and his wife, Susie, have two children, Kevin and Karen, and reside in Durham, North Carolina.

# Contents

# Chapter 1

# Introduction

## 1.1 INTRODUCTION

Temperature has a universal effect on life. All life processes depend on chemical reactions that, in turn, are dependent on temperature as based on the principles of the Arrhenius equation (Burton and Edholm, 1955). Toxicology is defined as the study of the adverse effects of chemicals on living organisms (Klaasen and Eaton, 1991). The interplay between temperature and the physiological response to toxic chemicals is the subject of this book.

The principle of the Arrhenius equation is based on thermal kinetics and states that the rate of chemical reactions increases exponentially with a rise in temperature (Figure 1.1). Most molecular, cellular, and physiological processes have a positive temperature coefficient, meaning their activity increases in a manner similar to that predicted by the Arrhenius equation. Thermal biologists often use $Q_{10}$ to describe the effects of temperature. The $Q_{10}$ of most biological processes ranges between 2 and 3, which equates to a doubling and tripling of the reaction rate with a 10°C increase in temperature (see Chapter 4 for discussion). Nerve conduction velocity, axonal transport, heart rate, cell division, and tissue metabolism are select examples of physiological processes with positive temperature coefficients (Table 1.1). There are notable exceptions to the general effect of temperature. For example, electrical resistance of excitable membrane and the height of action potentials increase in magnitude with cooling. These processes have a negative temperature coefficient, or a $Q_{10}$ that is greater than 0 but less than 1.0. Cellular and molecular mechanisms of toxicity also have positive temperature coefficients (Table 1.2). Processes such as receptor

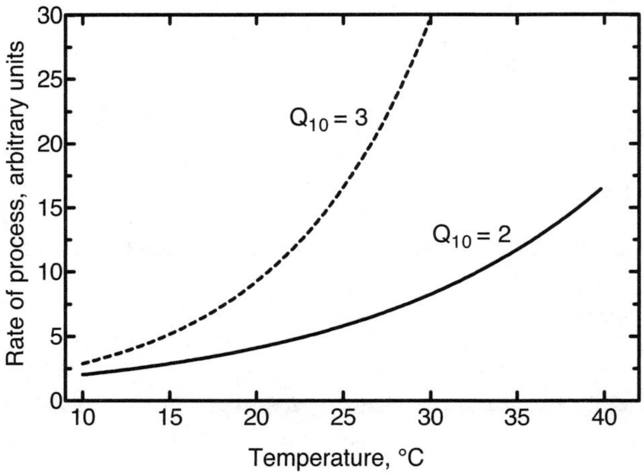


**Figure 1.1 Diagrammatic representation of the effect of temperature on the rate of a chemical reaction or physiological process. Theoretical functions show the effects of a doubling ($Q_{10}$ = 2) or tripling ($Q_{10}$ = 3) of the rate of the process with a 10°C increase in temperature. (Adapted from Schmidt-Nielsen, K. (1975). *Animal Physiology: Adaptation and Environment.* London: Cambridge University Press.)**

binding, lipid peroxidation, metabolic deactivation of a toxicant, and oxidative phosphorylation generally increase with temperature, but there are some notable exceptions. Toxic mechanisms such as pyrethroid-induced depolarization of nerve tissue and induction of some protective proteins exhibit a negative temperature coefficient (see Chapter 4). Altogether, one can see that toxicology is inexorably linked to temperature. The molecular mechanisms of toxicity are, like all life processes, subject to the Arrhenius temperature effect. This also includes those processes with a negative temperature coefficient that appear to "run uphill" and counter the laws of thermodynamics, but they can only be achieved with energy from chemical reactions that have a positive temperature coefficient.

The temperatures of all animals either conform to the temperatures of their environments or are regulated to be independent of the environment (Figure 1.2). The mechanisms of toxicity in temperature conformers and temperature regulators will be manifested in different ways depending on the environmental temperature. Regardless of the nature of the organism's thermoregulatory capabilities, there will be a range of physiological temperatures where the toxicity of a chemical is exacerbated or attenuated by a change in temperature (Figure 1.3). One would predict that the Arrhenius effects on mechanisms of toxicity would be minimized in species that are temperature regulators. However, some temperature regulators, especially small rodents, can lower their body temperature to attenuate

**Table 1.1 Effects of Temperature on Cellular and Physiological Processes**

| *Species* | *Parameter* | *Temperature Coefficient* | $Q_{10}$ *(temperature range)* |
|---|---|---|---|
| Dog | Heart rate | Positive | 1.52 (33–37°C) |
| Guinea pig | | | 1.32 (32–38°C) |
| Rat | | | 1.62 (35–38°C) |
| Frog | | | 2.1 (15–25°C) |
| Fish | | | 1.77 (12–25°C) |
| Rat | Tissue metabolism | Positive | 2.4 (30–39°C) |
| | Cerebral cortex | | 1.6 (30–39°C) |
| | Heart ventricle | | 2.16 (30–39°C) |
| | Liver | | |
| HeLa cells | Cell division | Positive | 10.7 (34-37°C) |
| Fish (gar) | Axonal transport (fast) | Positive | 2.2 (15–25°C) |
| | Axonal transport (slow) | | 2.8 (15–25°C) |
| Cat | Nerve membrane resistance | Negative | 0.49 (30–40°C) |
| Human | Action potential amplitude in nerves of arm and hand | Negative | 0.4 (21–31°C) |
| Rabbit | Nerve conduction velocity | Positive | 3.7 (20–40°C) |
| Guinea pig | | | 1.3 (30–40°C) |
| Frog | | | 1.5 (15–36°C) |

*Source*: Most data taken from Altman and Dittmer (1966) except for axonal transport (Cancalon, 1988) and nerve electrophysiology (Janssen, 1992).

the toxic effects of many chemicals. Temperature conforming species face relatively large changes in body temperature, and the interaction between temperature and mechanisms of toxicity should be profound. The effects of temperature on chemical toxicity can be as basic as diffusion across a cell membrane or as complex as an integrated thermoregulatory response. I have chosen to use the Arrhenius effect as a simple starting point to explain the fundamental processes of temperature and toxicology and to then move into a comprehensive explanation of these phenomena using an integrative, comparative, and environmental approach.

## 1.2 THE UNIQUE NATURE OF THERMOREGULATION

Biomedical researchers are often called upon to define when an insult triggers a so-called *significant physiological disturbance.* The unique nature of the thermoregulatory system makes it an ideal selection for such an evaluation. Temperature regulation is an autonomic process that is a hallmark of physiological homeostasis. When one considers the multitude

**Table 1.2 General Mechanisms of Toxic Chemicals**

| |
|---|
| Interference with normal receptor–ligand interactions |
| Neuroreceptors and neurotransmitters |
| Hormone receptors |
| Transport proteins |
| Interference with membrane functions |
| Excitable membranes |
| Ion flux |
| Membrane fluidity |
| Interference with cellular energy production |
| Oxidative phosphorylation uncouplers |
| Impairment in oxygen delivery |
| Inhibition in electron transport |
| Binding to biomolecules |
| Interference with enzyme function |
| Lipid peroxidation |
| Oxidative stress |
| Perturbation in calcium homeostasis |
| Cytoskeletal alterations |
| Phospholipase activation |
| Toxicity from selective cell loss |
| Hormonal imbalances |
| Birth defects |
| Genetic alterations |
| Cancer initiation/promotion |

*Source*: Modified from Klaassen, C.D. and Eaton, D.L. (1991). Principles of toxicology. In: *Casarett and Doull's Toxicology: The Basic Science of Poisons,* Amdur, M.O., Doull, J., and Klaassen, C.D., Eds., pp. 12–49. New York: Pergamon Press.

of parameters of the *internal milieu* that are regulated by homeostatic processes, body temperature comes to mind as one with a high degree stability. That is, in the evolution of eutherian mammals that range in body mass by over seven orders of magnitude, temperature regulation has been conserved. This is evident by the narrow limits of internal body temperatures ranging from 36 to 40°C among species. Other autonomic processes such as blood pressure, cardiac output, and respiration vary considerably within and between species. It is the nature of these processes to vary because the autonomic nervous system responds to the ever changing demands of the organism to deliver oxygen and nutrients and eliminate carbon dioxide and other waste products — these demands being intimately related to the organism's size, age, environment, and

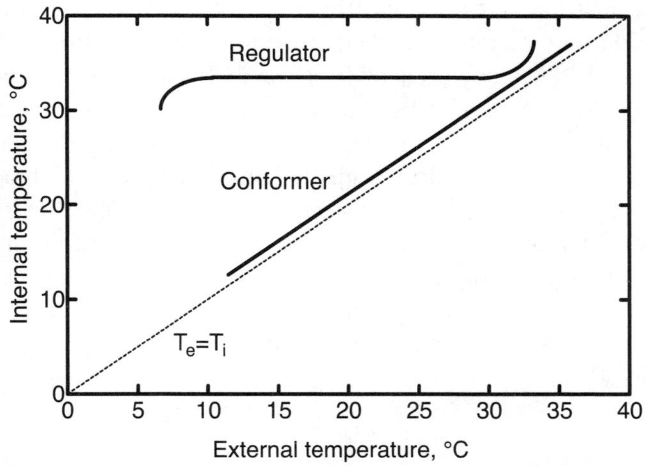


**Figure 1.2** All organisms can be classified as temperature conformers or temperature regulators. Conformers have an internal temperature equal to the external temperature. Regulators use energy to maintain an independent internal temperature over a selected range of external temperatures. The dashed line depicts uniform environmental ($T_e$) and internal temperature ($T_i$). (Adapted from Prosser, C.L. (1973). *Comparative Animal Physiology,* 3rd ed. Philadelphia: W.B. Saunders.)

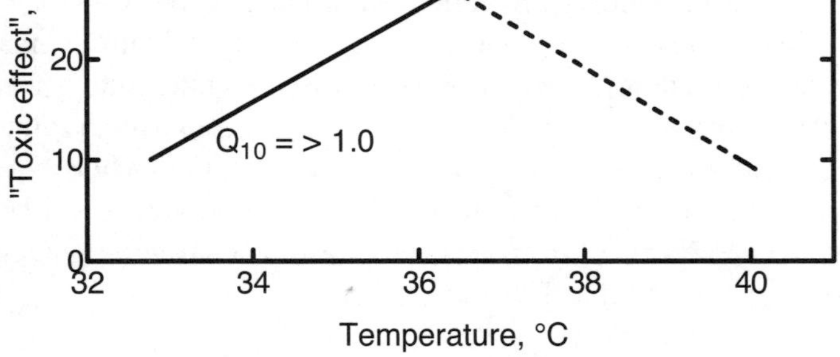


**Figure 1.3** Possible interactions between temperature and a cellular or molecular mechanism of toxicology. Most mechanisms have a $Q_{10} > 1$ and increase in activity with temperature. A few mechanisms have a negative temperature coefficient or $Q_{10} < 1$ and increase in activity with cooling.

many other factors. It is noteworthy that dysfunction of the heart and lungs is a leading cause of human morbidity and death and accounts (justifiably!) for the major area of funding in biomedical research.

With the exception of hibernation and torpor, body temperature is regulated at approximately the same level in a healthy homeothermic organism from soon after birth until the impending point of death. Dysfunction of the thermoregulatory system is rarely considered in the etiology of disease other than in circumstances of exposure to acute thermal stress. Among the parameters regulated by the autonomic nervous system, body temperature is one of the most stable, a characteristic that is essential for life by providing a stable thermal environment for all biochemical processes in the body.

## 1.3 WHY SHOULD TOXICOLOGISTS STUDY TEMPERATURE?

Temperature has apparently not been considered a significant factor in most toxicological studies. In a computerized search of multiple data bases from 1980 to 2003, this author found that only 0.5% of the more than 571,000 papers in toxicology reported measuring body temperature. So why should researchers in toxicology be concerned with temperature as a factor that might affect their particular biological endpoint? First, while body temperature is normally stable, altered regulation is seen upon exposure to toxicants, with the effects usually being more pronounced in smaller mammals. Second, while large mammals such as humans may regulate a stable core temperature, environmental stress will nonetheless exacerbate the physiological and behavioral responses to a toxicant. Third, by virtue of body temperature's stable nature, any environmental perturbation or insult that changes it should be considered as a biologically significant event. Tracking other autonomic parameters in subjects exposed to toxicants is fraught with inherent variability. Thus, one should look at temperature regulation as a hallmark of homeostasis and consider the significance when it is affected by a toxicant. To this end, one can use body temperature as a benchmark of toxicological exposure because a temperature change suggests a significant change in physiological homeostasis.

### *1.3.1 Temperature Is a Benchmark of Acute Toxicity in Rodents*

Rodents are the species of choice in most toxicological investigations. Acute toxicity testing in rodents has provided an extensive data base of potential hazards of environmental toxicants. In test batteries that screen for toxicants, body temperature is frequently used as a benchmark of overt toxicity (Moser, 1991, 1995; Tamborini et al., 1990). Generally, a decrease in body temperature in a test species was considered as a significant sign of toxicity, placed in the same class of sequelae as body weight loss, decreased appetite, and reduced motor activity. The problem with the use of body temperature in these evaluations is the manner in which it was measured in the rodent. Hand-held probes were typically used in past studies to measure the colonic or rectal temperature of a mouse or rat that had been extensively handled and manipulated in the process of a battery of behavioral and physiological evaluations. Some researchers have relied on temperature measurements in restrained animals to assess chemical toxicity. The thermoregulatory response of the rodent and other test species to these types of manipulation is profound (Chapter 7). In many of these studies, one finds the so-called baseline core temperature of the control rats to be at least 38°C, a value which is at least 1°C above the normal temperature (see Chapter 2).

This hyperthermic response obviates any subtle effects that a toxicant at low doses would have on body temperature, and thus limits the usefulness of body temperature as a sensitive benchmark of toxicity. This accounts for the common misconception that a significant change in temperature (e.g., hypothermia) is simply a gross indication of acute toxicity. However, the development of radiotelemetry to monitor core temperature and other physiological parameters in undisturbed animals provides toxicologists with a sensitive means of detecting changes in thermal homeostasis that occur from subtle toxic insults that would not be detected using conventional hand-held probe techniques (see Chapters 3 and 7 for more discussion).

### 1.3.2 Temperature Regulation as a Window to Autonomic Physiology

Body temperature in birds and mammals is regulated through behavior and motor responses of the autonomic nervous system, termed *thermoeffectors*. These motor systems are called upon to increase heat production or reduce heat loss to ensure a regulated body temperature in the face of ambient heat and cold stress as well as when heat is produced internally from exercise or during fever. When body temperature changes in the face of exposure to a toxicant or drug, it is in fact an indication of a dysfunction or change in activity of one or more behavioral and autonomic thermoeffectors. Changing the environmental temperature of exposure in an animal administered a toxicant allows one to discern if a particular thermoeffector such as metabolic thermogenesis, peripheral vasomotor tone, or evaporative water loss is affected. For example, a toxicant or drug that mediates constriction of skin blood flow would have little effect on body temperature in a cold environment because the peripheral vasculature would already be constricted. However, in a warm environment, skin blood flow is normally elevated to increase heat loss, and a toxic agent that causes peripheral vasoconstriction would lead to marked hyperthermia. Overall, ambient temperature can be modulated in a way to focus the effects of a toxicant on a particular thermoeffector (see Table 3.1).

### 1.3.3 Temperature-Dependent Processes

Since all biochemical and physiological processes are directly influenced by temperature, how can one be sure that a particular endpoint that is affected by the toxicant is not actually an indirect result of the toxicant changing body temperature? Many toxicologists and pharmacologists are cognizant of the temperature dependency and attempt to apply corrective measures to maintain a consistent thermal environment within the test subject. In many cases, the hypothermic effect of an agent is blocked by raising ambient temperature or by placing the animal on a heating pad. The issue is in fact more complex than just simple thermal kinetics. Many toxicants elicit a regulated reduction in body temperature, meaning that the subject activates heat-dissipating thermoeffectors and core temperature is regulated at a lower level. Preventing the drop in temperature by raising ambient temperature can actually exacerbate the toxicity of the chemical agent. I have found that many toxicologists are unaware of the potential effects of toxicant-induced changes in body temperature and are uncertain if preventing the change in temperature will alter the toxicity of the test compound. The temperature dependency issue is raised throughout this book and is given special emphasis in Chapter 4.

## 1.4 THREE APPROACHES TO STUDYING TEMPERATURE AND TOXICOLOGY

The effects of temperature on toxicological responses can be covered in many ways. In the preparation of this book, I have chosen to emphasize an integrative, comparative, and environmental approach, discussed in the following sections. Briefly, integrated thermoregulatory responses in temperature regulators and conformers are influenced by a variety of factors, with environmental temperature being one of the most critical (Figure 1.4). Changing the body temperature of a temperature-regulating or -conforming species will alter chemical toxicity by affecting the intake and tissue absorption as well as the molecular mechanisms of toxicity.

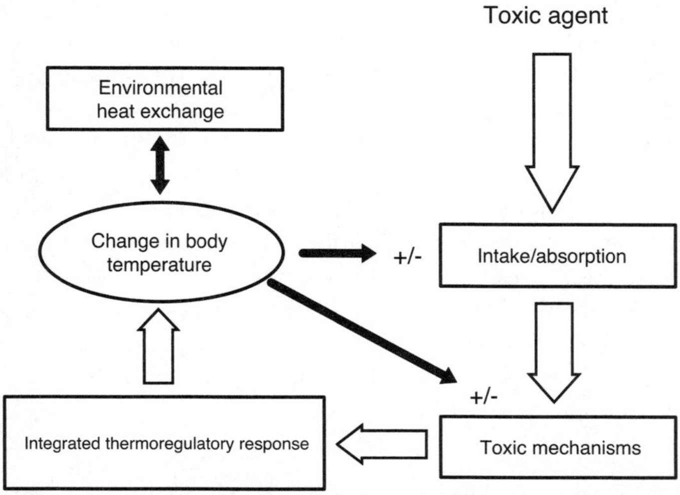


**Figure 1.4 Integrated thermoregulatory response of temperature regulators and conformers can affect chemical toxicity. A change in body temperature can alter the intake and/or absorption and molecular mechanisms of toxicity of a chemical. Ambient temperature is the most critical environmental factor that modulates the potential change in body temperature.**

### *1.4.1 Integrative Approach*

Integration in physiology typically implies studying normal or abnormal function at all levels of biological organization and not simply studying a fragment of a response at, for example, the subcellular level. An integrative approach is paramount in temperature regulation because this system can, with few exceptions, only be studied in the intact organism. Temperature regulation is ideally studied with a holistic view, taking into account the interaction between the environment, thermoreceptor and thermoeffectors function, feedback control, and a regulated core and skin temperature. To this end, a book on temperature and toxicology must be framed with a strong integrative approach.

Many researchers in toxicology have determined how the body temperature of humans, experimental animals, wildlife, and other organisms is affected by exposure to an array of pesticides and other toxic chemicals. Only a handful of these studies have provided an integrative approach to explain the thermoregulatory response to a toxicant. That is, a change in body temperature in a toxicological study was often considered to be the result of an acute dysfunction of homeostatic processes. This may be true for some chemicals, especially when given in high doses that approach lethal levels. However, it is important to consider the nature of the thermoregulatory response in terms of the coordination and integration of behavioral and autonomic thermoeffectors. Furthermore, since thermoregulatory responses are mediated through the cardiovascular and respiratory systems, an integrative approach means one should consider the interplay between the direct thermoregulatory response and the indirect effects on the heart and lungs, as well as effects on the kidney, gut, liver, dermis, and other organ systems. Age, nutritional state, environmental stress, and many other parameters also factor into the integrative thermoregulatory response. Finally, in an integrative approach one must consider the overall consequences of a thermoregulatory response. A major part of this book is devoted to understanding how the behavioral and autonomic responses to raise or lower body temperature affect the response to a toxicant.

### 1.4.2 Comparative Approach

Temperature and toxicology cannot be fully appreciated without a strong emphasis on the comparative physiological responses. Comparative physiology has provided pharmacologists and toxicologists with invaluable tools to study mechanisms of action of drugs and toxicants. Comparative animal physiology strives to study the ways in which diverse groups of organisms perform similar functions and respond to environmental stress (Prosser, 1973). The comparative physiologist uses the kind of organism as the experimental variable and emphasizes its evolutionary history of life in diverse environments. In this book, it is the response to diverse thermal environments that makes the comparative responses to toxic substances such a critical issue. Moreover, comparisons of thermoregulatory responses within the class of mammals are essential in the extrapolation of toxicological data from experimental animals to humans. Comparison of responses between classes of vertebrates as well as between vertebrates and invertebrates also provides a breadth of coverage that improves the assessment of molecular mechanisms of toxicity.

### 1.4.3 Environmental Approach

Environmental temperature is a primary factor limiting the growth, reproduction, and survival of all animal and plant life. A change in environmental temperature from ideal levels affects the physiological response to toxicants in different ways depending on whether the animal is a temperature conformer or temperature regulator. The temperature of a conformer varies with environmental temperature, and the efficacy of a toxicant will vary accordingly. Temperature regulators respond to heat and cold by activating thermoeffectors to maintain a stable core temperature. Depending on the severity of the thermal environment, the physiological and behavioral responses impart stress that alters the organism's thermoregulatory response to a toxicant. Environmental heat and cold stress places limits on the thermoregulatory system, thereby affecting the efficacy of a toxicant. Overall, the integrative and comparative aspects of toxicology and temperature are inseparable from the organism's *external thermal milieu*, including ambient temperature (i.e., air, land, or water), solar radiation, wind, and humidity. Other types of stresses also affect the thermoregulatory response to toxicants. Environmental physiology is thus a focal point to tie together the integrative and comparative thermoregulatory responses in all organisms exposed to toxic agents.

### 1.4.4 Presentation and Breadth of Coverage

It is my intention to provide the reader with a comprehensive, up-to-date review and analysis of the literature on temperature and toxicology. Over 500 references were used in the preparation of this book. The book focuses primarily on the response of rodents since they represent the primary test species in toxicological studies. The meager human data base, including emergency room case reports and epidemiology studies, is incorporated throughout the book.

It is important to comment on the selection of studies used to prepare the book. Much of the research in toxicology over the past 50 years, including studies in temperature regulation, utilized acute doses that approached or sometimes exceeded lethal levels. Many of the high-dose studies are cited and discussed in the book, including data on pesticides that are now outlawed in many parts of the world (e.g., DDT). Inclusion of these studies in this book might be questionable to toxicologists interested in the human health effects of environmentally relevant levels of pesticides and other contaminants. In spite of the high doses and acute nature of these studies, this author felt that this would be an ideal opportunity to present the work in a treatise with the idea that it may be useful information for future toxicologists. First, a thorough review may prevent unnecessary duplication of many of the acute dosing studies. Second, acute poisonings in humans and other species will nonetheless continue to be a key issue in toxicology and thermoregulation. Third, relatively high doses of pesticides and other toxicants continue to be used in experimental rodent studies because extrapolating thermoregulatory and other toxic effect data in rodents to humans generally involves an inverse relationship between body mass and sensitivity (see Chapter 5). Any thermoregulatory effects of relatively low doses of xenobiotics and toxins have also been included in the book.

Overall, this book is meant to provide an integrative, comparative, and environmental assessment of temperature and toxicology, from the molecular to whole-animal level. Chapter 2 provides a review of the basics of temperature regulation in homeotherms and poikilotherms that is written in sufficient detail for students in the biological sciences and biomedical researchers to grasp the concepts of thermal physiology and to interpret the data presented in the book. Chapter 3 is the largest chapter in the book, covering all aspects of the acute effects of toxicants on thermoregulation. This includes anticholinesterase pesticides, airborne pollutants, metals, alcohols, chlorinated hydrocarbons, and many other toxicants that affect thermoregulation in experimental mammals and humans. Chapter 4 addresses the data base of both whole-animal and *in vitro* studies that deal with effects of temperature on chemical toxicity. Chapter 5 deals specifically with the acute thermoregulatory response of rodents to toxic insult and how an integrative thermoregulatory response can affect their recovery and survival of the toxicant. This chapter addresses the problems of extrapolating the thermoregulatory effects in rodents to those of large mammals, including humans. The acute hypothermic response so commonly seen in rodents is infrequently observed in humans. In fact, humans often show a fever when exposed to a variety of toxicants. Hence, the febrile and hyperthermic responses to organophosphate insecticides, ethanol, metal fumes, and other toxicants are covered in Chapter 6. Environmental heat and cold stress is extremely relevant in human exposures to airborne pollutants and other toxicants. This subject matter along with the impact of psychological stress and restraint is covered in Chapter 7. Comparative physiological responses of invertebrates, fish, amphibians, and wildlife are addressed in Chapter 8. This chapter also focuses on the potential impact of endocrine-disrupting toxicants on the thermoregulatory system and their interaction with the hypothalamic–pituitary–thyroid axis. The study of heat shock proteins and other stress proteins is one key area where temperature regulation and molecular biology go hand-in-hand. That toxicants and thermal stress both cause the expression of stress proteins should be of interest to thermal physiologists and toxicologists with a cellular and molecular perspective. This area of study along with the impact of genetic variability is covered in Chapter 9. Finally, the thermoregulatory responses to natural toxins and venoms is covered in Chapter 10. Toxins and venoms have a profound effect on the health of humans, agricultural animals, and other species. Their effects on temperature regulation are remarkable but have rarely been reviewed in the literature.

# Chapter 2

# Principles of Temperature Regulation

## 2.1 INTRODUCTION

It was not until the advent of the small mercury thermometer in the 19th century that physiologists could begin to study thermoregulation. It was recognized that, among all animal life, relatively few species can be considered as true temperature regulators, meaning that they regulate their body temperature within narrow limits under a wide range of external and internal heat loads and heat sinks. *Temperature regulators* use autonomic and/or behavioral motor responses, termed *thermoeffectors*, to defend their body temperature against changes in heat gain and heat loss to the environment as well as the heat production from exercise. With some exceptions, all invertebrates are considered to be *temperature conformers*, meaning they lack any behavioral or autonomic mechanisms to regulate temperature independently of ambient temperature. The body temperature of most temperature conformers is almost always equal to that of the ambient conditions. All animal life is essentially capable of responding to thermal stimuli and eliciting a corrective motor response. Even unicellular organisms exhibit thermotropism and will move toward or away from a thermal stimulus.

## 2.2 TERMINOLOGY

Toxicologists desiring a better understanding of how temperature affects their particular biological endpoint will find a variety of terms in the literature to describe the thermoregulatory characteristics of a species. There is often overlap in definitions that can create confusion for both trained thermal physiologists and specialists in other fields of biology and medicine (Table 2.1). The study of temperature regulation is conventionally divided into two broad categories: *tachymetabolic* species, including birds and mammals, have a high basal metabolic rate; *bradymetabolic* species, including reptiles, amphibians, fish, and invertebrates, have a relatively low basal metabolic rate. Tachymetabolic species are also termed *endotherms*, because they regulate body temperature primarily through internal heat production derived from the sum of all metabolic processes. Many bradymetabolic species are *ectotherms*, meaning that body temperature is regulated behaviorally by modulating the heat transfer between the body and environment.

*Homeothermy* and *poikilothermy* are also frequently used terms that classify species into those that maintain their body temperature within a narrow and wide range, respectively. These terms often lose their strict definition depending on the environmental circumstances. This is especially relevant in small mammals dosed with drugs or toxicants. In the glossary of terms for thermal physiology (IUPS, 2001), homeothermy is defined as "the pattern of temperature regulation in a tachymetabolic species in which core temperature…is maintained within arbitrarily defined limits despite much larger variations in ambient temperature." Poikilothermy, defined as "large variability of body temperature as a function of ambient temperature in organisms without effective autonomic temperature regulation," is usually equivalent to temperature conformity.

**Table 2.1 Common Thermoregulatory Terms**

| *Term* | *Often Equated to* | *Phyla* | *Exceptions* |
|---|---|---|---|
| Homeotherm | Endotherm, tachymetabolic, temperature regulator, warm-blooded | Birds, mammals | Torpor in small rodents and birds |
| Poikilotherm | Ectotherm, bradymetabolic, temperature conformer, cold-blooded | Reptiles, amphibians, fish, invertebrates | Endothermy in honeybees and some fish and reptiles |
| Hibernator | | Mammals | |
| Torpor | | Birds, mammals | |
| Heterotherm | | Birds, mammals | |

Most species that are homeothermic are endothermic and tachymetabolic, but some endotherms will display poikilothermic tendencies. For example, some rodents and birds enter periodic torpor and allow body temperature to transiently drop to near ambient levels, but torpor is distinct from hibernation. As will be seen in Chapter 3, rodents dosed with toxic chemicals display a thermoregulatory response that some researchers define as poikilothermy. Some insect species are able to use endothermic mechanisms to regulate body temperature over a limited range of ambient temperature and thus are homeothermic.

*Heterothermy* is occasionally used to describe tachymetabolic species that display variations in temperature that are otherwise homeothermic. Primitive mammals such as monotremes and marsupials are generally considered as heterotherms. *Local heterothermy* is a useful term to describe a state where temperature of parts of the body comprising the thermal shell varies above or below normal. For example, extremely cold limbs brought on from exposure to a toxicant or drug would be considered a local heterothermic response (e.g., see Chapter 7).

## 2.3 HEAT BALANCE

Temperature regulation is fundamentally based on the body heat balance equation, which is a mathematical expression of the rate at which a subject generates and exchanges heat with its environment:

$$S = M - (W) - (E) - (C) - (K) - (R) \tag{2.1}$$

where $S$ is the rate of heat storage in the body (positive for an increase in body heat content), $M$ is metabolic rate, $W$ is work rate (positive for useful mechanical power accomplished; negative for mechanical power absorbed by the body); $E$ is evaporative heat transfer (positive for evaporative heat loss; negative for evaporative heat gain); $C$ is convective heat transfer (positive for heat transfer of heat to the environment; negative for transfer to the body); $K$ is conductive heat transfer (positive for heat transfer to the environment; negative for transfer to the body), and $R$ is radiant heat exchange (positive for heat transfer to environment; negative for transfer to body). The dimensions in Equation (2.1) are Watts (W), a measure of heat flow. However, the equation terms are often expressed in units normalized to surface area (Watts per square meter) or body mass (Watts per kilogram).

The explanation of heat balance can be simplified in a diagram relating a balance between the total sources of heat production and heat loss (Figure 2.1). The net heat storage ($\Delta S$) of a body is equal to the difference

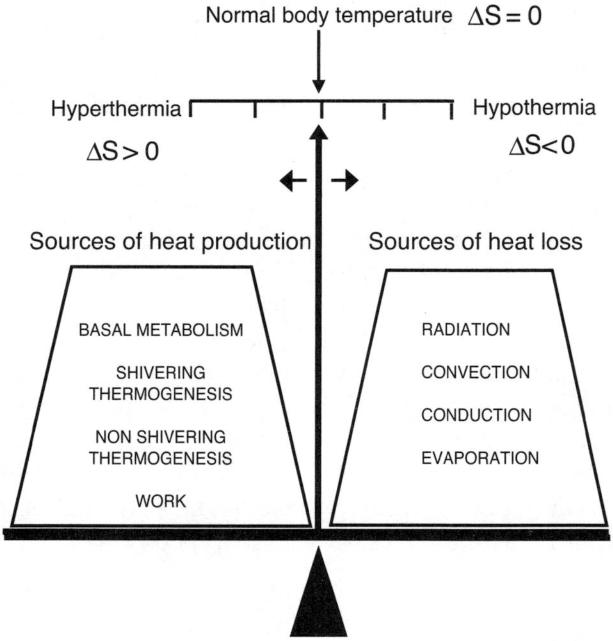


**Figure 2.1 Diagrammatic representation of the heat balance equation. A change in heat storage ($\Delta S$) occurs with an imbalance between the sums of heat production and heat loss.**

between all sources of metabolic heat production (basal metabolism, shivering and nonshivering thermogenesis, and heat from work) and the sum of the avenues of heat exchange. When body temperature is stable ($\Delta S$ = 0), heat production is equal to heat loss. When heat production exceeds heat loss, $S$ is positive and the subject becomes hyperthermic; when heat loss exceeds heat production, $S$ is negative and the subject becomes hypothermic. The routes of heat loss by evaporation, conduction, convection, and radiation are factors of greater or lesser importance depending on the species and ambient conditions (Table 2.2). If one pathway of heat loss is impeded, then heat must be dissipated by other avenues or the animal will quickly become hyperthermic. Or if heat loss through one avenue is accelerated, then heat loss through other avenues must be restricted or the animal will become hypothermic.

Under room temperature conditions, the majority of heat loss occurs through radiation, defined as the net heat transfer by electromagnetic radiation between two black surfaces. Under comfortable ambient conditions, humans dissipate approximately 67% of their total metabolic heat production by radiation (Table 2.2). Convection, the net heat transfer between a surface and a moving fluid (air or water), is conventionally

**Table 2.2 The Partitioning of Heat Loss of Nude Human Subjects as a Function of Ambient Temperature**

| *Room Temperature* | *Radiation* | *Convection* | *Evaporation* |
|---|---|---|---|
| Comfortable (25°C) | 67% | 10% | 23% |
| Warm (30°C) | 41% | 33% | 26% |
| Hot (35°C) | 4% | 6% | 90% |

*Source*: From Folk, G.E., Jr. (1974). *Textbook of Environmental Physiology.* Philadelphia: Lea and Febiger.

divided into natural and forced. Natural refers to the heat loss by convection that occurs under still air conditions and accounts for a small fraction of the total heat loss. Forced convection refers to the heat exchange that occurs when the medium of air or water moves around the body. There is a marginal area of still air with insulating properties around the skin of humans and other species. Any movement of the air from wind or movement by the animal disrupts this still air layer, accelerating convective heat loss. The wind chill factor is an example of an acceleration in heat loss by forced convection.

Conduction, the net heat transfer between a surface and a solid or stationary fluidic medium, is usually quite low because so little of the bare surfaces of mammals and birds comes in contact with substrate. Evaporation, the heat transfer driven by the vapor pressure gradient caused by the evaporation of water from a wet surface, accounts for approximately 20% of the total heat loss under comfortable ambient conditions. During heat stress, the ability to dissipate heat by radiation and convection is limited, thus making evaporation the principal avenue of heat loss.

## 2.4 THE THERMOREGULATORY SYSTEM

The approach used here to overview the thermoregulatory system will differ slightly from that used in the rest of the book. It is my intention to present and explain the thermoregulatory system in sufficient detail for toxicologists, pharmacologists, and other researchers who have a scientific background but little training in thermal physiology. Compared to the other chapters in this book, where specific literature citations are given for each topic, in this chapter a general reference is made for each topic by citing appropriate books and review papers.

There are many noteworthy contributions, including articles, books, and reviews on thermoregulation, that have been used in the preparation of this chapter. Some of the references were published decades ago but nonetheless provide accurate and thorough coverage of a particular field of study. Brief summaries of the literature references can be found in three volumes of review articles by eminent thermoregulatory researchers (Whittow, 1970). These volumes contain a wealth of information on the thermoregulatory responses of rodents, carnivores, primates, reptiles, amphibians, fish, and invertebrates. The books *Temperature and Life* (Precht et al., 1973) and *Comparative Animal Physiology* (Prosser, 1973) are excellent sources of information on temperature effects on animal and plant life. Clark and others compiled a massive survey of studies on the effects of drugs and other agents on thermoregulation that were published in the 1980s and 1990s (e.g., Clark and Lipton, 1985a, 1985b; Lipton and Clark, 1986).

Literature reviews over the past several decades provide thorough analyses on various aspects of the neural mechanisms of temperature regulation. Many recent articles by Boulant summarize the advances made on the neurophysiology of temperature regulation (Boulant et al., 1989; Boulant, 2000; also see Blatteis, 1998; Gordon and Heath, 1986; Wang and Lee, 1989). Bligh (1998) published a treatise of his years of work in temperature regulation that provides a thorough overview of the development of neural models. Advances in thermogenesis, including the function of brown adipose tissue, were brought forth by Himms-Hagen and many other investigators (Himms-Hagen, 1986, 1990). Temperature acclimation, including metabolic, hormonal, and neurological facets of adaptation, has also been a principal area of thermoregulatory research (e.g., Chaffee and Roberts, 1971; Brück and Zeisberger, 1990; Fregly, 1989; Horowitz, 2003). The field of environmental physiology is thoroughly reviewed in the textbook by G.E. Folk, Jr., and others (1998). Aging and temperature regulation in humans has been recently reviewed by Kenney and Munce (2003). Fever is, of course, the mainstay of thermoregulatory research (see Chapter 6, Introduction). Material from a book this author published on temperature regulation in rodents is also referenced throughout this chapter (Gordon, 1993). Advances in transgenics and the use of knockout mice to study responses to fever and other thermoregulatory processes were recently reviewed by Leon (2002).

### 2.4.1 Interspecies Body Temperatures

Temperature regulation in mammals and birds has evolved with the development of autonomic and behavioral mechanisms to regulate the balance between heat production and heat loss. Depending on the species, the temperature of the core (i.e., rectal, colonic, brain) is tightly regulated in the face of marked variations in ambient heat and cold stress. The core temperature is distinct from the *thermal shell*, which represents the skin and mucosal surfaces of the body engaged in heat exchange with the environment. The thermal shell includes tissues under the surfaces whose temperature may deviate from the core owing to heat exchange with the environment. The term *body temperature* is by definition an average of all the temperatures in the body (IUPS, 2001). However, most researchers using the term *body temperature* are usually making reference to the core temperature. For example, if one states in a study that a toxicant lowered body temperature, the assumption is almost always made that it was the core temperature that was measured. Actually, the core and shell temperatures were likely changing in a parallel fashion from the toxicant, but the implication of any statement of body temperature is really a reference to the temperature of the core.

Most mammals maintain a mean core temperature of 36 to 39°C; birds generally regulate their core temperature several degrees above that of mammals (Table 2.3). Humans, rats, hamsters, and mice happen to maintain approximately the same core temperature during the day (Table 2.3). Other mammals, such as cats, dogs, rabbits, sheep, and cattle, have a

**Table 2.3 Summary of Daytime Normothermic Core Temperatures Measured in the Rectum or Cloaca and Approximate Lower Critical (Lct) and Upper Critical (Uct) Temperatures for Selected Species of Mammals and Birds**

| *Species* | *Core Temperature (°C)* | *Thermoneutral Zone* | |
|---|---|---|---|
| | | *LCT (°C)* | *UCT (°C)* |
| Human | 37 | 24 | 31 |
| Rat | 36.6–37.5 | 28 | 34 |
| Hamster | 36.0–37.8 | 28 | 34 |
| Mouse | 36.0–37.6 | 26 | 34 |
| Guinea pig | 38.1–38.6 | 30 | 31 |
| Rabbit | 39 | 13 | 20 |
| Dog | 38–39 | 18 | 25 |
| Cat | 39 | 24 | 27 |
| Cattle, dairy | 38–39 | 5 | 16 |
| Goat | 38–39 | 20 | 26 |
| Sheep | 39 | 13 | 31 |
| Horse | 38 | | |
| Swine | 37–38 | 0 | 20 |
| Chicken | 41–42 | 19 | 29 |
| Pigeon | 43 | 20 | 30 |

*Source*: Most data from Altman and Dittmer (1966); data for rodents from Gordon (1993).

relatively high core temperature, about 2 to 3°C higher than that of most rodents and humans. The data in Table 2.3 represent approximate minimal daytime core temperatures collected from animals that presumably were not stressed prior to the measurement. All mammals and birds have a circadian rhythm of body temperature with an amplitude of approximately 1 to 2°C (for review, see Refinetti and Menaker, 1992). The circadian rhythm of body temperature represents a regulated oscillation in core temperature that is manifested in active as well as resting animals. That is, a distinct circadian temperature rhythm is apparent even when the effects of motor activity on core temperature are eliminated. Most rodents are nocturnal, and their core temperature is relatively low during the day, a time when the majority of toxicological studies are performed. Humans are obviously diurnal and generally active during the day when their core temperature is at its highest level. In extrapolation studies from laboratory rodents to humans, it is important to note that we often make comparisons between species in their nocturnal and diurnal state.

### 2.4.2 Thermal Homeostasis in the Unrestrained Rat

Because much of the toxicological data in this book focuses on rodents, emphasis is going to be placed on their thermoregulatory responses. Many scientists who are not well read in thermoregulation assume that the body temperature of rodents is unstable and not as well regulated as that of humans and other large mammals. Some may have developed this opinion from anecdotal information or based it on conclusions from studies of restrained or stressed rodents that can sustain marked temperature fluctuations (see Chapter 7). In fact, healthy rats and other rodents have well-developed thermoregulatory systems and are able to control body temperature over a wide range of ambient temperatures. For example, in a study from this laboratory, core temperature and heart rate (a reflection of metabolic rate) were monitored by radiotelemetry in rats of the Long-Evans strain housed individually on a wire-screen floor for 24 h at one of several ambient temperatures (Figure 2.2). By housing on a wire-screen floor, the animals were unable to burrow into bedding material, which would have altered their operative temperature. The core temperature was recorded at 5-min intervals, and stability at each ambient temperature was determined. It was shown that core temperature was tightly regulated during the day and night, varying by just 0.39°C, or ±1.3%, relative to the change in ambient temperature over a range of 15 to 30°C. Heart rate increased with a decreasing ambient temperature, reflecting the effects of cold exposure on cardiac output and metabolic demand. Only at ambient temperatures above 30°C did thermoregulatory control start to break down, and the rats developed mild hyperthermia.

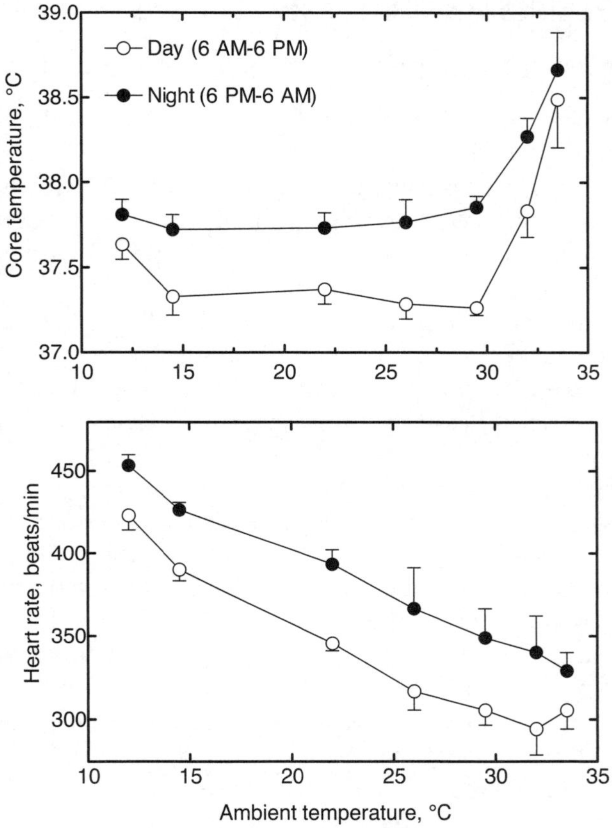


**Figure 2.2 Example of the thermal homeostatic capability of the Long-Evans rat exposed to a range of ambient temperatures for 24 h. Core temperature and heart rate monitored at 5-min intervals using surgically implanted radiotelemetry units (see Figure 3.1 for details). Note how heart rate increases proportionately with a reduction in ambient temperature. (Data modified from Yang, Y. and Gordon, C.J. (1996). *J. Therm. Biol.* 21: 353–363.**

Radiotelemetry provides an ideal means of monitoring core temperature in rodents without handling or disturbing the animals. Toxicologists and pharmacologists might consider a chemical agent that induces, for example, a 0.5°C decrease in core temperature as biologically insignificant. However, in terms of the stability of core temperature over a wide range of ambient temperatures as depicted in Figure 2.2, a deviation of 0.5°C accounts for more than 100% of the normal range of temperature variation. Hence, relatively small changes in core temperature in a rodent may well be considered to be biologically significant.

## 2.5 MECHANISMS OF TEMPERATURE REGULATION

### *2.5.1 Temperature Regulation as a Servo Control System*

One way to view thermoregulation is to compare it to the operation of a servo-loop regulated system (Figure 2.3A; Stolwijk and Hardy, 1974; Schmidt-Nielsen, 1975). In this fundamental model of regulation, an error signal ($S_e$) is generated by a comparator that sums the difference between a reference signal ($S_1$) and a feedback signal ($S_2$) that is a measure of the system's output. Controlling elements respond to the error signal and activate a corrective response to maintain the controlled system within certain limits. This servo-loop concept is an ideal way to develop a basic understanding of how the thermoregulatory system of homeotherms and poikilotherms responds to internal or external stresses that raise or lower body temperature.

The thermoregulatory system has four main components that make up the servo-loop: receptors that provide feedback, integrating and central processing neurons that provide a set-point and error signal, thermoeffectors that represent the controlling elements, and the controlled system, represented by the temperature of the core and shell. The role of these components in the servo-loop regulatory system is touched on briefly here and explained in more detail below. There are essentially two feedback loops: one loop to control heat retention and production and another loop for the control of heat loss. In the heat gain and retention system (Figure 2.3B), there is a theoretical set-point or threshold temperature that is compared with feedback from cold receptors in the skin and core. At temperatures below the set-point, an error signal is generated that actuates effectors to increase heat production and reduce heat loss. Homeotherms possess both autonomic and behavioral thermoeffectors; with few exceptions, poikilotherms must rely solely on behavioral mechanisms. The thermoeffector responses raise body temperature, thereby nullifying the feedback signal from the cold thermoreceptors. Similarly, in the heat loss system, internal or external heat loads activate warm receptors (Figure 2.3C). Each species has a particular threshold temperature for activating heat loss effectors. Autonomic and behavioral thermoeffectors are activated to reduce the heat load and nullify the feedback signal from the warm thermoreceptors.

In tachymetabolic species, the mean temperature of the thermal shell is the critical feedback signal to drive the thermoregulatory system under most environmental conditions. That is, the temperature of the core in these species is quite stable, even for rodents, as is evident by the rat's stable core temperature in Figure 2.2. Hence, tachymetabolic species rely on the change in temperature of the thermal shell to activate the appro-

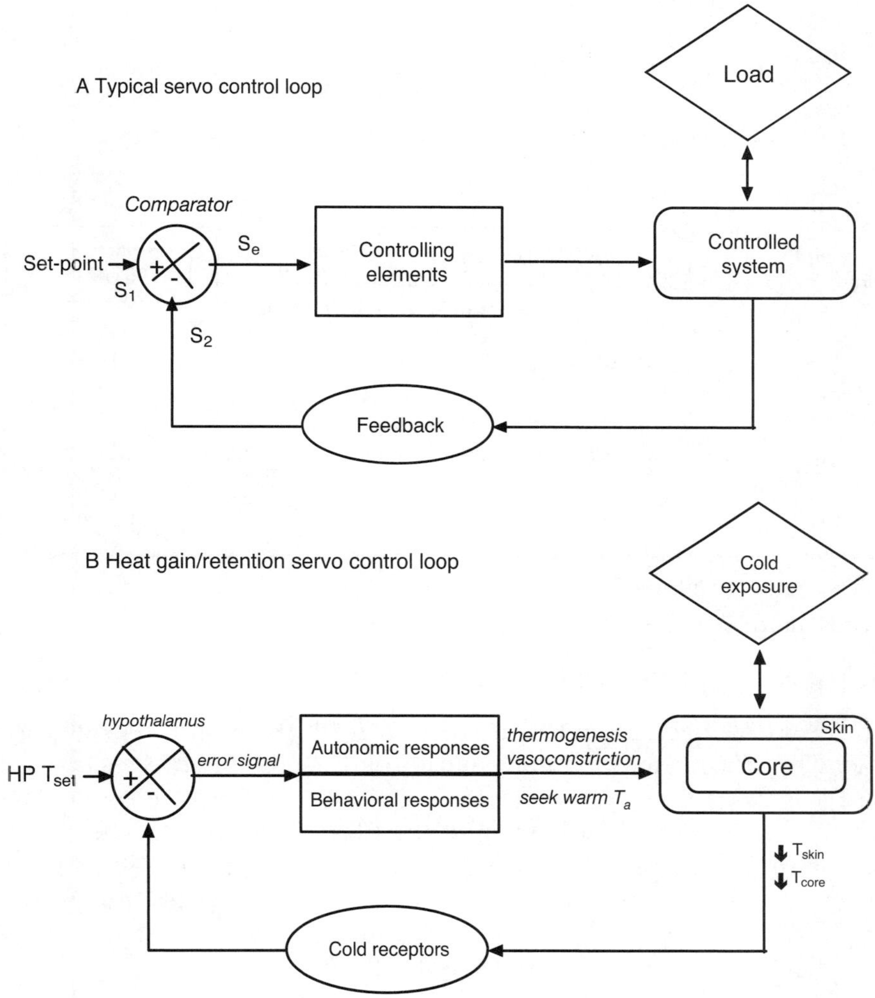


**Figure 2.3 (A) General concept of a servo-loop feedback control system. The analogy of the servo-loop feedback system is used to explain the regulation of thermoeffectors for heat gain (B) and heat loss (C) (see next page). For discussion of models, see Schmidt-Nielsen (1975) and Stolwijk and Hardy (1974).**

priate thermoeffectors to regulate a stable core temperature. When regulatory mechanisms begin to be overwhelmed, signals from the thermal core manifest strong feedback signals to prevent a further deviation in core temperature. Bradymetabolic, temperature-conforming species do not have the thermal shell and rely more on feedback from the core to regulate body temperature.

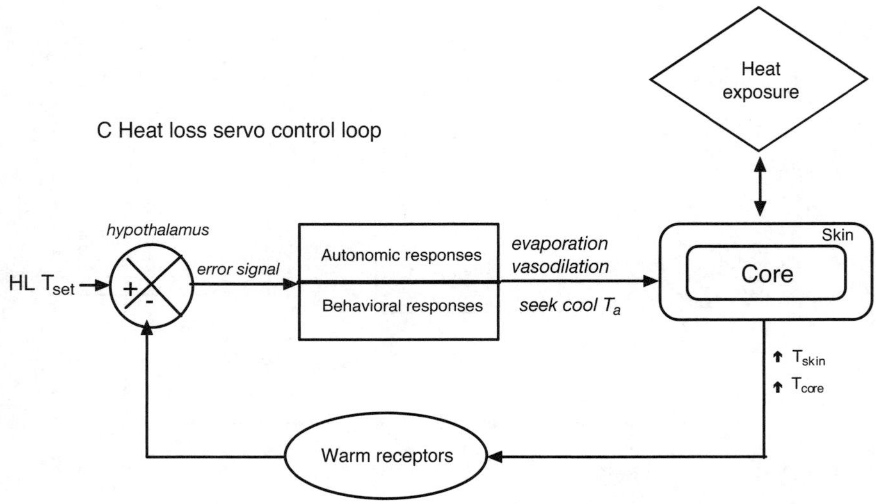


**Figure 2.3 (continued)**

## *2.5.2 Neurophysiological Mechanisms*

There has been a tremendous amount of work done on the neurophysiology of thermoregulation in mammals, birds, and other species over the past 50 years (for reviews see Boulant et al., 1989; Gordon, 1993; Boulant, 2000). Our current understanding of the CNS mechanisms of thermoregulation are based primarily on studies using neurophysiological, neurochemical, and neural lesioning techniques. Neurophysiological studies have demonstrated how temperatures in the skin and core are detected and then compared with a theoretical reference or set-point signal with the induction of corrective thermoeffector responses (Figure 2.4A, B). Thermal transduction occurs with activation of warm and cold receptors that respond with an increase in firing activity to heating and cooling, respectively. Recent studies have begun to unravel the identity of proteins in thermoreceptors that confer the properties of thermal sensitivity as well as their responsiveness to nonthermal chemical stimulants such as capsaicin and menthol (Patapoutian et al., 2003). Thermal receptor information is carried on C-fibers and Aδ-fibers through the spinothalamic and trigeminal afferent systems. Warm and cold thermoreceptors in the preoptic area and anterior hypothalamus (POAH) and other parts of the CNS also detect changes in temperature.

The thermal information from the skin and core is summed and integrated in the POAH and other sites in the CNS where warm- and cold-sensitive neurons undergo reciprocal inhibition (Figure 2.4B). The

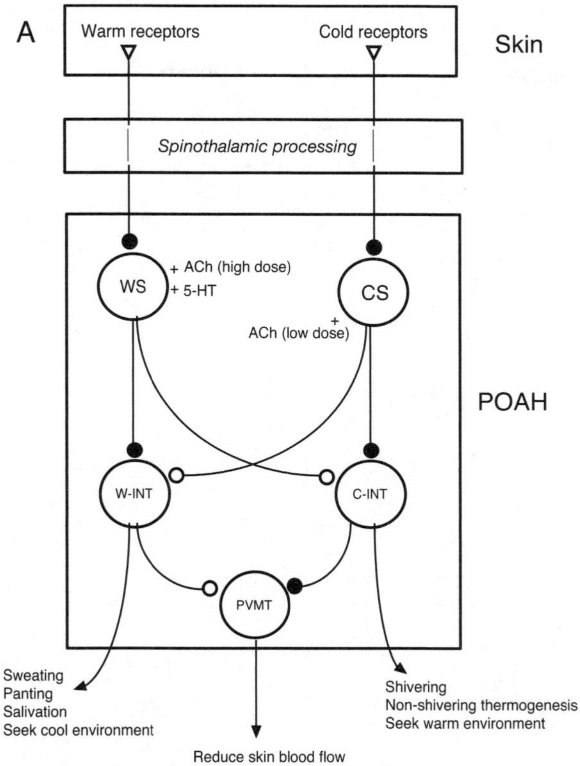


**Figure 2.4 (A) Basic neural circuit for regulation of body temperature. Thermoreceptor activity in the skin eventually passes into the POAH area, where thermal stimuli from skin and core are integrated, and appropriate effector signals are generated to control thermoeffectors for heat gain and heat loss. (B) (see next page) Pattern of temperature versus firing rate activity in cutaneous thermoreceptors, warm- (WS) and cold-sensitive neurons (CS), and integrating neurons in the POAH (CS-int, WS-int, PVMT). Note how reciprocal inhibition of CS-int and WS-int neurons leads to development of effector signals that have zero activity at the normal regulated core temperature of 37°C (Gordon and Heath, 1986; Gordon, 1993; Boulant, 2000; Bligh, 1998).**

concept of reciprocal inhibition of warm and cold neural pathways is based on neurophysiological studies showing that localized heating of neurons in the POAH will suppress activity of neurons that are stimulated by cooling the skin. Some neurons that are insensitive to changes in temperature of the CNS are affected by skin heating or cooling. The activity of some neurons in the POAH that is facilitated by skin heating is suppressed by POAH cooling. All together, one can visualize a network of warm-sensitive, cold-sensitive, and thermally insensitive neurons in the POAH as well as other locations in the CNS that behave

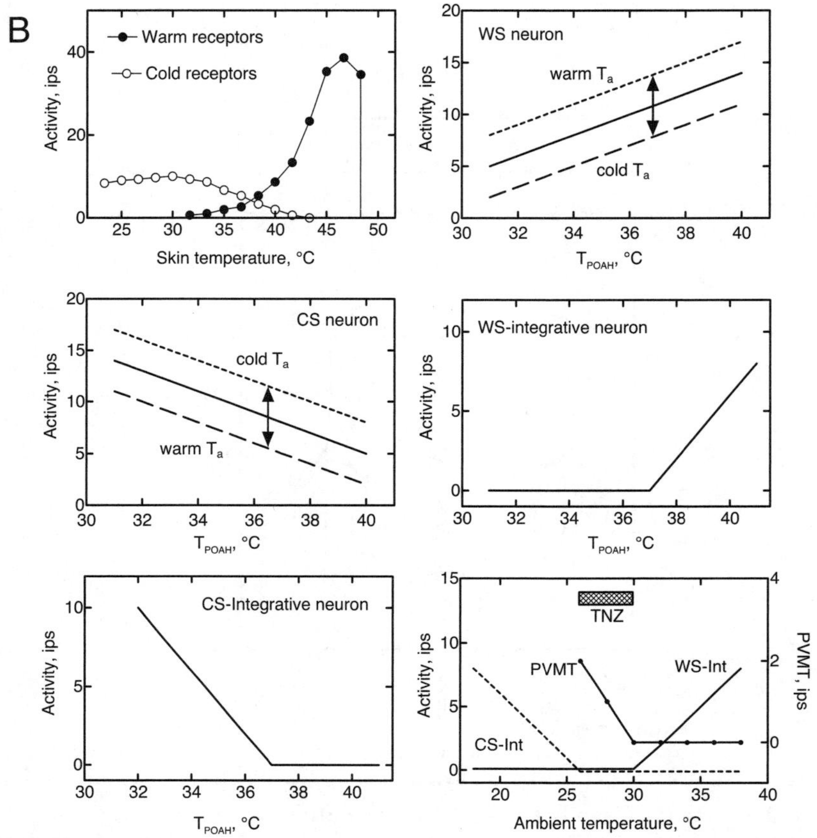


**Figure 2.4 (continued)**

as if they reciprocally inhibit the other and are driven by thermal input from the skin (Figure 2.4B).

Stereotaxic implantation of thermodes into the POAH area of a mammal or bird to locally heat or cool thermoregulatory centers was a mainstay of thermoregulatory research in the 1960s. These studies were the forerunners of the neurophysiological studies that confirmed the existence of the neural networks for thermoregulation (Hammel, 1968; Heath et al., 1972). Local heating of the POAH stimulates a heat loss response characterized by reduction in metabolic thermogenesis, peripheral vasodilation, and sweating or panting. The animal behaves as if it were hot and seeks cooler temperatures. The heat loss response is sufficient in magnitude that heat loss exceeds heat production and the subject becomes hypothermic. Local cooling of the POAH area elicits an increase in heat production and cutaneous vasoconstriction, and the subject seeks warmer ambient temperatures. Heat production exceeds heat loss, and the subject becomes hyperthermic.

Neurons in the POAH integrate thermal information as well as non-thermal stimuli that can have a bearing on thermoregulation. Blood pH, blood pressure, oxygen level, osmotic tonicity, and glucose levels can all influence the activity of POAH neurons in a manner that would be expected based on their thermoregulatory effects (Boulant and Silva, 1989). Integrative neurons in the CNS are generally 5 to 10 times more sensitive to a change in local temperature than to changes in skin temperature. That is, a 1°C reduction in brain temperature has about the same effect on a thermoeffector response as would a 5 to 10°C reduction in skin temperature. Under most circumstances, birds and mammals do not rely on a change in brain temperature to drive thermoeffectors because core temperature remains stable in the face of large changes in ambient temperature. However, rodents exposed to toxicants undergo marked reductions in core temperature, and the responsiveness of CNS thermal receptors is thus pertinent to consider. In birds, integration of thermal stimuli in the spinal cord is thought to take on more importance as compared to mammals (Simon, 1999).

### 2.5.3 Neurochemical Mechanisms

The CNS's control of body temperature involves a complex interaction between many neurotransmitters, modulators, and hormones. A basic understanding of these processes is essential, especially in cases where the neurochemical mechanism of a toxicant is known and thus allows one to speculate on the possible effects on thermoregulation. The work of Feldberg and Myers in the 1960s demonstrating specific thermoregulatory responses when adrenergic and serotonergic neurotransmitters were microinjected into the CNS was a landmark study that spurred innumerable studies on the neurochemical control of body temperature (see Myers, 1980). The reviews by W.G. Clark and others bear witness to the hundreds of studies on the responses of mammals, birds, and other species to neurotransmitters, peptides, drug agonists and antagonists, and other agents administered peripherally or directly into the CNS (Clark and Lipton, 1985, 1985a; Lipton and Clark, 1986; also see Wang and Lee, 1989). While there was a tremendous effort to use neurochemical techniques to understand thermoregulation, there remains today a considerable controversy. Indeed, one will find opposite thermoregulatory responses for the same neurochemical given by the same route of exposure to the same species (Table 2.4). This can be frustrating to the toxicologist and pharmacologist who would like to have clear cause-and-effect of a transmitter-mediated change in body temperature.

It is likely that much of the variability in these studies, especially in rodents, was a result of using restrained or otherwise stressed animals. In

**Table 2.4 Summary of the Thermoregulatory Effects of Principal Neurotransmitters, Modulators, and Hormones Injected into the CNS of Laboratory Rodents**

| *Agent* | *Species* | *Injection Site*[a] | *Response* |
|---|---|---|---|
| Acetylcholine | Hamster | IVT | Decrease |
| Acetylcholine | Hamster | POAH | Increase |
| Acetylcholine | Rat | POAH, IVT | Decrease |
| Dopamine | Mouse | IVT | Decrease |
| Dopamine | Rat | IVT | Decrease or increase |
| Norepinephrine | Mouse, hamster, rat | IVT, POAH | Decrease |
| Norepinephrine | Guinea pig | IVT, AH | Increase |
| Serotonin | Mouse | IVT | Decrease |
| Serotonin | Rat | IVT | Increase or decrease |
| Bombesin | Mouse, rat | IC, IVT | Decrease |
| Choleocysto-kinin | Mouse, rat | IVT | Decrease |
| Choleocysto-kinin | Guinea pig | IVT | Increase |
| β-endorphin | Mouse, rat | IVT | Increase |
| Neurotensin | Mouse, hamster, rat, guinea pig | IC | Decrease |

*Source*: Taken from Gordon, C.J. (1993). *Temperature Regulation in Laboratory Rodents.* New York: Cambridge University Press.

[a]IVT = intraventricular; AH = anterior hypothalamus; IC = intracisternal.

most of the work from the 1960s, radiotelemetry was unavailable, and core temperatures were measured using rectal probes or with implanted probes that were tethered to the subject. Quan and Blatteis (1989) found that the injection of minute volumes of control or norepinephrine solutions into the CNS caused transient neural damage that altered the thermoregulatory response to the injected neurotransmitter. Microinjection was a common method of assessing how a neurotransmitter affected the CNS control of body temperature. The pressure of the microinjection induced the synthesis of prostaglandin E, and a hyperthermic response ensued that masked an actual hypothermic effect of norepinephrine. All together, the neurochemical studies of temperature regulation have generally shown distinct patterns of how neurotransmitters operate in the CNS thermoregulator centers, but the studies are occasionally found to be contradictory and should be viewed with the aforementioned caveats in mind.

One relatively simple working model for the rat and mouse is composed of a heat dissipatory pathway that is stimulated by serotonin and a heat producing and conserving pathway stimulated by low levels of cholinergic stimulation but suppressed when synaptic levels of acetylcholine are excessive (Figure 2.4A). Norepinephrine may either stimulate or suppress these pathways. This model is useful for explaining the hypothermic effects of anticholinesterase pesticides that lead to stimulation of central and peripheral cholinergic pathways (see Chapter 3). Microinjection of muscarinic agonists or anticholinesterase agents either intraventricularly or directly into the POAH area leads to activation of heat dissipatory thermoeffectors and hypothermia. Small doses of acetylcholine injected into the POAH have been shown to induce hyperthermia. Microinjection of 5-hydroxy tryptamine into the CNS generally elicits a heat dissipatory response (for review, see Gordon, 1994).

Most of the responses summarized in Table 2.4 are based on studies performed in rodents maintained at room temperature. The thermoregulatory effects of microinjected neurotransmitters will depend in large part on ambient temperature. For example, amphetamine-related drugs that elicit a marked increase in brain serotonin will evoke a hypothermic response in rodents at standard room temperature (e.g., 22°C). This would lead one to conclude that 5-HT is involved in driving heat dissipating responses. However, when amphetamines are administered to rats maintained at thermoneutral temperatures or warmer, there is a peripheral vasoconstriction and a profound hyperthermia that can be lethal. Thus, the model such as that in Figure 2.4A has a limited usefulness and can only be used as a starting point when one wants to understand how a toxicant or drug might affect body temperature by modulating the activity of neurochemical pathways in the CNS.

## 2.6 SET-POINT: REGULATED VERSUS FORCED CHANGES IN BODY TEMPERATURE

The concept of a thermostat with a set-point temperature as depicted in Figure 2.3A is an extremely useful analogy for explaining how a drug or toxicant affects thermoregulation in homeothermic and poikilothermic species. Thermal physiologists define set-point as "the value of the regulated variable which a healthy organism tends to stabilize by the processes of regulation" (IUPS, 2001). When external or internal interferences tend to alter the regulated variable (i.e., body temperature), the resulting thermoeffector activities counter the alterations. In other words, if an organism uses its thermoeffectors to maintain its core temperature at 37.5°C, then it is assumed that its set-point or reference temperature is set at 37.5°C. The set-point for temperature regulation may change with certain endogenous and environmental stimuli such as fever, starvation, and dehydration (see Chapter 5). Moreover, the circadian or nychthemeral variation in core temperature is considered to be a result of a 24-h oscillation in the set-point temperature.

In this book, the effects of toxicants on body temperature are often discussed in terms of a potential change in the set-point. The set-point term has in fact generated much debate and confusion over the past several decades (IUPS, 2001; Kanosue et al., 1997). Many researchers, including this author, have relied on set-point terminology to describe phenomena with a connotation that there is an actual reference temperature in the CNS. While an actual reference temperature has never been shown to exist, the thermoregulatory system subjected to thermal stress, pyrogens, and other stimuli behaves in a manner suggestive of an operative set-point temperature. Reciprocal inhibition of warm, cold, and thermally insensitive neurons in a manner depicted in Figure 2.4A can provide a framework for the generation of a set-point temperature (see reviews by Bligh, 2001; Boulant and Silva, 1989; Gordon, 2001). To sum up, the set-point is in fact an analogy of a mechanical or electronic engineering control system. Its existence has not been proven, but it is nonetheless a useful way of explaining most thermoregulatory responses.

The set-point allows one to distinguish a toxicant that elicits integrated changes in thermoregulatory control from one that simply imparts deficits in thermoeffector function (Figure 2.5). The thermoregulatory system of homeotherms and poikilotherms attempts to maintain a core temperature ($T_c$) equal to the set-point temperature ($T_{set}$). This process is essentially continuous in homeotherms and intermittent in poikilotherms when options are available to behaviorally thermoregulate. In a thermoneutral environment for a healthy, tachymetabolic species, $T_{set}$ is equal to $T_c$ and thermoeffectors for heat gain and heat loss are balanced and at a minimal level of activity. The animal has a normothermic body temperature and selects an ambient temperature that is comfortable and associated with minimal energy expenditure. Normothermy (or cenothermy) means core temperature is controlled within ±1 S.D. of the range associated with normal resting, thermoneutral conditions.

Normothermic body temperature can increase or decrease in a forced or regulated fashion (Figure 2.5). An increase in the set-point, as occurs with a fever, means there is a transient period where $T_{set} > T_c$. The animal responds as if it were cold and selects warmer ambient temperatures and activates thermoeffectors to increase heat production (shivering and nonshivering thermogenesis) and reduce heat loss (peripheral vasoconstriction). Thermal physiologists view infectious fever as the cornerstone of a set-point elevation. Noninfectious agents can also increase the set-point, and this phenomenon

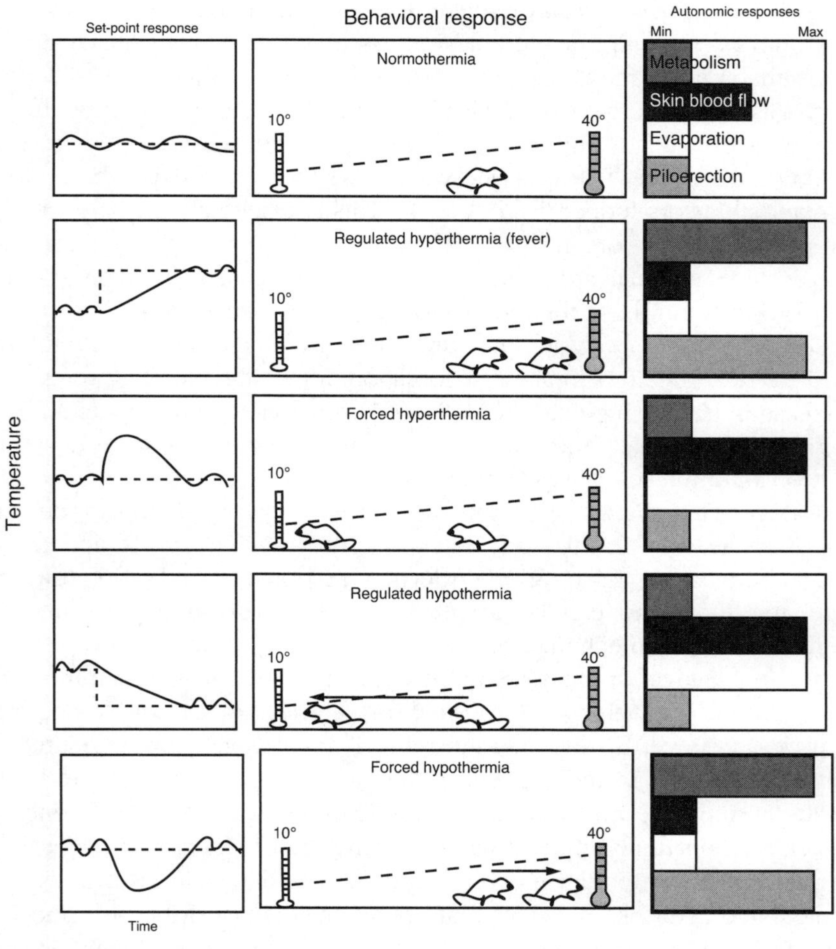


**Figure 2.5 Summary of behavioral and autonomic responses of a homeotherm when subjected to manipulation of body (solid line) and set-point temperature (dashed line): normothermia, regulated hyperthermia (fever), forced hyperthermia, forced hypothermia, regulated hypothermia. Modified from Gordon et al. (1988).**

is also termed *regulated hyperthermia*. As the hyperthermic response progresses, there is eventually an equaling of $T_{set}$ and $T_c$, and the animal reaches a steady state with an elevated body temperature. During *forced hyperthermia*, $T_c$ increases above $T_{set}$, as would occur by exposure to high ambient temperatures or by administering toxicants or drugs that stimulate metabolic thermogenesis but without affecting the CNS control mechanisms. During forced hyperthermia, thermoeffectors are activated to reduce heat gain and increase heat loss to lower body temperature. The animal seeks a colder environment to facilitate heat loss and lower body temperature to normal. *Forced hypothermia* refers to the state when $T_c$ is forced below $T_{set}$, as would occur during acute cold exposure or treatment with toxicants or drugs that impair metabolic thermogenesis without affecting CNS control mechanisms. The organism responds with an activation of thermoeffectors to minimize heat loss and increase heat production. A warmer environment is sought to reduce heat loss. *Regulated hypothermia* occurs when internal or external factors reduce $T_{set}$ below $T_c$. This is essentially the opposite of a fever because the organism feels hot and responds by seeking cooler temperatures and activating thermoeffectors to increase heat loss and reduce heat production. These thermoeffector responses persist until $T_c$ is equal to $T_{set}$ but at a lower body temperature. One also finds the term *anapyrexia* used to describe a pathological condition in which there is a regulated decrease in body temperature (IUPS, 2001). Anapyrexia and regulated hypothermia are essentially the same, but this author prefers the latter, especially in describing the responses to toxic agents.

Categorizing thermoregulatory responses into forced versus regulated responses is essential for understanding the mechanism of action of a toxic chemical or drug. If the thermal response can be identified as regulated then one can be assured that the toxicant is affecting CNS thermoregulatory mechanisms. A forced response could be mediated with or without activation of CNS pathways. In homeotherms, simultaneous measurement of behavioral thermal preference and body temperature provides a powerful tool to determine if the response is regulated. As will be discussed in more detail in Chapter 3, many toxicants and other insults administered acutely to rodents lead to a transient preference for cooler ambient temperatures as body temperature decreases. Measuring skin and body temperature can also be used to determine if the response is regulated. That is, an agent that elicits an increase in skin blood flow and hypothermia is likely to be a regulated response. However, it is possible that a marked increase in heat loss from peripheral vasodilation could change body temperature without affecting the set-point. Behavioral temperature preference is advantageous because it is relatively easy to monitor in undisturbed animals and, of all thermoeffectors, provides a rapid and most sensitive indication of a change in set-point (Gordon, 1993). Of course, behavioral temperature preference is really the only effector one can measure to assess if there is a change in the set-point in a poikilotherm.

## 2.7 THERMOEFFECTOR MECHANISMS AND THE THERMONEUTRAL ZONE

Measuring metabolic rate, evaporative water loss, and skin temperature (or skin blood flow) over a range of ambient temperature reveals a general pattern of thermoeffector activity that is typical for most tachymetabolic species (Figure 2.6). There is a range of ambient temperatures termed the *thermoneutral zone*, where metabolic rate is at or near basal levels. In this zone, temperature regulation is achieved by control of sensible heat loss, meaning without regulatory changes in metabolic rate or evaporative water loss. As ambient temperature decreases below the thermoneutral zone, the blood flow to the skin is minimal as a result of peripheral vasoconstriction. With further cooling, metabolism must increase above basal levels by shivering and nonshivering thermogenesis in order for heat production to match heat loss to the environment.

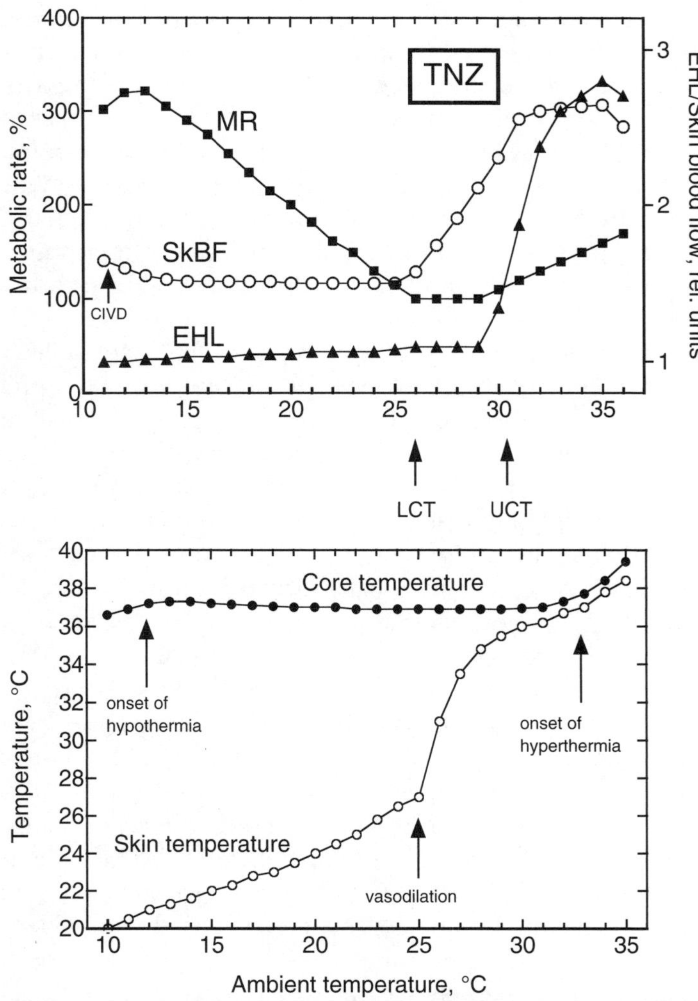


**Figure 2.6 General pattern of core and skin temperature and activity of autonomic thermoeffectors as a function of ambient temperature in a homeotherm. SkBF, skin blood flow; EHL, evaporative heat loss; MR, metabolic rate; LCT, lower critical temperature; UCT, upper critical temperature; TNZ, thermoneutral zone.**

The ambient temperature at which metabolic rate increases is termed the *lower critical temperature*. As ambient temperature decreases below the lower critical temperature, skin temperature falls passively but may increase with extreme cold exposure as a result of cold-induced vasodilation (CIVD). This is a protective response to keep exposed tissues from freezing. Eventually, the point is reached where metabolic rate cannot maintain the pace of heat loss, and the animal becomes hypothermic. As ambient temperature increases through the thermoneutral zone, skin blood flow increases and there is a disproportionate rise in skin temperature. At temperatures above the thermoneutral zone, evaporative heat loss mechanisms (i.e., panting, sweating, saliva grooming) are activated to maintain thermal balance. This ambient temperature is termed the *upper critical temperature*. It is also identified with the point where core temperature and metabolism begin to rise. At around the point of the upper critical temperature, skin temperature has increased to a level that is just below core temperature, reflecting maximal redistribution of warm blood from the core to the periphery. With further increase in ambient temperature, skin and core temperature parallel each other until the point of thermoregulatory failure. At this point, evaporative heat loss is ineffective and core temperature spirals upward leading to hyperthermic death.

The thermoneutral zone and the slope of the metabolism versus ambient temperature below the lower critical temperature represent the relative sensitivity of a homeotherm to cold. Rodents are small and have a relatively large surface area to mass ratio, meaning that they lose body heat faster and must rely more on a high metabolic rate rather than adjustments in peripheral vasomotor tone to thermoregulate below the lower critical temperature. Mice and rats have a lower critical temperature of 28 to 31°C, which is notably much warmer than the standard temperature for housing in most laboratory settings. That is, under standard conditions they are cold stressed and thermoregulate by maintaining a metabolic rate above basal levels. That rats are cold stressed at standard room temperatures is evident from the heart rate response presented in Figure 2.2. Heart rate at an ambient temperature of 22°C is about 13% higher than that at a thermoneutral temperature of 30°C.

The thermoneutral zone varies widely among species of mammals and birds (Table 2.3). The lower critical temperature of the rabbit is relatively low, reflecting this species' adaptation to cold environments. Rabbits are more susceptible to ambient heat stress as compared with rodents. Large agricultural mammals such as cattle have a relatively low upper critical temperature. In the summer, they can face a dire situation with dissipating excess heat, a problem that can be compounded by ingesting natural toxins in their feed that induce peripheral vasoconstriction (Chapter 10).

## 2.7.1 Metabolic Thermogenesis

The term *metabolic thermogenesis* is generally applied to situations where the animal's metabolism is utilized as a thermoeffector. In biological studies, metabolism is a universal term that describes the physical and chemical changes in living organisms. In thermal physiology, metabolism always refers to the transformation of chemical energy to work and heat (IUPS, 2001). The metabolic requirements of mammals and birds are conventionally divided into obligatory and facultative. Obligatory metabolism, the heat produced from basal metabolic processes, provides a sufficient amount of heat for thermoregulation in the thermoneutral zone. Facultative metabolism includes the heat from shivering and nonshivering thermogenesis and is called upon to meet increased energy demands during cold exposure. Hence, drugs or toxicants can effectively alter heat production by affecting obligatory, shivering, and nonshivering thermogenesis (Figure 2.7). A chemical-specific effect on shivering or nonshivering thermogenesis will not be apparent if the animal is housed at temperatures equal to or above thermoneutrality. An effect could also be difficult to observe if the animal is restrained, stressed, or subjected to any other kind of procedure that results in an elevated core temperature. That is, during stress-induced hyperthermia, the organism is in a heat stress situation and is expected to suppress heat production to lower body temperature. Obligatory metabolism also increases with heat stress. A

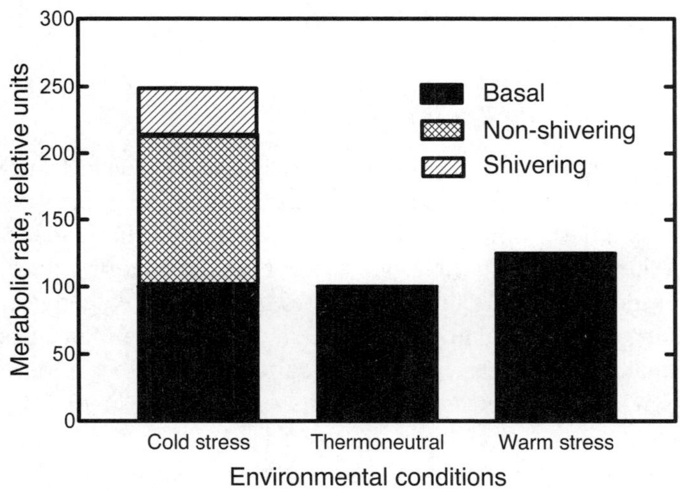


**Figure 2.7 Relative contribution of the three primary sources of heat production in rodents when housed in a warm, thermoneutral, or cold environment.**

chemical such as a blocker of oxidative phosphorylation leads to a decrease in metabolic rate at warm, cold, and thermoneutral temperatures.

### 2.7.1.1 Shivering Thermogenesis

Mammals exposed to cold or that are febrile will rely on shivering to supplement heat production. Shivering is uncomfortable and inefficient, and animals are unable to go about their natural behaviors when so much of their effort is being used to shiver. Shivering also disrupts the still air layer and thereby accelerates convective heat loss. With prolonged exposure to cold conditions, shivering in rodents is gradually replaced by nonshivering thermogenesis as the primary source of heat production.

There has been confusion in the literature over the thermoregulatory consequences of shivering versus toxicant-induced tremor. This is a pertinent issue to toxicologists because several classes of compounds induce a state of tremor. Toxicants such as DDT and chlordecone are tremorigenic. Acute exposure to these chemicals is associated with marked tremor, but the animals may nonetheless become hypothermic depending on ambient temperature (see Chapter 3). Tremor and shivering are unique muscular phenomena involving different patterns of synchronization and activation of muscular pathways. Cold-induced shivering elicits different frequencies of oscillation of skeletal muscles than does injection of tremorine, a chemical that elicits muscular tremor (Günther et al., 1983). Although it is not well understood, shivering is clearly a more effective means than tremor for generating heat for thermoregulation.

### 2.7.1.2 Nonshivering Thermogenesis

Nonshivering thermogenesis, defined as the heat produced from metabolic processes not involving contracting muscles, has been one of the most intensively studied topics in thermal physiology (Himms-Hagen, 1986, 1990). The key strategy in the development of nonshivering thermogenesis is the operation of metabolic pathways to accelerate oxygen consumption and produce heat, but without muscular contraction. Many tissues from cold-acclimated mammals, including heart and liver, have increased aerobic capacity and higher concentrations of enzymes involved in cellular respiration.

Recognized approximately 40 years ago, brown adipose tissue (BAT) is a key facet of nonshivering thermogenesis and has been one of the most intensely studied thermoregulatory processes (Himms-Hagen, 1990; Argyropoulos and Harper, 2002). Rodents rely on BAT as a major source of heat when exposed acutely and chronically to a cold environment.

BAT is found in other small mammals, including the orders insectevora (moles, shrews), lagamorphs (e.g., rabbit), and chiroptera (bats). It also occurs in appreciable amounts in the newborn of large mammals, including cow, goat, and humans. BAT also serves as an important source of heat in the arousal from hibernation and recovery from anesthetic-induced hypothermia, provides heat to elevate body temperature during fever, and is crucial in the development of thermoregulation from newborn to adult. With the exception of the guinea pig, newborn rodents are unable to shiver. Because of their small size and underdeveloped thermoregulatory mechanisms, young rodents rely on BAT as a major source of heat during cold exposure.

In spite of the pivotal role of BAT in thermogenesis in rodents, it is amazing to find that so little is known about how toxicants may affect BAT function. Toxic metals have been shown to interfere with BAT thermogenesis (Chapter 3). Other toxicants may affect BAT function in developing and adult rodents either directly or indirectly through the thyroid axis.

BAT is a unique structure that is packed with mitochondria and has a well developed sympathetic innervation. The release of norepinephrine from sympathetic terminals activates $\alpha$ and $\beta$ adrenergic receptors, stimulating lipase and the concomitant release of fatty acids as fuel for thermogenesis. A unique form of thyroxine 5'-deiodinase is found in BAT. This enzyme converts $T_4$ to $T_3$ during NE stimulation, and the binding of $T_3$ to BAT nuclear receptors leads to further expression of uncoupling protein (UCP). BAT is one of the most thermogenic tissues, capable of generating heat at a rate of 400 W/kg, or 80 times the basal metabolic rate of a rat. BAT occupies strategic locations in the rat, including interscapular, cervical, pericardial, intercostal, and perirenal regions. This presumably allows heat to be quickly transferred to these crucial anatomical regions when needed during cold exposure.

The thermogenic capability of BAT is attributed to a unique molecular adaptation that allows for uncoupling of oxidative phosphorylation in the mitochondria, resulting in a lowered ratio of ATP produced for each molecule of oxygen converted to water (Figure 2.8A, B). This is achieved by UCP, a proton-translocator that mediates a proton leak across the inner mitochondrial membrane. The proton leak reduces the electrochemical proton gradient in the mitochondrion and leads to uncoupling of oxidative phosphorylation. In other words, activation of UCP forces BAT to use more oxygen and produce greater quantities of heat. Over the past decade, the physiological role of UCP in uncoupling proton conductance has been discovered to occur in tissues other than BAT (Argyropoulos and Harper, 2002). There are several homologs of UCP that have been isolated in liver, skeletal muscle, and other areas in animals and humans. The uncoupling

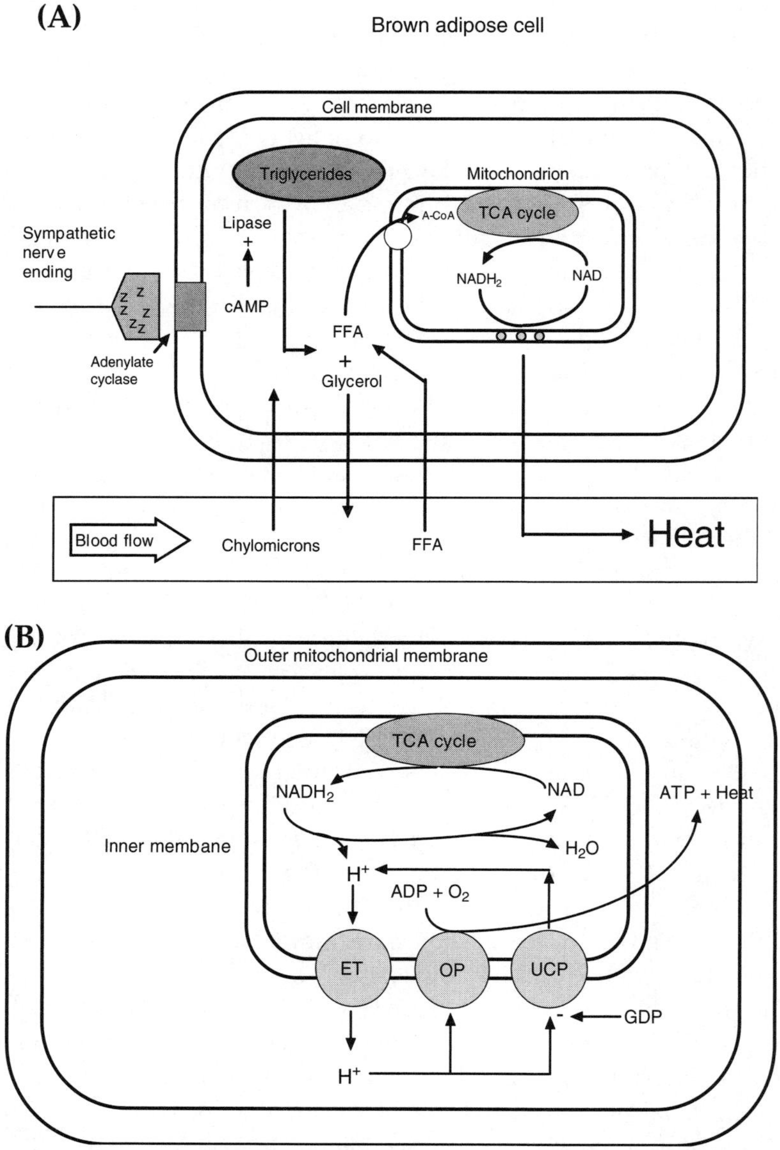


**Figure 2.8 Cellular (A) and subcellular (B) components responsible for heat production in brown adipose tissue (BAT). Diagram adapted from several sources (e.g., Himms-Hagen, 1990; Gordon, 1993).**

of oxidative phosphorylation in other tissues contributes significantly to the overall cellular metabolism, accounting for possibly 15 to 20% of the standard metabolic rate. In addition to energy metabolism in BAT and other tissues, UCPs have a role in reducing the production of reactive oxygen species and, hence, could have an important protective role in the protection of cells from toxicant exposure. In fact, metallothionein, a protein that binds heavy metals, is co-expressed with UCP in BAT and may serve an antioxidant protective role (see Chapter 3). The recent discovery of the ubiquitous nature of UCP and its homologs and their role in thermoregulation may play out as an important factor in the physiological response to toxicants.

### 2.7.2 Peripheral Vasomotor Tone

Control of skin blood flow in tachymetabolic species is an ideal thermoeffector because it requires insignificant amounts of metabolic energy. When housed at ambient temperatures within the thermoneutral zone, heat loss by radiation, convection, and conduction is controlled through adjustments in peripheral vasomotor tone, with no observable changes in metabolic rate. Sympathetic innervation to the precapillary sphincters and arteriovenous anastomoses (AVA) regulates the relative distribution of warm blood between the core and skin. AVAs are short channels that connect arterioles to venules and, when opened, shunt blood through the peripheral tissues allowing for a high rate of heat loss. Modulation of sympathetic tone is thought to be the main mechanism for the neural control of skin blood flow, but some species possess active vasodilatory mechanisms.

In view of the potential transdermal exposure to pesticides and other toxicants, it is relevant to consider how skin blood flow is modulated by heat and cold stress (see Chapter 7). Humans possess remarkable variation in skin blood flow that is modulated by thermoregulatory control. Overall skin blood flow can be as low as 150 to 200 ml/min in a cool environment and as high as 2,000 ml/min in a warm environment. It is important to note that the wide range in blood flow exceeds the tissues' metabolic requirements and reflects the demands of the thermoregulatory system to control heat exchange between the thermal core, shell, and environment. Indeed, the range of blood flow in some extremities that are inundated with AVAs can be more than 100-fold. A thorough analysis of the effects of ambient temperature on hand blood flow in three human subjects illustrates the profound changes in skin blood flow with temperature (Figure 2.9). Considerable variability is observed in the graph where blood flow is expressed as a function of ambient temperature; however, blood flow is highly correlated with skin temperature. Overall, there is a marked increase in blood flow when ambient temperature increases above 22°C or when skin temperature increases above 27°C. Humans possess a powerful cutaneous vasodilatory system to increase heat loss, a response that is mediated by the release of acetylcholine, which activates nitric oxide synthase mechanisms (Shibasaki et al., 2002). This active mechanism

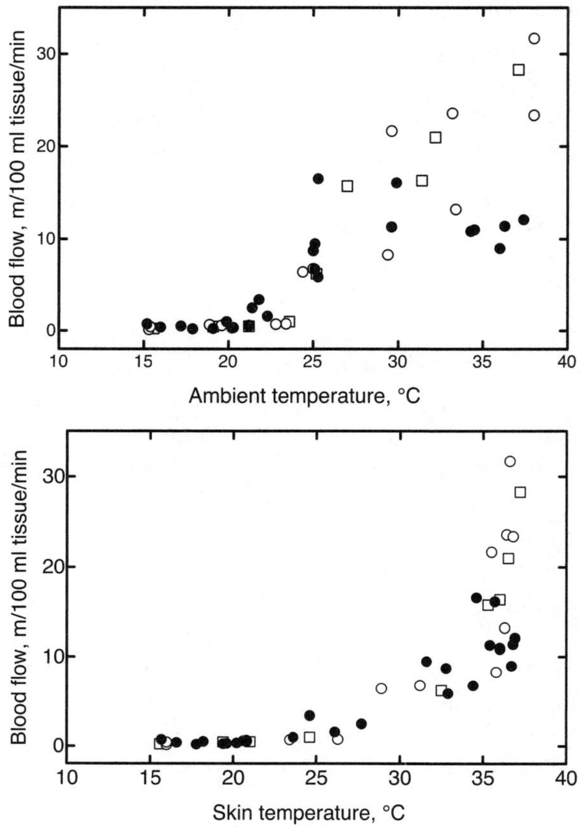


**Figure 2.9 Relationship between ambient temperature and skin temperature on hand blood flow in three human subjects exposed to a range of temperatures. (Data re-graphed from Forster, R.E., Ferris, B.G., and Day, R. (1946). *Am. J. Physiol.* 146: 600–609.)**

of vasodilation is important in view of the marked effects of anticholinesterase agents on skin blood flow and sweating and the potential dermal absorption of pesticides (see Chapters 6 and 7).

How peripheral vasomotor tone is used to thermoregulate depends on a species' body mass and unique adaptational characteristics. The role of body mass is discussed in more detail at the end of this section. Dry heat loss occurs most effectively from sparsely furred or bare surfaces that are well vascularized. For example, the inner surface of the rabbit ear is a main site for control of heat loss. The rabbit ear is highly vascularized and possesses well-developed sympathetic control of AVAs. The tail of the rat is long and slender with no fur and has well-developed control of blood flow. The tail accounts for about 7% of the rat's total surface area, and approximately 25% of the total heat production can be dissipated through the tail under ideal conditions (Gordon, 1993). Nearly all that is known about tail blood flow in the rat has been collected from studies in restrained animals. These studies have shown that blood flow to the tail is very low at ambient temperatures equal to standard room temperatures, similar to the functions described for skin blood flow in humans (Figure 2.9). As temperature is increased, there is a sudden increase in tail blood flow at a threshold ambient temperature of approximately 27 to 30°C. That is, within the thermoneutral zone for the rat, there is marked change in blood flow in a manner as depicted in Figure 2.6. Skin blood flow to the feet of the rat also shows a threshold, increasing at a critical temperature of approximately 22°C. The ears of guinea pigs, but not of mice and rats, are innervated with AVAs and serve as a site for the physiological control of heat loss.

Toxicologists should consider how the redistribution of blood flow when animals are subjected to heat or cold stress will affect the accumulation of a toxicant within a tissue or organ. The use of radioactive microspheres to measure organ and tissue blood flow has been an important tool in the study of the cardiovascular responses to thermal stress. A study of adult sheep serves as an ideal example to illustrate how heat stress can alter the distribution of blood flow to organs and tissues (Figure 2.10). When the sheep were exposed to a warm environment for several hours, blood flow was redirected to organs involved in heat dissipation. Heat stressed sheep pant and must increase blood flow to the diaphragm and respiratory muscles as well as to the nasal mucosa to dissipate heat by evaporation. Blood is drawn away from the digestive system (e.g., rumen), kidneys, and other organs during heating while blood flow to the brain remains unchanged. Blood is also directed to the skin of the ears and lower legs to enhance dry heat loss.

Assessing if the drug or toxicant causes vasodilation or vasoconstriction is a common approach in rodent studies, with the intent that the vasomotor response can be extrapolated to that of a human. To make comparisons in the vasomotor responses of rodents and humans, one must also consider the impact of body size. Phillips and Heath (1995) pursued a quantitative study of the impact of body mass on the role of peripheral vasomotor tone in thermoregulation. They noted that a small mammal such as a mouse or rat, by having a relatively large surface area to volume ratio, is an *ametabolic specialist*. That is, homeotherms of small size rely on changes in their metabolic rate as a primary means to regulate body temperature (also see Chapter 5). Large animals do not change metabolic rate as much with changes in ambient temperature and thermoregulate by controlling their surface temperature. The vasomotor index (VMI), a value reflecting the ability of an animal to regulate heat exchange through control of its skin temperature, was shown to

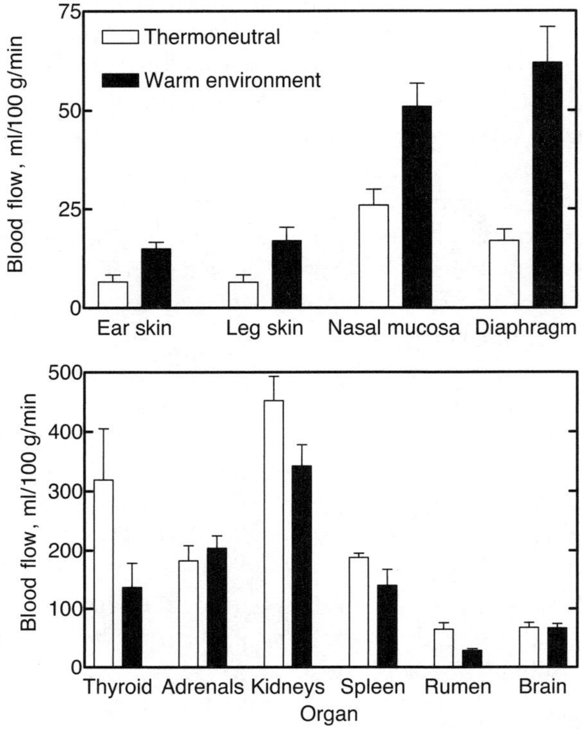


**Figure 2.10 Effects of exposure to a thermoneutral (19°C) or warm (40°C) environment on blood flow to selected organs in unanesthetized Merino wethers sheep weighing 21 to 32 kg. (Data from Hales, J.R.S. and Iriki, M. (1975). *Brain Res.* 87: 267–279.)**

be directly proportional to body mass. For example, the VMI of a 80-kg human was predicted to be more than sixfold greater than that of a 300-g rat. Could this mean that a vasomotor effect of a toxicant in a small mammal is magnified in a larger species that relies more on peripheral vasomotor tone to thermoregulate? Further work in comparative thermoregulatory responses to drugs and toxicants will help answer this question.

## *2.7.3 Evaporation*

The ability to dissipate heat by radiation, convection, and conduction is proportional to the difference between ambient and skin temperature. As ambient temperature increases above the thermoneutral zone, heat loss by evaporation of water becomes the principal mechanism to dissipate excess body heat. The latent heat of vaporization (λ) represents the quantity of heat that is absorbed from the skin or respiratory surface per gram of water that is evaporated. The latent heat of vaporization is inversely related to temperature (T) with the relationship

$$\lambda = 2{,}490.0 - 2.34 \times T \tag{2.2}$$

Thus, at 37°C, the vaporization of 1.0 g of water absorbs 2,403 J of heat. The evaporation of water occurs either passively or actively when thermoeffectors are called upon to dissipate heat. Passive water loss, also termed *insensible water loss* (now considered an outdated term), occurs from the diffusion of water through the skin and the loss in water with basal respiratory activity. Active, thermoregulatory water loss occurs when the animal pants, sweats, or applies moisture from saliva or urine onto the skin.

#### *2.7.3.1 Sweating*

Humans and other primates, cattle, horses, and sheep will sweat to dissipate heat by evaporation (Schmidt-Nielsen, 1964; Ingram and Mount, 1975). Sweat is produced by atrichial (eccrine) and epitrichial (apocrine) glands. Atrichial glands in humans are critical for thermoregulation, while epitrichial glands, found predominately in the axilla, are associated with sexual maturation and have thicker discharges not pertaining to thermoregulation. The general body surface of humans is richly innervated with atrichial glands that secrete copious amounts of hypotonic fluid and are under the control of the sympathetic nervous system. However, these efferent sympathetic pathways are unique in that acetylcholine is released from the nerve terminals. In view of the cholinergic sensitivity of atrichial sweat glands in humans, it is readily apparent why exposure to anticholinesterase insecticides and nerve gas agents leads to a profound sweating response (see Chapter 3). Species such as the goat and donkey appear to rely on circulating levels of epinephrine to drive sweating. It is also interesting to note that rodents have atrichial sweat glands in their foot pads, but they have no role in thermoregulation.

#### *2.7.3.2 Panting*

Some of the principal species that pant to dissipate heat by evaporation include sheep, dogs, rabbits, and cattle. Many other species, including rodents, increase breathing frequency when hot and dissipate heat by evaporation from the respiratory surfaces, but this is not panting. True panting animals are able to reduce their tidal volume to reduce the risk of respiratory alkalosis during panting and maintain a resonant ventilatory frequency that minimizes energy expenditure. For example, respiratory frequency of cattle increases from 40 breaths/min to 150 breaths/min when heat stressed. This is accompanied by a decrease in tidal volume from 0.95 to 0.5 l. Heat stressed sheep increase respiratory frequency from about 15 to over 100 breaths/min (Ingram and Mount, 1975). Birds do not sweat and therefore must rely on panting and gular flutter to dissipate heat by evaporation.

#### *2.7.3.3 Saliva Grooming*

Rodents neither sweat nor pant; they increase evaporative water loss by applying saliva to the fur and bare surfaces on the tail and scrotum. Saliva grooming and sweating achieve the same result of moistening skin to increase evaporative water loss but by completely different mechanisms of action. Normal grooming behavior accounts for about 7 to 8% of the total water loss by evaporation in rats maintained at room temperature. When exposed to acute heat stress, rats increase grooming behavior and are able to dissipate at least 90% of their total heat production by evaporation. This behavior can be maintained for several hours until dehydration ensues. Grooming to dissipate heat is an interesting integration of an autonomic and behavioral response. That is, to effectively groom, parasympathetic stimulation of salivary glands must occur concomitantly with grooming behavior in order to maintain a steady rate of evaporation. Moreover, there is similarity in the thermal stimulation of evaporative water loss in the rat and human. Thermal stimulation of salivation, like sweating, is under the control of cholinergic neurons. Salivary glands and sweat glands are activated by muscarinic cholinergic receptors and can be stimulated upon exposure to anticholinesterase agents. Pigs neither sweat nor pant but do wallow in mud when exposed to heat stress. Wallowing is akin to saliva spreading in that fluid is applied to the skin to maintain an effective rate of evaporative cooling when heat stressed.

### *2.7.4 Behavioral Thermoregulatory Effectors*

Considering all homeostatic processes that are regulated by the autonomic nervous system, one cannot fail to note that thermoregulation is unique in that it relies heavily on behavior as a predominant means of achieving regulation. The conscious sensing of temperature and utilization of behavioral mechanisms to create an optimal thermal environment are in nearly constant operation for humans and other species. The devotion of news media coverage to weather forecasts bears witness to the preoccupation that humans have with behavioral temperature regulation. Behavior certainly comes into play for other autonomic systems, but these responses are periodic and are usually activated when the parameter under control is subjected to a marked deviation. The sensations of hunger and thirst, for example, are behaviors that serve to regulate blood glucose and osmolarity, respectively.

Behavioral thermoregulation can be grouped into two categories: (1) natural, or the inherent behaviors displayed without the need for specialized apparatus, and (2) instrumental, which are the behaviors observed only with the use of specialized apparatus such as a temperature gradient or operant system. Natural thermoregulatory behaviors in rodents and other species are readily observed but difficult to quantify in a toxicological study. Individual animals assume a ball shape to restrict heat loss in the cold and stretch out to increase their surface area to dissipate heat. When housed in groups, rodents use natural behaviors to effectively control heat loss by huddling together in the cold and spreading out in the heat. Other more complex natural behaviors can provide an indication of the degree of thermal stress. For example, food hoarding and nest building are behaviors that can be affected by the relative amount of cold stress. Natural behaviors are especially crucial in developing rats and mice, which do not display homeothermy until approximately 14 days of age.

An array of laboratory instruments have been developed to quantify thermoregulatory behavior in rodents and other species (Gordon and Refinetti, 1993). Two principal devices, temperature gradients and operant systems, have been used to assess the effect of drugs and toxicants on temperature regulation. Each system has its advantages and disadvantages. The temperature gradient provides an environment that is conducive to natural thermoregulatory behavior to seek out an ideal thermopreferendum (Figure 2.11). The behavior can be maintained with little muscular effort, and this can be important in situations where a drug or toxicant may impair muscular activity. Temperature gradients are simple, large devices that do not allow for easy access for manipulating the animal. Very little training is needed for an animal in a temperature gradient, although it should be noted that rats require at least 6 h to adapt to a gradient, whereas mice adapt in less than 1 h (Gordon et al., 1991a). Operant systems provide a precise quantitative measure of a thermoregulatory behavior such as number of reinforcements per unit time, but the animals have to be well trained to use this system. Operant chambers are smaller and allow for easier access to manipulate the animals. Animals in operant systems must continually perform a motor task, and this may interfere with the true thermoregulatory behavior if the toxicant affects muscular function.

Animals exhibit thermotropic behavior when placed in a continuum of temperatures and will select a preferred temperature (i.e., thermopref-

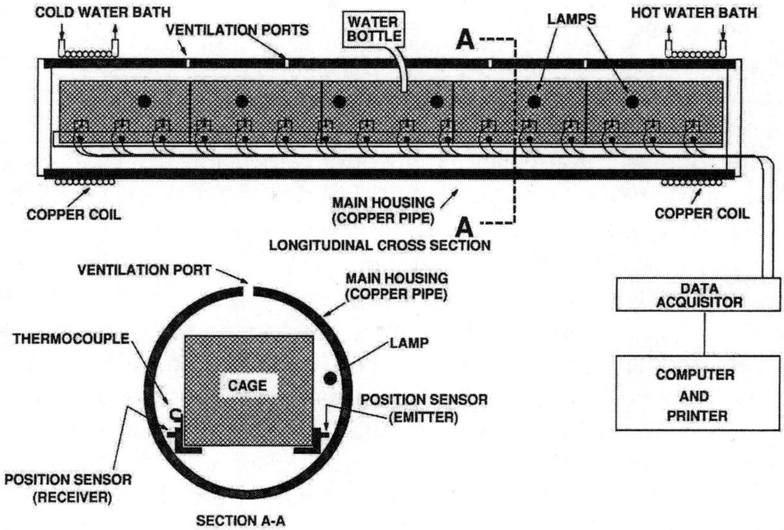


**Figure 2.11 Diagram of temperature gradient used to monitor selected ambient temperature in unrestrained rodent. (Modified from Gordon, C.J., Fogelson, L., Lee, L., and Highfill, J. (1991a). *Toxicology* 67: 1–14.)**

erendum or selected temperature). The thermopreferendum can refer to the skin, core, or substrate temperature. For homeotherms with a stable core temperature over a wide range of ambient temperature, the thermopreferendum is expressed in terms of the preferred or selected ambient temperature. The selected ambient temperatures of rodents and other homeotherms generally coincide with the thermoneutral zone (Table 2.3). That is, temperatures are selected that minimize energy expenditure. Many behavioral studies of rats suggest that they prefer temperatures well below thermoneutrality, but this is most likely attributed to improper study design where the rats were not allowed sufficient time to adapt to the novel environment of a temperature gradient. In fact, when allowed sufficient adaptation in a temperature gradient, rats display a clear circadian rhythm of selected temperature that is approximately 180° out of phase with the rhythm of core temperature (Figure 2.12). During the daytime when rodents are inactive and usually sleeping, they select temperatures of approximately 28 to 30°C. As the nocturnal phase approaches, core temperature undergoes a gradual then abrupt increase. Selected temperature falls from about 28 to 24°C and remains at cooler levels throughout the night. It is interesting to note that the selected temperature of the rat and other rodents is considerably higher than that of most animal vivariums (e.g., 22°C).

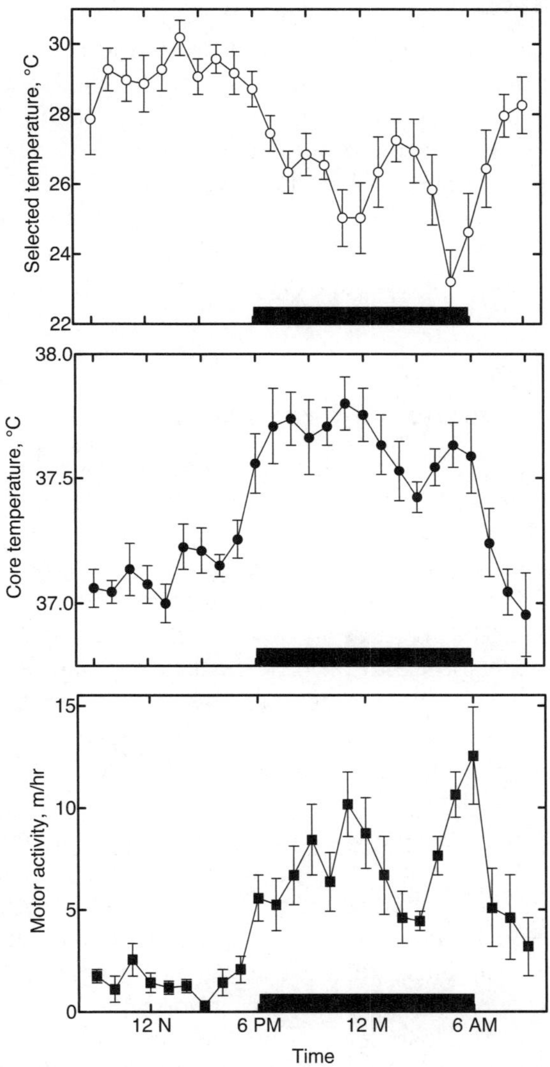


**Figure 2.12 Time-course of selected ambient temperature, core temperature, heart rate, and motor activity over a 24-h period in the unrestrained rat housed in a temperature gradient. (Data modified from Gordon, C.J. (1994).** ***Am. J. Physiol.*** **36: R71–R77.)**

Operant thermoregulatory systems are set up to motivate the rat to work for a reward of thermal reinforcement (Gordon and Refinetti, 1993). This is often done by housing the rat (generally shaved) in a cold chamber and training it to press a bar that activates a heat lamp. When housed in hot environments, rats can be trained to bar press for a jet of cool air or water spray. Operant thermoregulatory behavior in rats was used extensively in the past to assess the effects of drugs and a few toxic chemicals (see Chapter 3). The operant system is more amenable to the study of the effects of brain temperature and other manipulations on thermoregulatory behavior. For example, artificially warming the POAH area of a rat in an operant chamber will lead to a reduction in heat reinforcements; cooling the POAH accelerates heat reinforcements in the cold but attenuates cold reinforcements in a hot environment. Such an approach could be used to dissect the peripheral from the central thermoregulatory effects of a toxicant or drug.

## 2.7.5 Motor Activity: A Thermoeffector?

There is a misconception among some who are not trained in thermal physiology that rodents rely on their activity to generate heat for thermoregulation. This notion probably arose from the results of studies showing that wheel running activity and gross motor movement increased in magnitude as ambient temperature was reduced (for review, see Gordon, 1990, 1993). The increase in motor activity was considered to be a thermoregulatory response to produce heat to regulate body temperature in the cold. Toxicological studies may have reinforced the concept that motor activity is a thermoeffector because rodents were found to be inactive and hypothermic following acute exposure to a variety of chemicals.

The development of radiotelemetry to monitor temperature and motor activity has demonstrated that core temperature is influenced but not regulated by motor activity (Gordon and Yang, 1997; Honma and Hiroshige, 1978). The circadian temperature rhythm of the rat serves as an ideal example to illustrate the role of motor activity in temperature regulation. It is well known that core temperature and motor activity increase at night. The parallel waxing and waning of core temperature and motor activity would lead one to conclude that the heat produced from motor activity is responsible for the elevated core temperature at night. However, in a regression analysis of motor activity versus core temperature, Honma and Hiroshige (1978) showed that core temperature was elevated at night even when motor activity was equal to zero. Bursts in motor activity were directly correlated with transient elevations in core temperature in the rat, but this was not markedly dependent on ambient temperature (Gordon and Yang, 1997). Since motor activity is a frequently measured parameter in behavioral neurotoxicological evaluations, a better understanding of how an animal's activity may affect body temperature and related physiological processes is called for. Overall, the evidence suggests that motor activity is not a thermoeffector for basal thermoregulatory processes, but there is little known about whether motor activity has a thermoeffector function in animals exposed to toxicants.

## 2.8 POIKILOTHERMS

With few exceptions, fish, amphibians, reptiles, and invertebrates are poikilothermic. Their body temperature is essentially equal to ambient temperature; however, many poikilothermic species use their behavior to maintain a core temperature that is independent of ambient temperature. Poikilothermic vertebrates utilize neural networks of warm- and cold-sensitive neurons that are homologous to the pathways depicted in Figure 2.4A. Both peripheral and central thermal receptors have been identified in many species of poikilotherms.

Reptiles and amphibians are ectothermic and use behavior to regulate the absorption of solar radiation. Moving from sun to shade or modulating the amount of surface area exposed directly to the sun are effective mechanisms to control their body temperature (Precht et al., 1973; Prosser, 1973). In water, conductive heat exchange is approximately 23 times greater than in air. Hence, the body temperature of bradymetabolic fish and other species rapidly equilibrates with the water temperature, and there is generally little difference between the internal and external temperatures. Nonetheless, fish and other aquatic species use their behavior and seek water temperatures associated with their thermopreferendum (see Chapter 8). Some tadpoles have pigmented surfaces that absorb solar radiation. By orienting their position in shallow water, they can take advantage of solar radiation to thermoregulate by ectothermy. Endothermic capability is seen in some vertebrates, allowing them to display varying degrees of homeothermy. There are some fish (some species of tuna and shark) that possess counter-current heat exchange mechanisms that retain the heat produced by contracting muscles, allowing for the maintenance of an internal temperature that is warmer than that of the ambient water. Some reptiles such as the python are able to use muscular contractions to increase temperature above ambient temperature, a response that is critical in the incubation of a clutch of eggs.

Insects are thermotropic and can position themselves in a temperature gradient or orient to solar radiation to regulate their temperature. Some of the relatively large species of insects are endothermic. An abbreviated list of known endothermic insects includes species of wasps and bees, katydids, some moths, beetles, and butterflies (Heath and Heath, 1982). These species generate and conserve heat during flight and possess mechanisms of warming up prior to flight and during singing. This involves a combination of mechanisms of producing heat with muscles in the thorax and conserving heat loss. For example, the honeybee can maintain at least a 10°C difference between its thorax and the ambient temperature (cf. Figure 8.12).

# Chapter 3

# Acute Toxic Thermoregulatory Responses

## 3.1 INTRODUCTION

The regulation of normal body temperature is dependent on a balance between the organism's heat production and total sources of heat loss. This balance is maintained by an array of regulatory mechanisms, and homeothermic organisms are generally well adapted to maintain a normal core temperature in the face of marked changes in environmental temperature. A toxicant or drug will impart a change in body temperature when there is altered function of one or more components of the thermal homeostatic feedback loop (see Figure 2.3), including the thermal sensors, CNS integration and control, and thermoeffector function.

The majority of the studies on toxicology and temperature regulation involve short-term, acute responses in laboratory rodents. This chapter focuses on the integrative thermoregulatory responses to toxic insult, including the acute response of laboratory rodents, other mammals, and humans. In many toxicological studies, only body temperature was recorded, with no mention of how the changes in temperature were mediated. This chapter endeavors to explain the integrative thermoregulatory responses, meaning the homeostatic processes that are involved in mediating a change in body temperature. Toxicant-induced changes in body temperature that occur over several days, including febrile responses, are covered in Chapter 6.

## 3.2 GENERAL MECHANISMS

Whether a toxicant will cause an increase or decrease or have no effect on body temperature will depend largely on ambient temperature and the thermoeffector systems affected (Table 3.1). The potential changes in temperature, indicated by the number of arrows in each block of Table 3.1, is dependent on whether the animal is housed in a thermoneutral or relatively cool or warm environment. For example, a toxicant that blocks metabolic thermogenesis will manifest the most effective change in body temperature when exposure occurs at a relatively cool ambient temperature. A toxicant that induces peripheral vasoconstriction, thus restricting blood flow and heat loss from the skin, will have relatively minor effects in a cold environment because the animal is in a state of peripheral vasoconstriction but would lead to hyperthermia in a thermoneutral or warm environment. However, an agent that causes peripheral vasodilation would be mostly ineffective in a warm environment because skin blood flow is already elevated and an additional vasodilatory action should have little effect on total heat loss. A block of salivation in rodents or sweating in humans would have little effect in the cold but would lead to dramatic hyperthermia if the blocking agents are administered in a warm environment.

If the toxicant impairs one thermoeffector without affecting CNS thermoregulatory control, then one would expect the animal to utilize other thermoeffectors to maintain thermal homeostasis. For example, if skin blood flow was stimulated in a cold environment, then metabolic thermogenesis could increase to counter the increased heat loss. Stimulation of metabolism in a warm environment such as occurs by exposure to uncoupling agents is accompanied by a marked increase in evaporation (see Chapter 6). Overall, these are idealized situations, and toxicants generally affect the function of more than one thermoeffector system.

In addition to ambient temperature, it is also important to consider the species when attempting to predict how a change in thermoeffector function will affect the control of body temperature. For example, species such as rodents with a relatively small body mass and large surface area to body mass ratios rely mostly on metabolic thermogenesis to thermoregulate, whereas peripheral vasomotor tone becomes more critical in species with large body mass (see Chapter 2). Hence, one must be cautious in extrapolating a potential thermoregulatory effect from rodent to human or human to rodent. That is, an agent that modulates vasomotor tone in humans may have little effect on the thermoregulation of a rodent. Of course, many other species-specific differences can hamper the extrapolation of toxicological data (see Chapter 5).

**Table 3.1 Relative Hyperthermic (↑) and Hypothermic Effects (↓) of a Toxicant or Drug that Stimulates or Blocks Thermoeffectors at Cool, Thermoneutral, and Warm Ambient Temperatures[a]**

| | *Thermogenesis (increased)* | *Skin Blood Flow (increased)* | *Evaporation (increased)* | *Thermogenesis (blocked)* | *Skin Blood Flow (blocked)* | *Evaporation (blocked)* |
|---|---|---|---|---|---|---|
| Cool temperature | ↑ | ↓↓ | ↓ | ↓↓↓ | | |
| Thermoneutral temperature | ↑↑↑ | ↓ | | ↓↓ | ↑↑↑ | ↑ |
| Warm temperature | ↑↑↑↑ | | | ↓ | ↑↑↑↑ | ↑↑↑↑ |

[a]Number of arrows indicates relative magnitude of change in core temperature.

## 3.3 METHODS FOR MONITORING BODY TEMPERATURE

Until recently, investigations into the thermoregulatory effects of toxicants were made by measuring colonic or rectal temperature. This technique is accurate for single time point measurements of core temperature of species such as the rat and mouse. However, the stress from handling, repeated measurement, and/or restraint of animals to repeatedly measure colonic temperature poses a myriad of problems (see Chapter 7).

Measuring colonic temperature in a rodent imparts stress, leading to an elevation in temperature that can persist for several hours after the initial measurement. Such results obscure the true effects of a toxicant on body temperature. These methods also limit the sensitivity of body temperature as a biomarker, meaning that higher doses of toxicant are required to induce a detectable effect.

The advent of radiotelemetry has revolutionized the study of thermoregulation and other fields of physiology (Figure 3.1A). With telemetry, relatively small radiotransmitters with thermosensors are implanted into the abdominal cavity at least one week prior to testing. The temperature of the abdominal cavity is an accurate representation of the core body temperature. In the example shown in Figure 3.1B, the core temperature is continuously monitored in awake, unrestrained rats for one day before and for several days after oral administration of two doses of the organophosphate insecticide chlorpyrifos. Telemetry allows one to clearly observe the hypothermic response as well as the subtle elevation in daytime temperature during recovery from toxicant exposure. Such responses would not be apparent in rodents subjected to repeated colonic probing or restraint. The acute hypothermic response to this insecticide and other toxicants is discussed in more detail below, and the delayed hyperthermia is discussed in Chapter 6.

## 3.4 HYPOTHERMIA: A COMMON RESPONSE IN RODENTS

Hypothermia is the most frequently observed thermoregulatory response of mice, rats, and other relatively small mammals when they are administered acute doses of xenobiotics. The hypothermic response is discussed here in general terms and then in more detail with discussions of specific categories of toxicants.

Hypothermic responses to toxicants in experimental rodents were observed in the early part of the 20th century (for reviews of early

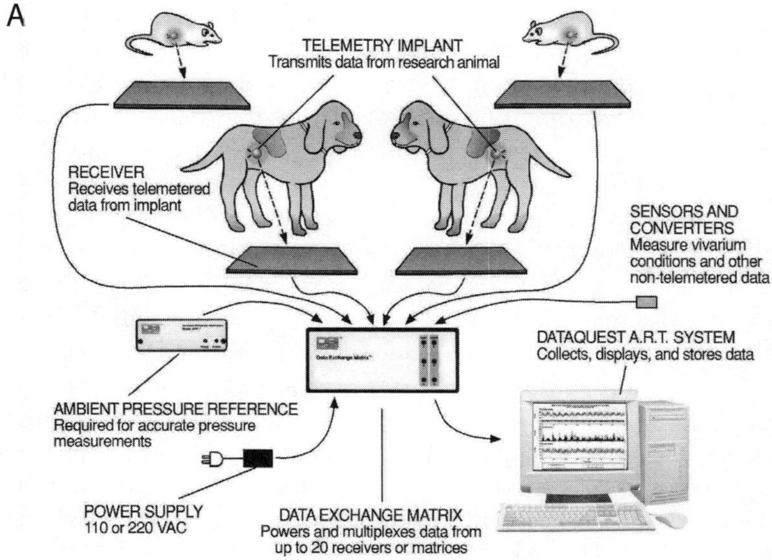


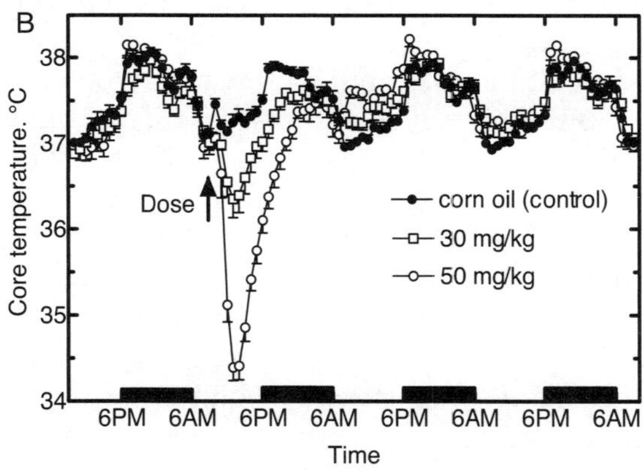


**Figure 3.1 (A) Diagram of a radiotelemetry system used to monitor body temperature, blood pressure, and other physiological parameters in rodents and other species. Drawing courtesy of Data Sciences International, St. Paul, MN. (B) Example of recording of body temperature from unrestrained and awake Long-Evans rats (males) before and after dosing with corn oil vehicle and two doses of the organophosphate insecticide chlorpyrifos. (Data modified from Gordon, C.J. and Mack, C.M. (2001). *Toxicology* 169: 93–105.)**

literature, see Fuhrman, 1946; Doull, 1972; Gordon et al., 1988). In many of these older studies, it was found that the hypothermic efficacy of a toxicant or drug was dependent on dose, route of exposure, species, and especially ambient temperature. Thermal stability of small rodents at temperatures below the thermoneutral zone is dependent in large part on the maintenance of a high metabolic rate. Healthy rodents are generally capable of maintaining a constant body temperature at relatively cold ambient temperatures for long periods provided they are given an adequate supply of food. Any agent that impairs the ability to maintain a constant rate of heat production will result in a rapid reduction in core temperature.

The hypothermic efficacy of a toxicant is generally proportional to the difference between the animal's lower critical temperature and ambient temperature of exposure. In the examples shown in Figure 3.2, which typify the responses to structurally diverse toxicants (i.e., metals, solvents, and anticholinesterase agents), exposure leads to rapid (i.e., 1 to 2 h after treatment) hypothermic response at temperatures below thermoneutrality. Occasionally, a hyperthermic response is observed when toxicants are administered at ambient temperatures above thermoneutrality. It is also important to note that, because of their smaller size, mice are more labile and generally show a greater decrease in body temperature. The majority of these toxicological studies have been performed at relatively cool temperatures because hypothermic effects were more readily detectable in rodents using the conventional technology of colonic probes.

It was thought that the toxicant-induced hypothermia was attributable to a dysfunction in thermoregulatory control. That is, the rodent exposed to the toxicant was considered to have sustained damage to its CNS control of body temperature or thermoeffectors responsible for heat generation and heat retention. This conclusion was based on the simple fact that toxicant-induced hypothermia was greater and more prolonged when exposures occurred at cool ambient temperature. If the toxicant simply blocked metabolic thermogenesis, then it would have been reasonable to expect a hypothermic response that was proportional to the decrease in ambient temperature (Table 3.1). However, it has been recognized over the past 20 years that the hypothermic response to many toxicants is mediated by an integrated thermoregulatory response of behavioral and autonomic thermoeffectors.

## 3.5 THERMOREGULATORY RESPONSE TO TOXICANTS

Most of what is known about the toxicology of thermoregulation is based on data from studies on pesticides, solvents, and other toxic chemicals that are used in agricultural and manufacturing applications. Discussed in

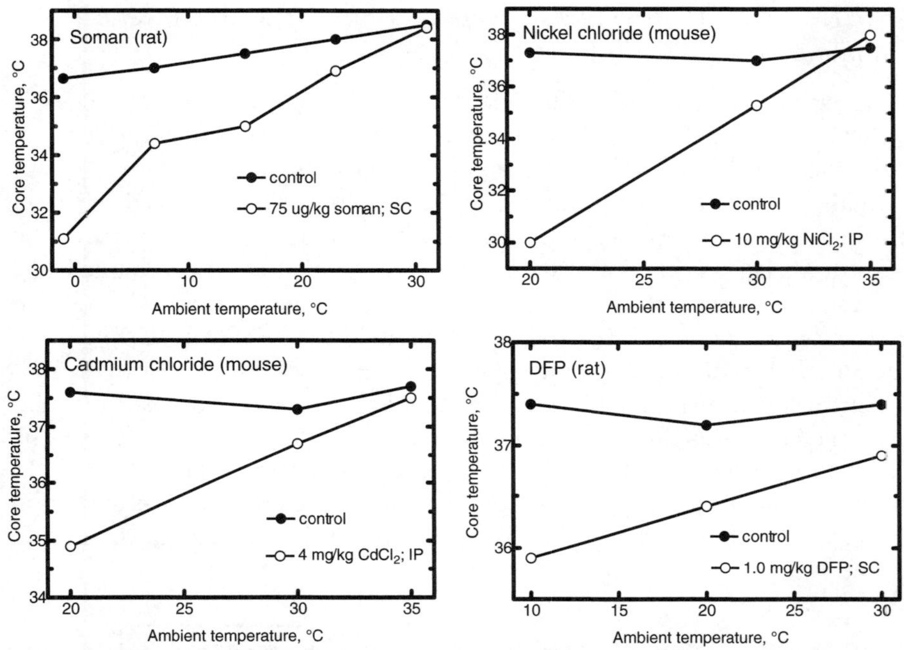


**Figure 3.2 Examples of how ambient temperature affects the thermoregulatory efficacy of a variety of toxicants, including soman (Wheeler, 1989), nickel chloride, cadmium chloride (Gordon and Stead, 1986), and diisopropyl fluorophosphate (DFP) (Gordon et al., 1991b). Core temperatures measured 1 to 2 h after exposure.**

the following sections are the thermoregulatory effects of selected groups of toxicants, including the anticholinesterase agents, chlorinated hydrocarbons, metals, airborne pollutants, and alcohols. While several of these toxicants have been banned from use for many years (e.g., DDT), it is nonetheless important to address the mechanisms of action of these compounds.

### *3.5.1 Anticholinesterase Agents*

The anticholinesterase (anti-ChE) agents were one of the first classes of chemicals to be studied for their specific effects on thermoregulation (Baetjer and Smith, 1956; Meeter, 1969; Meeter et al., 1971; Gordon, 1994). The anti-ChE insecticides, including organophosphate and carbamate compounds, are a major source of the pesticides used throughout the world. Understanding the toxicology of these agents is also important because of their use in the formulation of nerve gas agents with potential use as weapons of mass destruction.

### *3.5.1.1 Correlation between Hypothermia and Cholinesterase Inhibition*

The primary mechanism of toxicity of anti-ChE agents is the inhibition of acetylcholinesterase (AChE) activity. The anti-ChE agents that are pertinent to toxicology are the organophosphate- and carbamate-based compounds. Organophosphate agents form a covalent bond with the active site of AChE. The organophosphate–AChE bond is considered irreversible but can dissociate slowly with time. Carbamate agents form a reversible bond. In theory, a toxicological effect such as a change in body temperature will occur when AChE activity is inhibited to the level at which the accumulation of acetylcholine in synapses exceeds the rate of hydrolysis. With sufficient accumulation of acetylcholine, there is stimulation of cholinergic synapses in the peripheral and central nervous systems. This overstimulation of cholinergic pathways leads to a variety of acute sequelae that are characteristic of cholinergic poisoning, including excess salivation, reduced forelimb strength, tremor, miosis, and reduced motor activity (Figure 3.3).

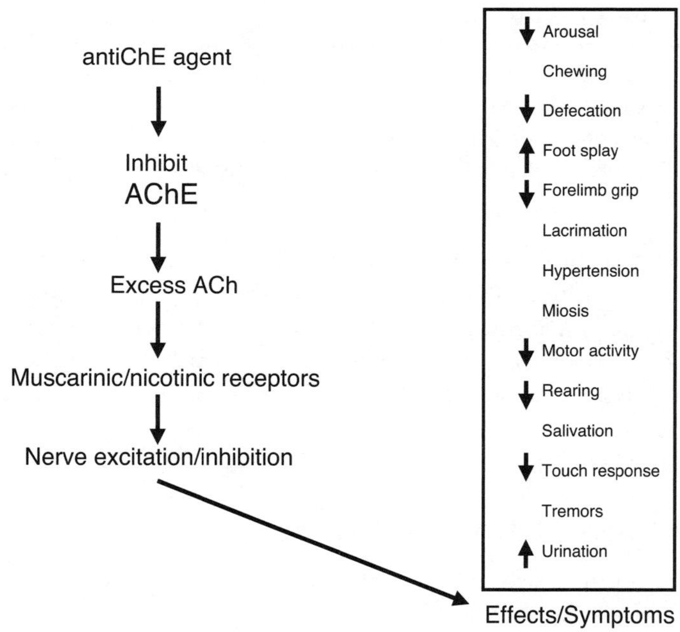


**Figure 3.3 General mechanism of action of anticholinesterase agents. Inhibition in acetylcholinesterase (AChE) activity leads to cholinergic stimulation. Stimulation of heat loss pathways leads to hypothermia (see Gordon, 1994; Ballantyne and Marrs, 1992).**

Hypothermia is a well-known benchmark of toxicity in rodents exposed to anti-ChE–based insecticides (Ballantyne and Marrs, 1992; Moser, 1995). For example, in a summary of studies on the thermoregulatory effects of anti-ChE agents (Gordon, 1994), 22 of 25 studies on organophosphates listed a hypothermic response within 24 h after exposure. Likewise, among the studies of carbamates, hypothermia was reported in 15 of 16 studies (also see Tables 3.2 and 3.3). Compared to many of the sequelae of cholinergic stimulation listed in Figure 3.3, hypothermia is an ideal toxicological parameter of anti-ChE exposure because it can be quantified in terms of a magnitude of change as well as a duration of effect. The other sequelae of the cholinergic crisis listed are not as easy to measure in a quantified manner in the undisturbed animal. The absolute change in core temperature is generally proportional to the magnitude of the dose of a toxicant, but the temperature change between doses at a given time point may be indiscernible. However, the integration of the change in temperature with time, termed the temperature index, is an ideal means of quantifying the thermoregulatory effects of anti-ChE and other toxic agents (Clement, 1991; Gordon and Mack, 2001).

During the early stages of acute exposure, the inhibition of AChE is proportional to the degree of cholinergic stimulation. The human health risk assessment of anti-ChE insecticides has utilized the inhibition in plasma or serum AChE activity as a threshold to limit exposure to these insecticides. To this end, it is important to characterize the threshold inhibition in AChE activity in the brain and peripheral tissues that is associated with a physiological change such as hypothermia. In a survey of five studies of the rat exposed to various organophosphate agents, an inhibition in brain AChE activity of 82.5% was associated with a −3.0°C reduction in core temperature (Gordon and Fogelson, 1993). Clement (1991) showed that the appearance of hypothermia in mice treated with sarin occurred when AChE inhibition in the hypothalamus exceeded 52%. In fact, it appears that for rodents tested at room temperature, an inhibition in brain AChE of 50% was the approximate threshold for eliciting a hypothermic effect (Gordon, 1994). Interestingly, carbamates lower core temperature with less inhibition in brain AChE activity as compared to organophosphates. For example, a 22% inhibition in brain AChE activity is associated with a 3.5°C reduction in core temperature measured 60 min after administration of physostigmine (Maickel et al., 1991). However, in nearly all of these studies, core temperature was measured with colonic probes at standard room temperature.

While the inhibition of brain AChE activity is the crucial facet of the neurotoxic effects of anti-ChEs, such a measurement is obviously impossible to make in human subjects, and indirect methods must be utilized.

**Table 3.2 Effects of Acute Exposure to Organophosphate Insecticides and Related Agents on Body Temperature[a]**

| *Species* | *Chemical* | *Dose* | *$T_a$, °C* | *Time (min)* | *$\Delta T_c$, °C* | *References* |
|---|---|---|---|---|---|---|
| Mouse | Disulfoton | 10 mg/kg (IP) | 25 | 240 | –2.0 | Costa & Murphy, 1983 |
| Mouse | Parathion | 40 mg/kg (IP) | 1 | 60 | –2.9 | Ahdaya et al., 1976 |
| Mouse | Parathion | 40 mg/kg (IP) | 27 | 60 | NC | Ahdaya et al., 1976 |
| Mouse | Sarin | 130 μg/kg (SC) | 22 | 180 | –7.9 | Clement, 1991 |
| Rabbit | Chlorphenvinphos | 33 mg/kg (IP) | RT | 180 | –1.7 | Gralewicz & Socko, 1997 |
| Rat | Chlorpyrifos | 30 mg/kg (PO) | 22 | 240 | –0.9 | Gordon & Mack, 2001 |
| Rat | Chlorpyrifos | 10 mg/kg/day (PO) | 22 | 5 days | –1.0 | Maurissen et al., 2000 |
| Rat | Diazinon | 200 mg/kg (PO) | 22 | 240 | –0.9 | Gordon & Mack, 2003 |
| Rat (f) | Diazinon | 200 mg/kg (PO) | 22 | 180 | –0.8 | Gordon & Mack, 2003 |
| Rat | Diisopropyl flurophosphate | 1.0 mg/kg (SC) | 20 | 120 | –0.8 | Gordon et al., 1991 |
| Rat | Fenthion | 75 mg/kg (PO) | 22 | 90 | –1.6 | Moser, 1995 |
| Rat | Paraoxon | 1.0 mg/kg (IP) | 22 | 240 | –1.9 | Coudray-Lucas et al., 1981 |
| Rat | Parathion | 7.0 mg/kg (PO) | 22 | 120 | –0.8 | Moser, 1995 |
| Rat | Soman | 125 μg/kg (SC) | 23 | 120 | –2.9 | Wheeler, 1989 |
| Rat | Tributyl-phosphoro-trithioate (DEF) | 200 mg/kg (IP) | 22 | 150 | –4.7 | Ray, 1980 |
| Rhesus monkey | Sarin | 30.6 μg/kg (IM) | 25 | 240 | –1.7 | Craig et al., 1959 |

[a]Data in this and following tables give the time (t) to reach a maximal change in core temperature ($\Delta T_c$) for a given dose, ambient temperature ($T_a$), and route of exposure. Parentheses after dose indicate route of exposure (PO, oral; IP, intraperitoneal; SC, subcutaneous; IV, intravenous); RT, room temperature; NC, no change.

**Table 3.3 Effects of Acute Exposure to Carbamate Insecticides and Related Agents on Body Temperature**

| *Species* | *Chemical* | *Dose*[a] | $T_a$, °C[b] | *Time (min)* | $\Delta T_c$, °C | *References* |
|---|---|---|---|---|---|---|
| Guinea pig | Physostigmine | 0.24 mg/kg/hr (SC) | 23 | 24 h | −1.8 | Lim et al., 1989 |
| Mouse | Carbaryl | 50 mg/kg (IP) | 27 | 60 | −0.9 | Ahdaya et al., 1976 |
| Mouse | Physostigmine | 0.4 mg/kg (IP) | 22 | 30 | −4.2 | Bhat et al., 1990 |
| Mouse | Propoxur | 10 mg/kg (SC) | 23 | 60 | −1.8 | Kobayashi et al., 1988 |
| Patas monkey | Physostigmine | 0.4 mg/kg (PO) (every 30 min for 3 h) | 35 | 120 | −0.4 | Avlonitou & Elizondo, 1988 |
| Pig | Physostigmine | 5 µg/kg/min (IA) | RT | 60 | 0.5 | Stemler et al., 1990 |
| Rat | Aldicarb | 0.7 mg/kg (PO) | 22 | 90 | −1.5 | Moser, 1995 |
| Rat | Carbaryl | 75 mg/kg (PO) | 22 | 180 | −2.6 | Gordon & Mack, 2001 |
| Rat | Methomyl | 5 mg/kg (SC) | 22 | 60 | −2.0 | Gupta et al., 1994 |
| Rat | Physostigmine | 0.5 mg/kg (SC) | 23 | 60 | −2.1 | Maickel et al., 1988 |
| Squirrel monkey | Physostigmine | 8.0 mg/kg (PO) | RT | 150 | −1.6 | Rupniak et al., 1992 |

[a]Parentheses after dose indicate route of exposure (PO, oral; IP, intraperitoneal; SC, subcutaneous; IV, intravenous).
[b]RT, room temperature.

The inhibition of serum or plasma ChE activity is often used as an index of potential inhibition of brain AChE activity in humans who may be exposed to anti-ChE. In the rat exposed to DFP, a significant hypothermic response was associated with an inhibition in serum ChE activity of 54% (Gordon and Fogelson, 1993). With other organophosphates, plasma or serum ChE activity can plummet to below 25% of normal before one sees a significant hypothermic effect. A 5-min exposure to vapors of DFP in the mouse causes core temperature to dip by over 4°C within 30 min, and it remains depressed for approximately 4 h (Scimeca et al., 1985). This was associated with a 66% inhibition of brain AChE activity that remained depressed in spite of a full recovery of core temperature and motor coordination.

One endeavor of risk assessment is to identify the threshold for a toxicological effect such as hypothermia. It would seem that the threshold AChE inhibition for induction of hypothermia could be reduced if the animals were subjected to colder temperatures and more sensitive methods were used to monitor core temperature (i.e., telemetry). Most rodent studies are performed at ambient temperatures that are comfortable for humans (~22°C) but happen to be approximately 6°C below the rat's lower critical temperature. In the data listed in Tables 3.2 through 3.8, the doses for a given decrease in core temperature would most likely be reduced if the animals were exposed to colder temperatures when exposed to the toxicants.

#### *3.5.1.2 Integrated Thermoregulatory Responses*

Heat loss thermoeffector pathways in rodents are driven with muscarinic pathways located in CNS thermoregulatory centers. Stimulating these pathways, either by CNS or systemic injections of cholinomimetic agents or by administering anti-ChE agents, leads to a stimulation of the heat loss pathways and a hypothermic response. Meeter and colleagues first determined that the control of skin blood flow plays a critical role in the development of hypothermia following exposure to organophosphates (Meeter, 1969). The fall in body temperature following exposure to organophosphates such as sarin, DFP, and chlorpyrifos occurred concomitantly with an increase in tail skin temperature of the rat (Meeter, 1969; Gordon and Fogelson, 1993; Gordon et al., 2002). The tail of the rat is a critical site for the regulation of dry heat loss (see Chapter 2).

The effects of organophosphates on vasomotor control are integrated with other thermoeffector systems and are tightly associated with the prevailing ambient temperature. At a standard laboratory temperature of 22 to 24°C, DFP elicited a marked elevation in tail skin temperature and a moderate decrease in metabolic rate. However, at a colder temperature

**Table 3.4 Effects of Acute Exposure to DDT and Other Chlorinate Hydrocarbons on Body Temperature**

| *Species* | *Chlorinated Hydrocarbon* | *Dose*[a] | $T_a$, °C[b] | *Time* | $\Delta T_c$, °C | *References* |
|---|---|---|---|---|---|---|
| Mouse (f) | DDT | 1250 mg/kg (IP) | 27 | 24 h | –0.8 | Ahdaya et al., 1976 |
| Rat | α-chlordane | 300 mg/kg (PO) | 24 | 4 h | –3.4 | Hrdina et al., 1974 |
| Rat | α-chlordane | 5 mg/kg/d (IM) | RT | 45 days | 1.1 | Hrdina et al., 1972 |
| Rat | Chlordecone | 50 mg/kg (IP) | 22 | 3 h | –1.2 | Cook et al., 1987 |
| Rat | Chlordecone | 75 mg/kg (PO) | 23 | 4.5 h | –1.7 | Swanson & Woolly, 1982 |
| Rat | p,p = DDT | 75 mg/kg (PO) | 22 | 12 h | 1.3 | Hudson et al., 1985 |
| Rat | p,p = DDT | 60 mg/kg (PO) | 22 | 24 h | 0.6 | McDaniel & Moser, 1997 |
| Rat | p,p = DDT | 5 mg/kg/day (IM) | RT | 45 days | 0.5 | Hrdina et al., 1972 |
| Rat | p,p = DDT | 25 mg/kg/day (IM) | RT | 45 days | 1.3 | Hrdina et al., 1972 |
| Rat | Dieldrin | 50 mg/kg (PO) | 23 | 4–6 h | –2.6 | Swanson & Woolly, 1982 |

[a]Parentheses after dose indicate route of exposure (PO, oral; IP, intraperitoneal; IM, intramuscular).
[b]RT, room temperature.

**Table 3.5 Effect of Acute Exposure to Inhaled and Intrathecally Instilled Toxicants on Body Temperature**

| *Species* | *Toxicant* | *Dose*[a] | $T_a$, °C[b] | *Time* | $\Delta T_c$, °C | *References* |
|---|---|---|---|---|---|---|
| Guinea pig | Diphenylmethane-4,4'-diisocyanate | 7.8 ppm | 25 | 3 h | –3.0 | Thorne et al., 1987 |
| Mouse | Chloroform | 2,000 ppm | 24 | 3.5 h | –4.0 | Gearhart et al., 1993 |
| Mouse | DFP | Vapors | 21–24 | 0.5 h | –4.4 | Scimeca et al., 1985 |
| Mouse | Formalin | 15 ppm | RT | 2 h | –3.1 | Jaeger & Gearhart, 1982 |
| Mouse | Ozone | 2 ppm | 22 | 2 h | –6.7 | Watkinson et al., 1996 |
| Rat | Carbon monoxide | 810 ppm | RT | 4 h | –1.3 | Mullin & Krivanek, 1982 |
| Rat | Ethanol | 14,900 ppm | RT | 4 h | –1.1 | Mullin & Krivanek, 1982 |
| Rat | Formalin | 15 ppm | RT | 2 h | –0.5 | Jaeger & Gearhart, 1982 |
| Rat | Ozone | 1.0 ppm | 18–20 | 2 h | –3.6 | Watkinson et al., 1993 |
| | | 1.0 ppm | 30–32 | 2 h | –0.9 | |
| Rat | Ozone | 0.8 ppm | 23 | 3 h | –1.3 | Mautz & Bufalino, 1989 |
| Rat | ROFA* | 2.5 mg (IT) | 22 | 2.6 h | –2.2 | Campen et al., 2000 |
| | | 2.5 mg (IT) | 10 | 2.6 h | –3.5 | |
| Rat | Trichloroethane | 6,100 ppm | RT | 4 h | –1.6 | Mullin & Krivanek, 1982 |
| Rat | Toluene | 8,567 ppm | RT | 50 min | –1.0 | Rebert et al., 1989 |
| Rat | Toluene | 3,100 ppm | RT | 4 h | –2.0 | Mullin & Krivanek, 1982 |

[a]Parentheses after dose indicate route of exposure (PO, oral; IP, intraperitoneal; SC, subcutaneous; IV, intravenous).
[b]RT, room temperature.

*ROFA, residual fly oil ash particulate matter.

**Table 3.6 Effects of Acute Exposure to Metals and Related Agents on Body Temperature**

| *Species* | *Chemical Agent* | *Dose*[a] | $T_a$, °C[b] | *Time* | $\Delta T_c$, °C | *References* |
|---|---|---|---|---|---|---|
| Rabbit | Arsenic trioxide | 3 mg/kg (IV) | RT | 60 min | −0.6 | Kawaguchi & Tsutsumi, 1982 |
| Rat (f) | Arsenic trioxide | 6 mg/kg (IP) | 25 | 2 h | −2.7 | Norris & Elliott, 1945 |
| Rat | Nickel chloride | 250 µmol/kg (SC) | 20 | 1.5 h | −3.0 | Hopfer & Sunderman, 1988 |
| Rat | Nickel chloride | 2.0 mg/kg (IP) | 23 | 60 min | −2.1 | Watanabe et al., 1990 |
| Mouse | Cadmium chloride | 4.0 mg/kg (IP) | 20 | 60 min | −2.7 | Gordon & Stead, 1986 |
| Mouse | Cadmium chloride | 2.0 mg/kg (IP) | 22 | 30 min | −4.0 | Martinez et al., 1993 |
| Mouse | Cobaltous chloride | 25 mg/kg (IP) | 23 | 30 min | −3.7 | Burke, 1978 |
| | | 25 mg/kg (PO) | 23 | 30 min | −2.1 | |
| | | 25 mg/kg (SC) | 23 | 30 min | −2.4 | |
| | | 25 mg/kg (IV) | 23 | 30 min | −3.9 | |
| Mouse | Nickel chloride | 10 mg/kg (IP) | 20 | 60 min | −7.3 | Gordon & Stead, 1986 |
| Mouse | Sodium selenite | 30 umol/kg (SC) | 20 | 30 min | −3.8 | Watanabe & Suzuki, 1986 |
| Mouse | Lead acetate | 25 mg/kg (IP) | 22 | 30 min | −2.5 | Martinez et al., 1993 |
| Mouse | Triethyltin | 6 mg/kg (IP) | 20 | 60 min | −6.0 | Gordon et al., 1984 |

[a]Parentheses after dose indicate route of exposure (PO, oral; IP, intraperitoneal; SC, subcutaneous; IV, intravenous).
[b]RT, room temperature.

**Table 3.7 Effects of Acute Exposure to Alcohols and Selected Organic Solvents on Body Temperature**

| *Species* | *Chemical Agent* | *Dose*[a] | $T_a$, °C | *Time* | $\Delta T_c$, °C | *References* |
|---|---|---|---|---|---|---|
| Mouse | Ethanol | 3 g/kg (IP) | 20 | 60 min | –5.2 | Gordon & Stead, 1986a |
| | | 3 g/kg (IP) | 30 | 60 min | –2.2 | |
| Mouse | Sulfolane | 400 mg/kg (IP) | 20 | 60 min | –2.9 | Gordon et al., 1986 |
| Rabbit | Sulfolane | 400 mg/kg (IP) | 10 | 180 min | –0.5 | Mohler & Gordon, 1988 |
| Rabbit | Sulfolane | 1000 μg (IVT) | 15 | 120 min | 0.47 | Mohler & Gordon, 1989 |
| Rat | Ethanol | 4 g/kg (IP) | 21 | 5 h | –2.9 | Gallaher & Egner, 1987 |
| Rat | Ethanol | 4 g/kg (PO) | 8 | 60 min | –2.3 | Myers, 1981 |
| | | 4 g/kg (PO) | 22 | 60 min | –1.6 | |
| | | 4 g/kg (PO) | 36 | 60 min | 1.1 | |
| Rat | Carbon disulfide | 400 mg/kg (IP) | 22 | 60 min | –1.3 | Herr et al., 1992 |
| Rat | Sulfolane | 400 mg/kg (IP) | 15 | 60 min | –1.6 | Gordon et al., 1984a |
| | | 400 mg/kg (IP) | 25 | 60 min | –1.5 | |

[a]Parentheses after dose indicate route of exposure (PO, oral; IP, intraperitoneal; SC, subcutaneous; IV, intravenous).
[b]RT, room temperature.

**Table 3.8 Effects of Acute Exposure to Formamidines, Pyrethroids, and Miscellaneous Agents on Body Temperature**

| *Species* | *Chemical Agent* | *Dose*[a] | $T_a$, °C[b] | *Time* | $\Delta T_c$, °C | *References* |
|---|---|---|---|---|---|---|
| Mouse (f) | Paraquat | 30 mg/kg (IP) | 22 | 24 h | –2.8 | Cagen et al., 1976 |
| Dog | Amitraz | 100 mg/kg (PO) | RT | 390 min | –1.5 | Hugnet et al., 1996 |
| Rat | Amitraz | 100 mg/kg (IP) | 22 | 1 h | –2.2 | Moser, 1991 |
| Rat | Cismethrin | 20 mg/kg (PO) | 22 | 1–2 h | –1.1 | Gilbert et al., 1989 |
| | Deltamethrin | 10 mg/kg (PO) | 22 | 1–2 h | 0.7 | |
| | Fenvalerate | 20 mg/kg (PO) | 22 | 1–2 h | –0.4 | |
| Rat | Cypermethrin | 20 mg/kg (PO) | 22 | 90 min | 0.9 | McDaniel & Moser, 1993 |
| | Cypermethrin | 60 mg/kg (PO) | 22 | 180 min | –1.5 | |
| | Permethrin | 75 mg/kg (PO) | 22 | 120 min | 0.75 | |
| | Permethrin | 150 mg/kg (PO) | 22 | 240 min | 1.7 | |
| Mouse | Dinitrophenol | 25 mg/kg (SC) | 22 | 50 min | 1.3 | Anari & Renton, 1993 |
| Rat | Dinitrophenol | 6–12 mg/kg (IV) | RT | 35 min | 1.2 | Takehiro et al., 1979 |
| Rat | Dinitrophenol | 6 mg/kg x 4@20 min (IV) | 15 | 120 min | 1.1 | Saiki & Mortola, 1997 |
| Rat | Triadimefon | 300 mg/kg (IP) | 22 | 30 min | –1.6 | Moser & MacPhail, 1989 |
| Rat (f) | Triadimefon | 300 mg/kg (IP) | 22 | 30 min | –2.8 | Moser & MacPhail, 1989 |

[a]Parentheses after dose indicate route of exposure (PO, oral; IP, intraperitoneal; SC, subcutaneous; IV, intravenous).
[b]RT, room temperature.

of 10°C, DFP injection resulted in a reduction in tail skin temperature along with a marked reduction in metabolic rate and hypothermia. That is, the hypothermic response in the cold was mediated primarily by a decrease in heat production, and the rat appeared to restrict heat loss from the tail to prevent an excessive hypothermic response. At a thermoneutral temperature of 30°C, tail vasodilation was ineffective to dissipate much additional heat, and metabolic rate could not be lowered below basal levels. Hence, the rat was unable to lower body temperature as much as in the thermoneutral environment, and DFP elicited an increase in evaporative water loss, presumably as an additional measure to increase heat loss and lower body temperature. This illustrates the balance between three thermoeffectors to achieve a hypothermic response under a wide range of ambient temperatures (Gordon et al., 1991b).

The response of the autonomic thermoeffectors to anti-ChE chemicals suggests that these agents elicit a regulated hypothermic response (see Chapter 2). The mechanism of action of these and other toxicants on temperature regulation can be better understood if their behavioral thermoregulatory responses can be monitored before and after administration of the toxicant. Behavioral thermoregulatory responses corroborate the findings of Meeter et al. (1971) and clearly show that the set-point for temperature regulation is reduced in rats exposed to organophosphates. The time-course of selected ambient temperature and core temperature in the rat monitored by telemetry exemplifies the regulated hypothermic response induced by administration of chlorpyrifos (Figure 3.4). When dosed with a control vehicle (corn oil), there was a transient decrease in selected temperature that reflects a heat dissipatory response from the stress of handling and injection (Figure 3.4). When dosed with chlorpyrifos, selected ambient temperature decreased from 30 to 25°C and the behavioral response preceded a 2.5°C decrease in core temperature. At the nadir of the decrease in core temperature, selected temperature increased rapidly, a response that presumably facilitated the recovery of core temperature. It is important to note that in the temperature gradient the rat has the option of selecting ambient temperatures as warm as 36°C. If the rat simply moved to a temperature range that was slightly above the thermoneutral zone, the hypothermic effects of chlorpyrifos would have been blocked. Administration of the organophosphate DFP also induced an abrupt selection for cooler temperatures that occurred concomitantly with a decrease in core temperature (Gordon, 1994a, 1997). In fact, most of the toxicants to be discussed in this chapter elicit a regulated hypothermic response in rats and mice (Table 3.9).

Without information on the behavioral thermoregulatory response, one could only conclude that exposure to anti-ChEs as well as other toxicants induced dysfunction in thermoregulation and an impairment in defense

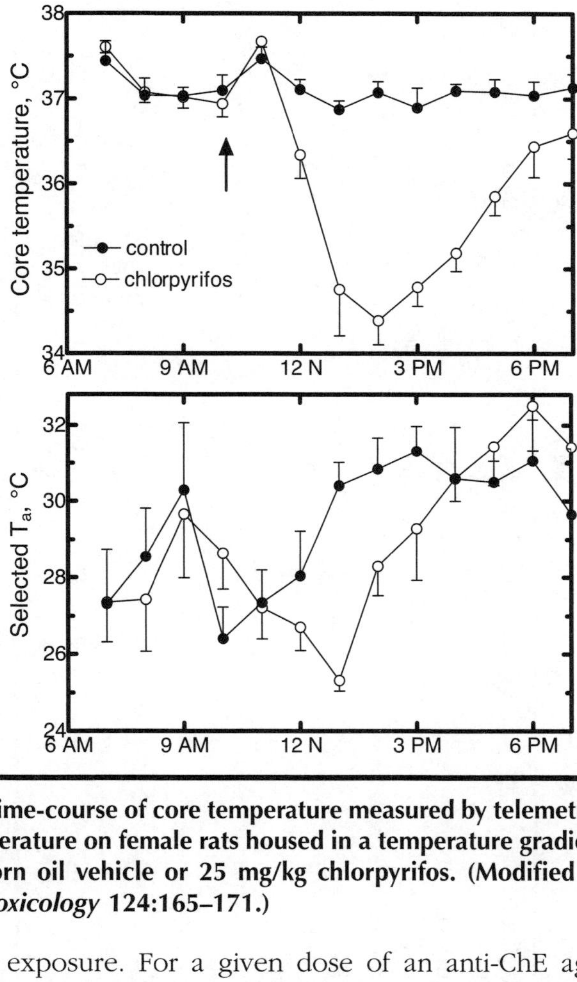


**Figure 3.4 Time-course of core temperature measured by telemetry and selected ambient temperature on female rats housed in a temperature gradient and administered the corn oil vehicle or 25 mg/kg chlorpyrifos. (Modified from Gordon, C.J. (1997). *Toxicology* 124:165–171.)**

against cold exposure. For a given dose of an anti-ChE agent, there is essentially a linear fall in core temperature with a reduction in ambient temperature, providing further support that body temperature decreases because of partial failure of heat gain and heat conserving mechanisms (Wheeler, 1989; Meeter, 1969). However, the coordinated response of autonomic and behavioral thermoeffectors to lower body temperature essentially defines a regulated hypothermic response. That the rat seeks colder temperature and increases skin blood flow allows one to conclude that the thermoregulatory set-point is reduced.

### *3.5.1.3 CNS Mechanisms*

The stimulation of CNS muscarinic pathways appears to be a primary cause of the acute hypothermic response elicited by anti-ChE insecticides.

**Table 3.9 Summary of Effects of Toxicants on Selected Ambient Temperatures ($ST_a$) and Core Temperatures ($T_c$) When Animals Are Housed in a Temperature Gradient. Response Type Indicates if Thermoregulatory Change Is a Regulated or Forced Change in Body Temperature[a]**

| *Species* | *Toxicant* | *Dose*[a] | $ST^b_a$ | $T_c$ | *Response Type* | *References* |
|---|---|---|---|---|---|---|
| Mouse | Cadmium chloride | 2 mg/kg (IP) | ↓ | ↓ | Regulated | Gordon & Stead, 1986 |
| Mouse | Chlordimeform | 60 mg/kg (IP) | ↓ | ↓ | Regulated | Gordon et al., 1985a |
| Mouse | 2,4-DNP | 20 mg/kg (IP) | ↓ | ↑ | Forced | Pertwee & Tavendale, 1979 |
| Mouse | Ethanol | 2.6 g/kg (IP) | ↓ | ↓ | Regulated | O'Connor et al., 1989 |
| Mouse | Nickel chloride | 10 mg/kg (IP) | ↓ | ↓ | Regulated | Gordon & Stead, 1986 |
| Mouse | Sodium selenite | 30 µmol/kg (SC) | ↓ | ↓ | Regulated | Watanabe & Suzuki, 1986 |
| Mouse | Sulfolane | 400 mg/kg (IP) | ↓ | ↓ | Regulated | Gordon et al., 1986 |
| Mouse | Triethyltin | 6 mg/kg (IP) | ↓ | ↓ | Regulated | Gordon et al., 1984 |
| Rat | Chlordimeform | 60 mg/kg (IP) | ↑ | ↓ | Regulated* | Gordon & Watkinson, 1988 |
| Rat (f) | Chlorpyrifos | 25 mg/kg (PO) | ↓ | ↓ | Regulated | Gordon, 1997 |
| Rat | DFP | 1.5 mg/kg (IP) | ↓ | ↓ | Regulated | Gordon, 1994a |
| Rat | Ethanol | 3.0 g/kg (PO) | ↓ | ↓ | Regulated | Gordon et al., 1988a; Briese & Hernandez, 1996 |
| Mouse | Lead acetate | 100 mg/kg (IP) | ↓ | ↓ | Regulated | Gordon et al., 1987 |
| Rat | Sulfolane | 800 mg/kg (IP) | NC | ↓ | Regulated* | Gordon et al., 1985 |

*Animal preferred warmer temperatures than controls but not warm enough to offset hypothermic effects of the toxicant.

[a] Parentheses after dose indicate route of exposure (PO, oral; IP, intraperitoneal; SC, subcutaneous; IV, intravenous).
[b] NC, no change.

Co-administration of muscarinic antagonists such as scopolamine and atropine blocks most of the hypothermic response elicited by anti-ChE agents (Gordon and Grantham, 1999; Maickel et al., 1991; Meeter and Wolthuis, 1968). It would appear that noncholinergic pathways and ambient temperature may affect which neurotransmitter systems are operative in the thermoregulatory responses of anti-ChEs. For example, Maickel et al. (1991) found that the efficacy of atropine and scopolamine to block physostigmine-induced hypothermia was attenuated with decreasing ambient temperature. When temperature was shifted above or below thermoneutrality, specific neurotransmitter systems were activated or suppressed, and this modulation could well alter the relative activity of cholinergic systems in animals exposed to anti-ChEs. Organophosphates led to a significant turnover of norepinephrine in the hypothalamus, suggesting that a noradrenergic pathway is also involved in mediating the hypothermic and hyperthermic effects (Coudray-Lucas et al., 1983). Administration of prazocin, a peripheral α-adrenergic antagonist, exacerbates the hypothermic response to soman and physostigmine in mice (Clement, 1993). It is interesting to note that the organophosphate DEF (S,S,S,-phosphorotrithioate) induces a profound hypothermic response but is a weak inhibitor of AChE activity (Ray, 1980). DEF-induced hypothermia is not affected by administration of cholinergic antagonists. Thus, while the cholinergic pathways are very important in the acute hypothermic response to anti-ChEs, it is clear that other neurochemical pathways are operative in the mediation of anti-ChE–induced hypothermia (see Gordon, 1994, for review).

### 3.5.2 Chlordecone

There has been considerable research on the thermoregulatory mechanisms of chlordecone (kepone), a chlorinated hydrocarbon insecticide. In spite of the presence of intense tremor, chlordecone and related agents induce an acute hypothermia in the rat (Table 3.4). Although not well studied, the heat production resulting from tremor is apparently inconsequential to the rat's overall heat balance. In fact, the pattern of muscle contraction in shivering is distinct from tremor, and one should not necessarily equate tremor with a thermoregulatory effector response. The frequency patterns of electromyograms in mice, rats, and rabbits administered tremorine, a chemical that elicits Parkinsonia-like tremor, differ markedly from the electromyograms when subjected to acute cold exposure (Günther et al., 1983; see Chapter 2).

Hsu et al. (1986) determined that the hypothermic response to chlordecone in the restrained rat maintained at an ambient temperature of 8 or 22°C was a result of an inhibition of metabolic thermogenesis. In addition, there was no indication that chlordecone affected peripheral vasomotor mechanisms. It is important to emphasize here that the restrained rat is unable to shiver in cold environments and is more susceptible to hypothermia (also see Chapter 7). When the rat was tested under unrestrained conditions, chlordecone induced a prolonged hypothermic response that was associated with an elevation in tail skin temperature, but there was no effect on metabolic thermogenesis (Swanson and Woolley, 1982). Chlordecone had no effect on metabolic rate when rats were treated at a temperature of 22°C, whereas a rapid and sustained reduction in metabolic rate was observed when the rat was maintained at a cold temperature of 10°C (Cook et al., 1987). Moreover, when the rats were allowed to behaviorally thermoregulate in a temperature gradient, chlordecone elicited a preference for cooler temperatures. The preference for cooler temperatures and peripheral vasodilation suggest a regulated hypothermic response in the rat during the first 6 h after systemic exposure to chlordecone.

Chlordecone and dieldrin are related in chemical structure, but chlordecone is tremorigenic while dieldrin is a convulsant. Dieldrin also induced a prolonged hypothermic response concomitant with an increase in tail skin temperature (Swanson and Woolley, 1982). Interestingly, there was a significant increase in core temperature 1 to 4 days after exposure to chlordecone (Cook et al., 1987; Swanson and Wooley, 1982). The delayed hyperthermic state was associated with a normal tail skin temperature, suggesting that the hyperthermic response is regulated (see Chapter 6). That is, if the CNS control of body temperature was unaffected during the hyperthermic state, an increase in tail skin temperature would be expected as a response to dissipate excess heat.

#### *3.5.2.1 CNS Mechanisms*

The acute hypothermic and delayed hyperthermic effects of chlordecone may be a result of direct effects of the toxicant on specific CNS loci. Specific thermoregulatory effects of chlordecone can be elicited by localized administration into the CNS. For example, infusions of small amounts of chlordecone (40 and 320 μg) into the lateral and third ventricles elicited a moderate hyperthermic response that persisted for at least 24 h after injection (Figure 3.5A). However, injection into the intracisternal space elicited a hypothermic response (Figure 3.5B). The hypothermic response to intracisternal injection of chlordecone was associated with an increase in tail skin temperature (Figure 3.5C). An increase in heat loss by tail vasodilation that precedes the reduction in core temperature provides evidence of a CNS mechanism to explain the hypothermic effect of chlordecone when it is administered systemically. The hypothermic response to central chlordecone appears to be mediated by activation of

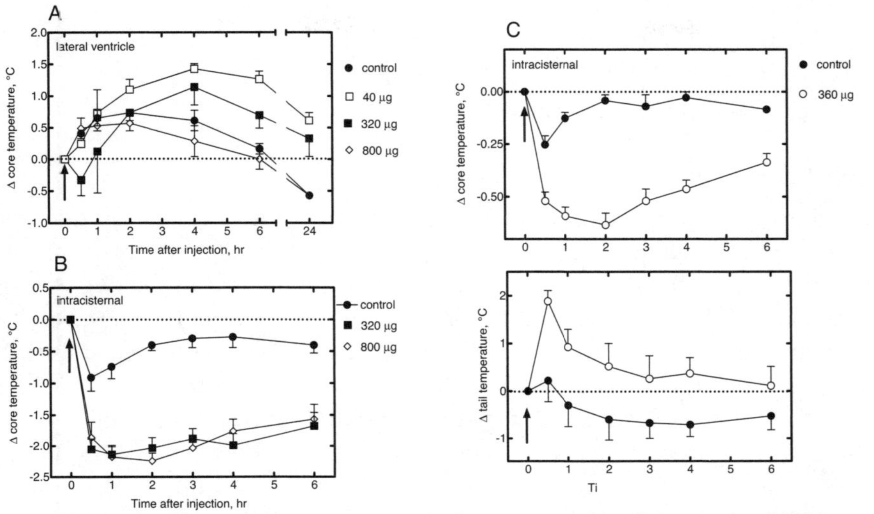


**Figure 3.5 Hyperthermic and hypothermic responses to CNS administration of chlordecone. (A) Infusion into the lateral ventricle elicits a hyperthermic response that persists for at least 24 h. (B) Infusion into the intracisternal space elicits a hypothermic response. (C) Intracisternal infusion of chlordecone accompanied by tail vasodilation. Note that the control vehicle consists of ethylene glycol. (Data from Cook, L.L., Edens, F.W., and Tilson, H.A. (1988). *Neuropharmacology* 27: 871–879.)**

adrenergic pathways because pretreatment with 6-hydroxydopamine, an agent that induces degeneration of adrenergic neurons in the CNS, effectively blocked the hypothermic response to chlordecone (Cook et al., 1988a). Selective adrenergic antagonists also blocked the hypothermic effects of chlordecone. Overall, the hypothermic and delayed hyperthermic effects of chlordecone may be explained by selective toxicity within the CNS. A CNS-mediated hyperthermic response appears to be manifested upon recovery from the acute hypothermic effects of chlordecone.

### *3.5.3 Airborne Toxicants*

Studying the thermoregulatory responses to airborne pollutants and other inhaled toxicants has been a challenge because of the difficulty in monitoring body temperature while maintaining the animal in specialized systems for exposure to the airborne agent. The thermoregulatory effects of many inhaled toxicants should be viewed with caution because of the common use of restraint. In many studies, rats and other rodents have to be restrained to achieve nose-only exposure to the toxicant (Mautz, 2003; Narciso et al., 2003). However, the advent of radiotelemetry has allowed for marked advances in this field because the animals can be exposed to the airborne pollutants while being monitored undisturbed without restraint or anesthesia.

### *3.5.3.1 Ozone*

Commonly studied air pollutants such as ozone, carbon monoxide, and a variety of volatile organic solvents induce hypothermic responses in rodents when they are exposed at ambient temperatures below thermoneutrality (Table 3.5). Vaporized organic agents such as formalin, trichloroethane, and toluene cause hypothermia at relatively high concentrations, but the mechanism of action has not been well studied. The toxicity of ozone on thermoregulation in the mouse and rat has been characterized using radiotelemetry (Watkinson et al., 1993, 1995, 1996). The hypothermic effects of ozone on rodents are dependent on the prevailing environmental temperature. For example, at a temperature of 18 to 20°C, exposure to 0.37 ppm ozone led to a 1°C decrease in core temperature in the unrestrained rat within 2 h; 1.0 ppm ozone resulted in a 3.5°C reduction in core temperature. Raising ambient temperature to 30 to 32°C reduced the hypothermic efficacy of ozone with a decrease of just 0.9°C after 2 h of exposure to 1 ppm (Watkinson et al., 1993). The core temperature of the mouse appears to be even more sensitive with a 6.7°C decrease following 2 h of exposure to 2 ppm ozone when maintained at 21 to 23°C (Watkinson et al., 1996).

The thermoregulatory responses to ozone as well as other pollutants in the rat are associated with marked effects on the cardiovascular system that may be dependent on the change in body temperature (Watkinson et al., 1993). Radiotelemetric monitoring of heart rate and core temperature reveal how a decrease in heart rate precedes the hypothermic response (Figure 3.6). Heart rate decreased by approximately 50% within 1 h after the start of ozone exposure at a temperature of 18 to 20°C. When the rats were maintained at 30 to 32°C, exposure to 1.0 ppm ozone resulted in a 35% reduction in heart rate, but the hypothermic response was minimal compared to that in the cool environment. These data exemplify how the reduction in heart rate is not a simple result of hypothermia because the bradycardia precedes the hypothermic response and it persisted in a cool and thermoneutral environment. The change in body temperature undoubtedly has a role in the cardiovascular response, but the effects are more complex than simple thermal kinetics as outlined in Chapter 4. It is interesting to note that guinea pigs exposed to 1 ppm ozone showed no discernable changes in core temperature or heart rate but did sustain pathological damage to the lungs in a similar degree to that observed in rats (Campen et al., 2000). The acute hypothermic

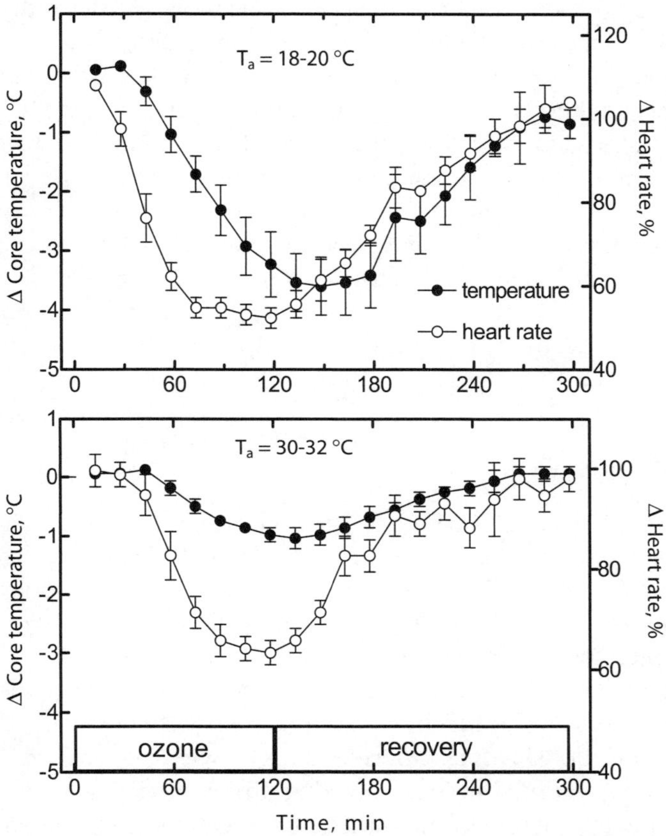


**Figure 3.6 Time-course of core temperature and heart rate after exposure to 1.0 ppm ozone in rats maintained in a cool or thermoneutral environment. (Data modified from Watkinson, W.P., Aileru, A.A., Dowd, S.M., Doerfler, D.L., Tepper, J.S., and Costa, D.L. (1993). *Inhal. Toxicol.* 5, 129–147.)**

response to ozone exhibited by the rat and mouse is protective because it lowers their metabolism and rate of intake of the toxicant (see Chapter 4). The lack of hypothermic effect in the guinea pig may exacerbate its sensitivity to ozone.

The mechanism of ozone-induced hypothermia is unresolved. There is little known about how ozone affects autonomic and behavioral thermoeffectors to effect a decrease in core temperature. Mautz and Bufalino (1989) measured oxygen consumption, core temperature, and a variety of respiratory parameters in rats exposed to ozone and found that the metabolic rate of the restrained rat began to decrease within 60 min after exposure to 0.8 ppm ozone. A hypothermic response to this level of ozone was observed after 100 min of continuous exposure. Overall, exposure to 0.8 ppm ozone led to a 25% reduction in metabolic thermogenesis, which is likely to be a major cause of the hypothermic response. Ozone also affects thyroid function, causing significant reductions in circulating levels of $T_3$ and $T_4$ following 24 h of exposure to 1 ppm (Clemons and Wei, 1984). Otherwise, there is apparently little known about how other thermoeffectors such as peripheral vasomotor tone and behavioral thermoregulation contribute to the effects of ozone on body temperature. Ozone causes marked damage to the alveolar epithelium, resulting in free radical formation and a severe inflammatory response in alveolar fluids (Wiester et al., 1996; Campen et al., 2000). Little if any ozone penetrates into the blood and CNS, and it is possible that the pulmonary inflammation from ozone activates afferent neurons, leading to hypothermic and bradycardic responses, but no mechanism has been elucidated.

### *3.5.3.2 Carbon Monoxide*

The hypoxemia resulting from exposure to carbon monoxide places limits on metabolic thermogenesis, resulting in hypothermia at relatively cool ambient temperatures in rats (Gautier and Bonora, 1994). Carbon monoxide poisoning and hypoxia have distinct effects on the thermoregulatory system. For example, both hypoxia (i.e., atmosphere of 10 to 14% oxygen) and carbon monoxide (0.03% in air) reduce oxygen consumption and decrease core temperature in the rat; however, hypoxia was shown to inhibit both shivering and nonshivering thermogenesis, whereas carbon monoxide blocked nonshivering but not shivering thermogenesis. The differences appear to be attributable to the direct effects of hypoxia on chemoreceptor function, whereas carbon monoxide interacts with a different mechanism of action on the control of respiration (Gautier and Bonora, 1994).

Considering the large number of poisonings from carbon monoxide each year, it is amazing that there is so little known about its thermoregulatory effects. There have apparently been no studies on the behavioral thermoregulatory responses to carbon monoxide. Understanding the thermoregulatory effects of carbon monoxide might improve the methods for its treatment in human poisonings (see Chapter 5). Benignus (1994) predicted that little if any hypothermia would be expected in humans exposed to the levels of carbon monoxide that cause profound hypothermia in the rat. However, endogenous release of carbon monoxide within the CNS has recently been discovered as a novel pathway in the mediation of fever (see Chapter 6). This being the case, it would seem that studies on the toxicology of carbon monoxide should focus on relatively low levels of exposure. Future work using radiotelemetry would provide an ideal means of improving our understanding of how thermoregulation is affected by carbon monoxide. In regard to the potential febrile effects, it is interesting to note the results of an old clinical case report on the carbon monoxide poisoning of an 18-

year-old male who exhibited a core temperature of 39.2°C at 3 h after admission (Craig et al., 1959; see Chapter 5).

#### 3.5.3.3 Particulate Matter

The pathological effects of a class of pollutants referred to as particulate matter (PM), commonly found in various sources of urban air pollution, have been intensely studied. PMs with a diameter of <10 μm are of greatest concern and have been found to induce a marked inflammatory response in the lung, characterized by leukocyte infiltration, release of cytokines, and formation of free radicals in the alveoli (Li et al., 1997). Inhaled PMs are thought to be associated with increased human morbidity and mortality; concomitant exposure to thermal stress may exacerbate the toxicity of PMs (see Chapter 7).

Exposure of rats to residual oil fly ash (ROFA), a highly soluble sample of PMs, leads to a marked hypothermic response (Campen et al., 2000a). Rats under light halothane anesthesia were given an intrathecal instillation of 2.5 mg ROFA while core temperature and heart rate were recorded by radiotelemetry. Core temperature decreased from 37 to 35°C and then exhibited an apparent recovery (Table 3.5). However, approximately 24 to 36 h after ROFA exposure, core temperature again decreased by approximately 1°C and remained depressed for several hours. Similar to ozone, the hypothermia is exacerbated by exposure to cooler temperatures and is associated with a severe bradycardia. ROFA is a highly heterogeneous substance containing a variety of metals including nickel, vanadium, and iron. Hence, there has been an interest in determining if some components of the PM are more toxic than others. In one study, inhalation exposure to nickel (1.3 to 2.1 mg/m$^3$) but not vanadium (1.7 mg/m$^3$) over a 4-day period led to hypothermia and bradycardia in rats (Campen et al., 2001). In addition, mixtures of nickel and vanadium appeared to have a synergistic effect in terms of a reduction in body temperature and heart rate. Like ozone, PMs apparently do not cross into the circulation, and it is not clear how the peripheral effects of these toxicants lead to such a rapid reduction in body temperature. Future studies to assess the effects of PMs on thermoeffector function will lead to a better understanding of their mechanism of action on thermoregulation.

### 3.5.4 Metals

#### 3.5.4.1 Body Temperature and Metabolic Rate

The thermoregulatory effects of heavy metals, metallic salts, and metals conjugated to organic moieties have been well studied (Table 3.6). Exposure to many types of metals leads to disruptions in metabolic thermogenesis and alters the CNS control of body temperature in rodents. Salts of metals including mercury, cadmium, cobalt, nickel, copper, zinc, lead, magnesium, and others elicit an acute hypothermic response when administered to rats and mice. Telemetric recordings of core temperature in the rat show an acute hypothermic response to 100 or 250 µmol/kg nickel chloride with recovery over 4 h after dosing (Hopfer and Sunderman, 1988). During recovery from nickel chloride exposure, there was a marked disruption in the circadian temperature rhythm for several days along with a notable increase in the daytime core temperature. This delayed hyperthermia is similar to that of the fever observed in the recovery phase from exposure to anti-ChE agents (see Chapter 6).

There is a direct relationship between the $LD_{50}$ of a metallic salt and the threshold dose to reduce metabolic rate or body temperature in the mouse maintained at a relatively cool ambient temperature of 15°C (Figure 3.7). Using a least-squares multiple regression to calculate a no-effect dose, the threshold dose for lowering core temperature averaged 8.6% of the $LD_{50}$ for 11 metal salts (range 0.5 to 39%). This illustrates that by housing rodents at cold temperatures that challenge but do not overwhelm their capacity to maintain a normal core temperature, their sensitivity to metals and other toxicants can be exacerbated. Such a technique can be useful for defining the lowest dose of efficacy for a toxicant.

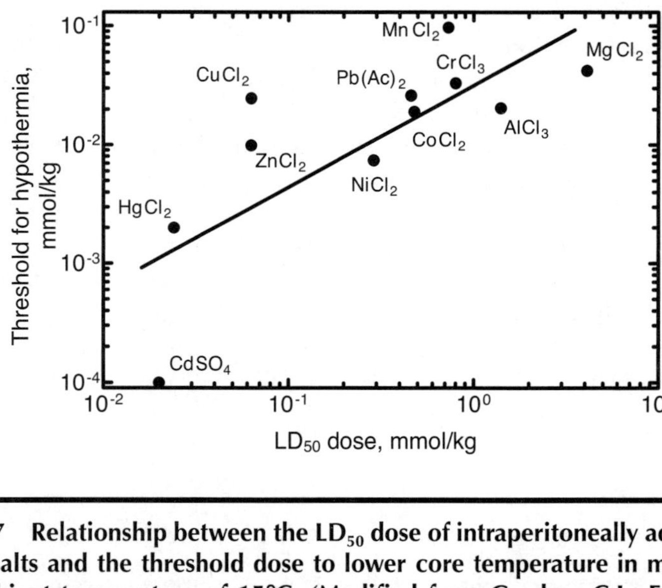


**Figure 3.7 Relationship between the $LD_{50}$ dose of intraperitoneally administered metallic salts and the threshold dose to lower core temperature in mice housed at an ambient temperature of 15°C. (Modified from Gordon, C.J., Fogelson, L., and Highfill, J.W. (1990). *J. Toxicol. Environ. Health* 29, 185–200.)**

It is interesting to note that many of the metal salts that slowly cross the blood–brain barrier are nonetheless capable of eliciting a rapid hypothermic response when administered parenterally. In one of the few comparisons of the effects of route of exposure on thermoregulation, intravenous and intraperitoneal administration of cobaltous chloride were about twice as effective in lowering the core temperature of the mouse when compared to oral or subcutaneous exposure (Table 3.6). Instillation of some metals into the trachea also elicits a profound hypothermic response as a result of a pulmonary inflammatory response (see Section 3.5.3.3). On the other hand, humans exposed to metal fumes such as zinc, copper, lead, and others can develop pulmonary inflammation and a rise in core temperature termed *metal fume fever* (see Chapter 6). Overall, the peripheral mechanisms by which metals can induce hypothermia or hyperthermia are not well understood.

### *3.5.4.2 Brown Adipose Tissue*

Some heavy metals have been used as tools to study the mechanisms of brown adipose tissue (BAT) thermogenesis. Nonshivering thermogenesis in BAT and liver appears to be a sensitive target for some metals. Cadmium and zinc have both been found to block BAT thermogenesis in the rat (Noli et al., 1998; Rebagliati et al., 2001). BAT thermogenesis and the activity of key enzymes for BAT function were significantly reduced when measured *in vitro* 24 h after a single exposure to cadmium chloride in cold-exposed rats (Figure 3.8). Cadmium was also shown to directly inhibit BAT thermogenesis *in vitro* (Noli et al., 1998). Rats injected with cadmium and maintained at 22°C were found to have cadmium in the BAT at a concentration of $1.7 \pm 0.9$ μg/g of tissue. However, when the rats were given the same dose of cadmium but exposed to an ambient temperature of 4 °C, cadmium concentration increased to $15.6 \pm 6.1$ μg/g tissue. It appears that the marked increase in blood flow to BAT during cold exposure could exacerbate the toxic effects of some metals. That is, there is a greater risk of thermoregulatory dysfunction during cold exposure because of the combined effects of increased blood flow to BAT, more sequestering of cadmium, and thus, greater inhibition of nonshivering thermogenesis.

Heavy metals appear to affect thermogenesis through disruption of the thyroid axis. Cadmium and zinc were shown to directly block BAT thermogenesis by inhibiting the conversion of $T_4$ to $T_3$, an essential step in the maintenance of BAT function (see Chapter 2). Zinc has been found to decrease the activity of 5'-deiodinase in rat liver (*in vitro* and *in vivo*), which should lead to normal levels of $T_4$ but reduced levels of $T_3$ (Pondal et al., 1995). In addition, it was shown that cadmium reduces serum levels

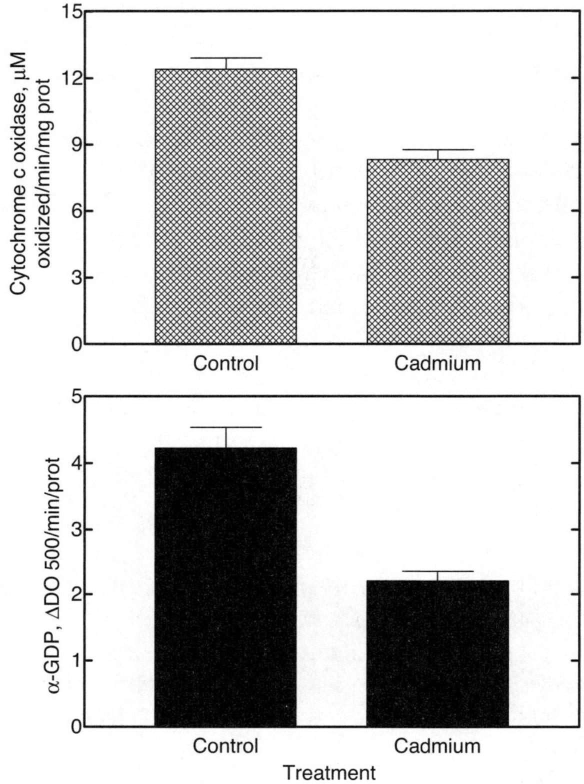


**Figure 3.8 BAT mitochondrial cytochrome c oxidase and α-glycerophosphate dehydrogenase (α-GDP) enzyme activity in rats measured 24 h after treatment with cadmium chloride (220 µg/100 g, intraperitoneally) and exposed to an ambient temperature of 4°C. (Data modified from Noli, M.I., Pavia, M.A. Jr., Mignone, I.R., Brignone, J.A., Hagmuller, K., and Zaninovich, A.A. (1998). *Bull. Environ. Contam. Toxicol.* 61:31–37.)**

of $T_3$ and $T_4$. Dysfunction in the hepatic de-iodination of $T_4$ would lead one to expect marked impairment in the ability to increase heat production via nonshivering thermogenesis during cold exposure. In summary, some metals have been shown to impair metabolic thermogenesis during cold exposure by directly blocking BAT thermogenesis as well as indirectly through reductions in circulating levels of $T_4$ and/or $T_3$.

The toxicity of metals may have a unique link to BAT function through the expression of metallothionein. Metallothionein is a cysteine-rich protein that has a high affinity for heavy metals. The protein is thought to be involved in the detoxification of metals and is found in high levels in blood, liver, and other organs following metal exposure. Metallothionein is also expressed in BAT of mice and rats when they are cold-exposed

(Beattie et al., 2000). Both uncoupling protein (UCP) and metallothionein are co-expressed in BAT with exposure to cold and following administration of norepinephrine and isoproterenol. This would suggest that the synthesis of these proteins is mediated by activation of the sympathetic nervous system. It is well known that UCP is critical for BAT thermogenesis (Chapter 2). UCP and metallothionein also appear to protect cells against oxidative stress. Although not well studied, metallothionein may participate in the control of BAT thermogenesis. It follows that cold exposure–induced expression of metallothionein in BAT may alter the sensitivity of rodents to some metals. Cold and warm acclimation in rodents affects their sensitivity (i.e., change in $LD_{50}$) to some metals, but there has been no attempt to link this altered sensitivity to BAT function (see Chapter 4).

### *3.5.4.3 Autonomic and Behavioral Effects*

The effects of metals on autonomic and behavioral thermoeffectors have been relatively well studied. Subcutaneous administration of sodium selenite led to a marked decrease in metabolic rate and hypothermia when mice were housed at an ambient temperature of 20°C (Watanabe and Suzuki, 1986). Raising ambient temperature to 33°C blocked the hypothermia, resulting in an increase in metabolic rate (Figure 3.9A). When mice were given the option of selecting their thermal environment by housing in a temperature gradient, selenite administration resulted in a marked preference for cooler temperatures, an effect that persisted for at least 90 min (Figure 3.9B). Thus, like many other toxicants discussed in this chapter, the mouse dosed with sodium selenite prefers a relatively cool environment that leads to a reduction in metabolic rate and hypothermia. The toxicity of sodium selenite, as indicated by lethality, is markedly increased when the mouse is maintained in a warm environment where it cannot lower its body temperature (Chapter 4).

Nickel, cadmium chloride, and lead acetate elicit an acute hypothermic response that, like that caused by sodium selenite, is mediated by both behavioral and autonomic thermoeffectors (Gordon and Stead, 1986; Gordon et al., 1987; Martinez et al., 1993). In the mouse, nickel and cadmium chloride led to a rapid reduction in metabolic rate, preference for cooler temperatures, and hypothermia when housed in a temperature gradient (Christensen and Fujimoto, 1984; Gordon and Stead, 1986). The temperature gradient is a relatively simple means of assessing behavioral thermoregulatory responses in rodents; however, more complex operant systems have also been used to detect thermoregulatory effects of metals as well as many types of drugs. Watanabe et al. (1990) assessed the effects of nickel chloride in rats maintained in an operant setting where they were trained to press a lever to obtain

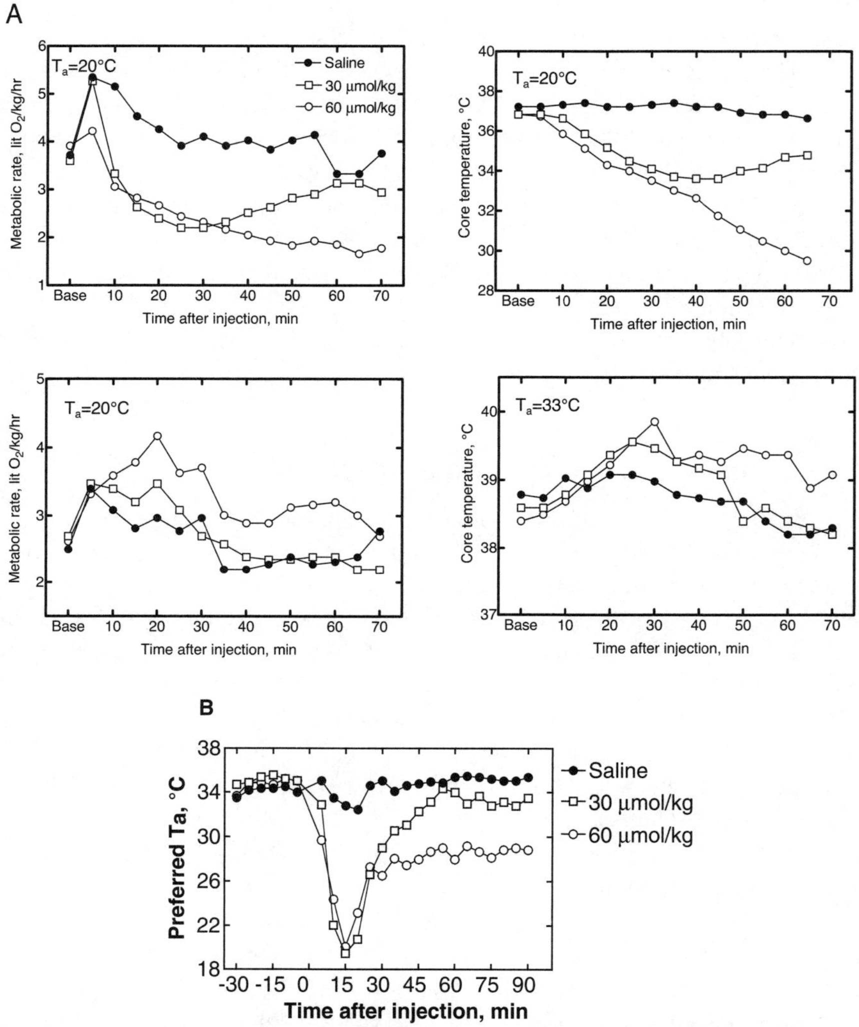


**Figure 3.9 (A) Time-course of core temperature and oxygen consumption of mice housed at an ambient temperature of 20°C and selected temperature of mice housed in a temperature gradient and injected with either saline or 30 μmol/kg sodium selenite (B). (Data modified from Watanabe, C. and Suzuki, T. (1986). *Toxicol. Appl. Pharmacol.* 86: 372–379.)**

heat while exposed to a cold temperature. A nickel dose of 2 mg/kg caused a transient hypothermia at an ambient temperature of 21°C. Exposure to nickel also led to a reduction in bar pressing behavior for heat. The operant responses combined with the effects of nickel on body temperature suggest, as with other toxicants, that nickel elicits a reduction in the set-point for temperature regulation.

In one of the few studies where skin temperature has been measured, Burke (1978) found that the marked hypothermic effects of cobaltous chloride in the mouse were accompanied by a ~2°C reduction in tail skin temperature. The decrease in tail skin temperature was apparently a passive response caused by the lowering of core temperature, and it would appear that there was no active attempt to dissipate heat by peripheral vasodilation. The marked decrease in heat production in mice dosed acutely with metals such as cobalt chloride (Gordon et al., 1990) at relatively cool ambient temperatures is probably more than sufficient to effect a rapid drop in core temperature without the need for an increase in heat loss. Peripheral vasodilation would be expected to be operative as the ambient temperature approaches thermoneutrality and in situations where metabolic thermogenesis does not undergo such marked reductions. Moreover, since the role of peripheral vasomotor tone in thermoregulation increases with body mass (see Chapter 2), the lack of a response in the mouse may not be unexpected.

The hypothermic effect of cadmium chloride and lead acetate is directly correlated with brain levels of the metals when measured 90 min after intraperitoneal (IP) injection (Martinez et al., 1993). The mouse brain is much more permeable to cadmium than lead, and the hypothermic effect of cadmium is correlated with a striking increase in brain levels of cadmium relative to lead (Martinez et al., 1993) Raising the ambient temperature to 35°C resulted in a complete block of the hypothermic effects of lead acetate but not cadmium chloride, although the hypothermic efficacy of cadmium was markedly reduced at the warmer temperature. Raising the temperature from 22 to 35°C had no effect on the deposition of lead and cadmium in the brain when measured 90 min after injection; however, ambient and body temperature did affect the long-term deposition of heavy metals in the brain (see Chapter 4). The ability of cadmium to induce hypothermia at cool as well as relatively warm temperatures suggests an active role of this metal to stimulate heat loss pathways in mice.

#### *3.5.4.4 Primate and Human Studies*

The hypothermic response to nickel has been studied with the expectation of the possible development of a model for understanding human responses. Nickel-induced hypothermia was observed more than a century ago in humans and has been reported in electroplating workers accidentally ingesting nickel solutions (Sunderman et al., 1988). Mild hypothermia (–0.6°C) was also reported in adult women infused intravenously with magnesium sulfate (Parsons et al., 1987). In a study on the effects of lifetime exposure to cadmium ingestion in the Rhesus monkey, body temperature and many other physiological parameters were measured repeatedly in males that ingested cadmium chloride at doses of up to 100 μg per gram of food (i.e., 100 ppm) for a period of 9 years (Masaoka et al., 1994). However, no significant effects on body temperature were observed, although other toxic manifestations such as reduced body weight gain were noted. Lifetime exposure studies such this one are rare. In spite of the lack of effect of cadmium on body temperature, it would be of interest to assess such a prolonged treatment of a heavy metal on thermoeffector responses under conditions of heat and cold stress.

### 3.5.4.5 Neural Mechanisms

Because cholinergic pathways in the CNS of mouse and rat mediate a heat loss–hypothermic response, Burke (1978) hypothesized that the hypothermic effects of cobaltous chloride were attributable to central cholinergic stimulation. However, pre-injection of cholinergic antagonists, including atropine and hexamethonium, had no effect on the hypothermic effects of cobaltous chloride. On the other hand, chlorpromazine pre-treatment was effective in blocking about 50% of the hypothermic effect of cobaltous chloride. Interestingly, administration of serotonergic and noradrenergic drugs to block the effects of serotonin or deplete brain levels of norepinephrine had little effect on the hypothermic effects of cobaltous chloride in the mouse (Burke and Brooks, 1979; Burke et al., 1983). Pre-treating the rabbit with α-receptor adrenergic blocking agents resulted in complete inhibition of the hypothermic effects of intravenously administered arsenic ($As_2O_3$), suggesting that the hypothermic response is mediated by the hypothalamic release of norepinephrine (Kawaguchi and Tsutsumi, 1982).

Direct injection of small quantities of toxicants into the CNS can help resolve the mechanism of action of the toxicant on thermoregulation. Intracerebroventricular (ICV) injection of microgram quantities of cadmium chloride in the mouse (Christensen and Fujimoto, 1984) and cobaltous chloride (Burke, 1978; Burke and Brooks, 1979) in the mouse and rat was shown to elicit a hypothermic response. This would suggest that the hypothermic effect of these metals, when injected peripherally, is a result of small amounts penetrating the CNS and activating thermoregulatory pathways to effect a heat loss response. One possible mechanism by which some metals could affect the CNS control of body temperature is through their effects on calcium homeostasis. A balance between the levels of $Na^+$ and $Ca^{2+}$ is considered a critical mechanism in the regulation of normal body temperature (Myers, 1982). An increase in the ratio of $Ca^{2+}$ to $Na^+$ in the CNS results in hypothermia. Because nickel and cadmium are divalent cations and may cross cell membranes through $Ca^{2+}$ channels and compete with $Ca^{2+}$ receptors, it was postulated that parenteral exposure to nickel, cadmium, and possibly other divalent cations elicits a

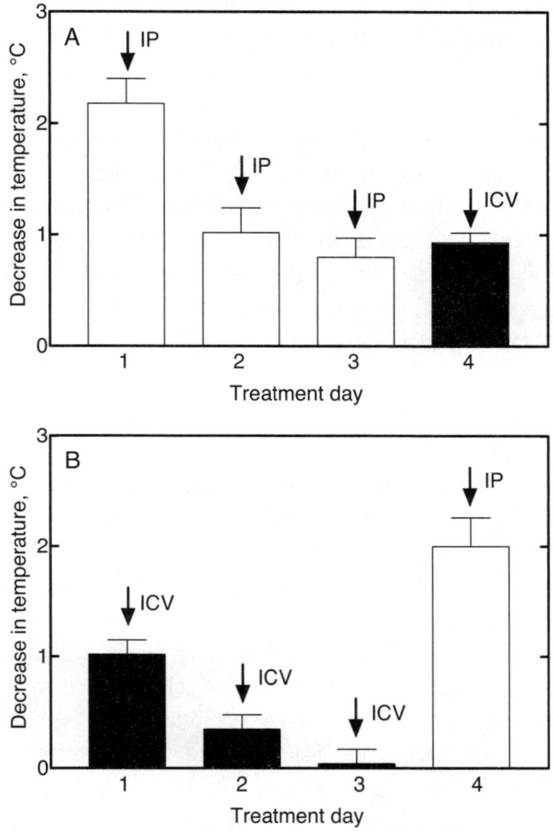


**Figure 3.10 Tolerance to hypothermic effect of systemic and CNS administration of cadmium chloride in the mouse. (A) Hypothermic response decreases with each daily IP (intraperitoneal) administration of 0.62 mg/kg cadmium chloride for days 1, 2, and 3. Intracerebroventricular (ICV) administration of 0.1 μg cadmium chloride on day 4 is similar in magnitude to response given on day 3. (B) Hypothermic response decreases with daily ICV injections of 0.1 μg cadmium chloride for days 1, 2, and 3. Hypothermic response on day 4 to IP administration of 0.62 mg/kg cadmium chloride is similar to response given on day 1. (Modified from Christensen, C.W. and Fujimoto, J.M. (1984). *Gen. Pharmacol.* 15: 263–266.)**

hypothermic response through mimicking excess calcium release in the CNS (Hopfer and Sunderman, 1988).

Christensen and Fujimoto (1984) performed an experiment on mice suggesting that calcium balance in the CNS thermoregulatory centers does not solely explain the hypothermic effects of cadmium (Figure 3.10A, B). Tolerance to the hypothermic effects of intraperitoneal dosing with cadmium chloride in the mouse was apparent after just two days of dosing, but tolerant mice still displayed a normal hypothermic response when cadmium was given by an ICV route. Mice could also be made tolerant to the hypothermia with daily ICV dosing of low doses of cadmium, as indicated by a negligible change in body temperature on the third day of ICV dosing. However, mice that were tolerant to repeated ICV injections of cadmium nonetheless displayed a marked hypothermic effect to an intraperitoneal injection of cadmium. If the hypothermia was mediated solely by CNS pathways, then one would expect that a tolerant animal would show an attenuated response to a systemic injection. These experiments imply that there are separate peripheral and central mechanisms involved in the manifestation of cadmium-induced hypothermia.

### *3.5.4.6 Organotins*

The organotins are potent agents that are extremely neurotoxic and cause lesions in the central and peripheral nervous system. These agents are also potent inhibitors of oxidative phosphorylation. Trimethyltin (TMT) and triethyltin (TET) lead to acute hypothermic responses in rats and mice (Gordon et al., 1984; Gordon and O'Callaghan, 1995). Mice injected with TET undergo a rapid reduction in metabolic rate when housed at relatively cool temperatures and become hypothermic. An approximate 50% decrease in metabolic rate was observed within 10 min after mice were dosed with 8 mg/kg TET and housed at a temperature of 24°C. When allowed to behaviorally thermoregulate, mice dosed with TET preferred cool temperatures and allowed their core temperature to decrease. This regulated hypothermic response may afford protection to the neurotoxic effects of TET and possibly other organotins (see Chapter 4). The rat also displays a profound hypothermic response to TET and TMT (Leow et al., 1980; Gordon and O'Callaghan, 1995).

While systemic injections of 5 to 10 mg/kg TET result in an acute hypothermic response with recovery in approximately 24 h, ICV injection of 100 μg TET resulted in transient hypothermia for 30 min followed by a prolonged increase in core temperature persisting for several hours (Leow et al., 1980). Such a response is unique when compared to the hypothermic effects of centrally injected cadmium and cobalt. Why are the initial thermoregulatory responses to systemic and central injection of TET opposite? Systemic TET may disable thermoeffectors, especially thermogenesis and vasomotor control, thus preventing the animal from mounting a hyperthermic response. The organotins appear to elicit a delayed hyperthermic response that is similar to that seen with anticholinesterase agents (see Chapter 6). In one study using colonic probes to measure core temperature, rats injected intravenously with 8.0 mg/kg TMT showed a 0.5°C elevation in core temperature by 72 h after exposure (Gordon and O'Callaghan, 1995).

The hyperthermic response to organotins is likely to exacerbate their neuropathic effects in the hippocampus and other regions of the CNS (see Chapter 4).

### 3.5.5 Alcohols

Ethanol and other short-chain alcohols elicit acute hypothermic responses in rodents at ambient temperatures below thermoneutrality (Table 3.7). Ethanol is discussed in some detail here because, while it is the most heavily abused drug, it is also considered a neurotoxicant. It is also one of the few toxicants for which there is a plethora of toxicological and pharmacological data in both experimental animals and humans. In addition, there is considerable interest in understanding the potential interactions between chronic ethanol consumption and susceptibility to environmental toxicants. The mechanisms of toxicity to ethanol are comparable in some cases to other toxicants. Hence, the ethanol literature can be used to better understand the extrapolation of toxicological data from experimental animals to humans (see Chapter 5).

The thermoregulatory effects of ethanol have been well studied in experimental animals and humans (for review, see Kalant and Le, 1983). Myers (1981) concluded that ethanol was a "poikilothermic" agent. That is, following oral administration in rats, body temperature decreased upon exposure to cold temperatures but increased when the rats were exposed to warm temperatures exceeding thermoneutrality. It was concluded that the rat's body temperature behaved in a poikilothermic-like manner, akin to that of lower vertebrates when exposed to a range of ambient temperatures (Chapter 2). However, poikilotherms still can thermoregulate when given behavioral options to seek a source of heat, whereas alcohol-induced poikilothermia connotes a state of uncontrolled thermoregulation. Moreover, it has been discovered that ethanol, like many other neurotoxicants, elicited a regulated hypothermic response in mice and rats as characterized by vasodilation of tail blood flow (Crawshaw et al., 1997), a preference for cooler temperatures in a temperature gradient, and a concomitant reduction in body temperature (Gordon and Stead, 1986; Gordon et al., 1988a; O'Connor et al., 1989; Briese and Hernandez, 1996). Using a novel technique of subjecting mice to cyclic changes in ambient temperature, Crawshaw et al. (1997) developed an index of thermoregulatory disruption in two strains of mice. They found that the regulated hypothermic response to ethanol was greater in the C57BL/6J strain as compared to the DBA/2J strain, whereas the disruption in thermoregulatory control was greater in the latter strain. Genetic variations provide an ideal means of studying the mechanisms of action of a toxicant by using thermoregulatory endpoints (see Chapter 9).

### 3.5.6 Mechanisms

One key mechanism of the toxicity of ethanol and other alcohols appears to be related to the ability to disrupt membrane permeability. The presence of ethanol in cell membranes expands the volume and disrupts the lipid bilayer structure, resulting in altered permeability (Chen and Engel, 1990). Altered membrane permeability to $Ca^{2+}$ and the increased neuronal flux of $Ca^{2+}$ is thought to be critical in triggering ethanol-induced changes in body temperature (Rezvani et al., 1990). Intraventricular (IVT) infusion of EGTA, a $Ca^{2+}$ chelator (Myers, 1980) or verapamil, a $Ca^{2+}$ channel antagonist, blocked ethanol-induced hypothermia (Rezvani et al., 1986). The potency of a short-chain alcohol to affect temperature regulation was shown to be closely related to its membrane partition coefficient (Mohler and Gordon, 1991). That is, the more an alcohol can partition into the membrane fraction, the greater its expected narcotic-like potency and ability to alter thermoregulation. McCreery and Hunt (1978) showed that the effective dose of an alcohol to induce ataxia was inversely dependent on the membrane partition coefficient. In a study including methanol, ethanol, and other short-chain alcohols, there was a highly significant inverse relationship between the threshold dose to lower core temperature in rats maintained at an ambient temperature of 22°C and the alcohol's membrane coefficient (Figure 3.11). This relationship points to a common mechanism of action of these alcohols, namely, membrane disruption and increased permeability, to alter body temperature in the rat.

#### *3.5.6.1 Human Responses*

Ethanol is one of the few toxicants for which the thermoregulatory effects in humans have been studied under controlled conditions. Fellows et al. (1984) assessed the effects of oral dosing with 0.5 g/kg ethanol on forearm and hand skin blood flow, metabolic rate, and core temperature of adult males maintained at ambient temperatures of 21 and 30°C. Ethanol was more effective at inducing an increase in skin blood flow at 30°C as compared to 20°C but had no effect on metabolic rate. In spite of an increase in heat loss, core temperature was essentially unaffected with a decrease of just 0.18°C. However, Hrbek et al. (1985) found a 0.7°C decrease in the axilla temperature (a measure of core, but it should be interpreted with caution) 2 h after ingestion of 0.9 g/kg ethanol in a group of male and female adults. Ethanol doses of 0.3 and 0.6 g/kg were ineffective at changing core temperature. The hyperthermic effects of ethanol in human subjects are discussed in Chapter 6.

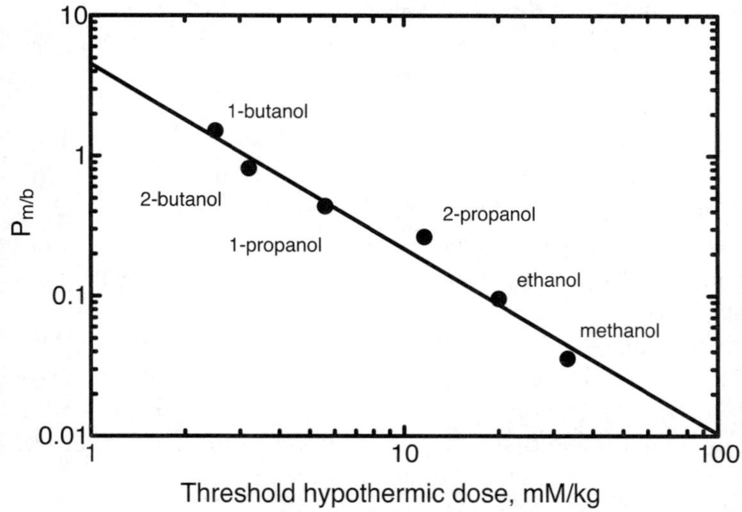


**Figure 3.11 Correlation between the threshold dose of alcohols for inducing hypothermia in the Fischer 344 rat and the membrane/buffer partition coefficient ($P_{m/b}$). (Modified from Mohler, F.S. and Gordon, C.J. (1991). *J. Toxicol. Environ. Health* 32: 129–139.)**

## *3.5.7 Organic Solvents*

In addition to the wealth of data on ethanol, acute parenteral and respiratory exposure to organic solvents has been found to elicit hypothermic responses (Table 3.7). This laboratory studied the thermoregulatory responses of the mouse, rat, and rabbit to sulfolane, an industrial solvent that results in a variety of neurotoxic sequelae (Gordon et al. 1984a, 1986). In the mouse and rat, systemic administration of sulfolane elicited a regulated hypothermic response characterized by a decrease in core temperature, reduction in metabolic thermogenesis, and preference for cooler ambient temperatures. In one of the few thermoregulatory studies of a toxicant in the rabbit, it was found that subcutaneous sulfolane administration led to a reduction in core temperature, no change in metabolic rate, and an increase in ear skin temperature. Although there was no information on the behavioral thermoregulatory response of the rabbit to sulfolane, the increase in ear skin temperature could indicate a peripheral vasomotor response to increased heat loss. Thus, in mouse, rat, and rabbit, acute sulfolane led to an apparent regulated hypothermic response. Compared to the acute hypothermic response, IVT administration of small quantities of sulfolane in the rabbit led to a slight hyperthermic response (Mohler and Gordon, 1989). Direct injections into the preoptic area had no effects on body temperature. Hence, the acute hypothermic response to sulfolane may be mediated by a peripheral mechanism of action.

It is also interesting to note the impact of body mass on the hypothermic response to sulfolane. A sulfolane dose of 400 mg/kg in the mouse, rat, and rabbit resulted in a decrease in core temperature of −2.9, −1.6, and 0.5°C, respectively, when measured over a period of 1 to 3 h after injection (Table 3.7). That is, the hypothermic efficacy of sulfolane was reduced with increasing body mass. Since the hypothermic response to sulfolane is protective in the mouse (see Chapter 4), the diminished hypothermia with increasing size could be an important factor in its comparative toxicity. The impact of body mass on the thermoregulatory response to a toxicant is crucial in extrapolation of toxicological responses from experimental animals to human and is discussed in more detail elsewhere (see Chapter 5).

### 3.5.8 Formamidines

The thermoregulatory effects of this class of insecticides have been studied in a variety of species. Formamidine insecticides, such as chlordimeform and amitraz, have been shown to elicit hypothermic responses in mouse, rat, and dog (Table 3.8). Measurements of behavioral thermoregulation and skin temperature in the rat and mouse suggest that chlordimeform elicits a regulated hypothermic response. In the anesthetized rat, chlordimeform elicited a rapid increase in tail and foot temperature, indicative of a peripheral vasodilatatory response to increase heat loss (Gordon and Watkinson, 1988). When housed in a temperature gradient, the mouse administered 75 mg/kg chlordimeform (IP) underwent a 4 to 6°C reduction in selected temperature below that of controls, a response that was maintained for at least 2 h after injection and was associated with a marked hypothermia (Gordon et al., 1985a). The rat treated with 60 mg/kg chlordimeform and housed in a temperature gradient exhibited a 14% reduction in metabolic rate and remained in a relatively cool area of the gradient. Overall, the behavioral and autonomic responses resulted in a 2°C decrease in core temperature (Gordon and Watkinson, 1988).

The behaviorally and autonomically mediated decrease in core temperature lessens the cardiovascular toxicity and lethality of chlordimeform. As the rat increased blood flow to the tail and foot following IP injection of 60 mg/kg chlordimeform, there was a precipitous decrease in heart rate and blood pressure. However, the moderate hypothermic response improved the likelihood of survival (see Chapter 4). The mechanism of the formamidine agents to elicit a hypothermic response is thought to be a result of their $\alpha_2$-agonistic properties. Formamidine-based agents selectively bind to $\alpha_2$ receptors in the brain (Costa et al., 1989). Pharmacological agents have been used to pinpoint the mechanism of action on temperature regulation. Administration of clonidine, an $\alpha_2$-agonist, elicits hypothermia (Livingston et al., 1984), whereas yohimbine, an $\alpha_2$-antagonist, is generally effective in blocking the hypothermic effects of formamidine compounds (Boyes and Moser, 1988; Hugnet et al., 1996).

## 3.6 TOXICANTS ELICITING HYPERTHERMIA

Relatively few toxicants elicit an acute hyperthermic response in rodents and other test species when housed under standard housing temperatures (~22°C). Rodents are well adapted to regulate body temperature when subjected to cold stress but have relatively poor defense against heat stress (Chapter 2). A chemical or drug may have no effect on body temperature at relatively cool temperatures but may lead to a hyperthermic response if exposure occurs at warm temperatures exceeding thermoneutrality. Unlike the preponderance of toxicants that appear to induce a regulated hypothermic response by directly affecting the CNS' control of body temperature, there is little evidence of toxicants that elicit an acute or rapid hyperthermic response via an increase in the set-point. However, many toxicants elicit a delayed hyperthermic response that is similar to a fever and develops by one or more days after exposure (Chapter 6).

### *3.6.1 DDT*

DDT is an organochlorine compound that is tremorigenic and a convulsant and is one of the few pesticides that can elicit an acute and apparently lethal increase in the core temperature (Hudson et al., 1985). DDT induces a marked hyperthermic response in the rat but apparently not in the mouse (Table 3.4). Not surprisingly, the thermoregulatory effects of DDT are dependent on ambient temperature. When adult rats are given a lethal dose of DDT (1,000 mg/kg) and housed at 22°C, tail skin temperature begins to increase prior to any change in core temperature by 3 to 5 h after injection (Woolley, 1973). There is then a precipitous decrease in tail skin temperature that is concomitant with a marked elevation in core temperature. At the point of death, around 6 h after dosing, core temperature is 40 to 42°C, while tail skin temperature is nearly below pretreatment levels. If the DDT-treated rat is placed in a cold room maintained at 3 to 4°C, core temperature falls dramatically, reaching near lethal levels within a few hours (Woolley, 1973). Hence, in spite of the intense tremorigenic activity of this agent, DDT-treated rats in the cold are unable to thermoregulate against a cold stress that is well tolerated by untreated rats. As with other tremorigenic compounds such as kepone, the heat from tremor in rats dosed with DDT appears to contribute little to their overall heat balance.

The rise in core temperature following a lethal dose of DDT is associated with elevations in heart rate and respiratory rate (Henderson and Woolley, 1970). Interestingly, the 10-day-old rat pup was unable to mount a hyperthermic response to high doses of DDT but instead displayed an acute hypothermia (4°C decrease). The rat pup survived for approximately twice as long as the adult when given lethal doses of DDT. It is important to note that homeothermic capacity in mouse and rat is attained at 14 days of age and then improves gradually over the next month (for review, see Gordon, 1993) The lower critical temperature of the 10-day-old rat is 6°C higher than that of the adult. Hence, when tested at standard room temperature, the rat pup is subjected to marked cold stress, and a hyperthermic response would be unlikely. In fact, the relatively cold environment of the rat pup is likely responsible for the hypothermia and prolonged survival time to acute DDT (see Chapter 4).

The vasoconstriction of tail blood flow at the onset of hyperthermic death in rats treated with DDT may be mediated by changes in 5-hydroxytryptamine serotonin (5-HT) turnover in the CNS. DDT-induced hyperthermia and tremor are associated with increased CNS turnover of 5-HT along with elevations in excitatory amino acids, including aspartate and glutamate (Hudson et al., 1985). The hyperthermic effect of DDT was associated with increases in the level of 5-hydroxyindoleacetic acid (5-HIAA), an indicator of enhanced turnover of 5-HT (Peters et al., 1972). Pre-administration of drugs that block the synthesis of 5-HT, such as p-chlorophenylalanine and 6-fluorotryptophan, attenuated the hyperthermic effects of DDT (Peters et al., 1972). It is interesting to note that amphetamine-type drugs are also robust mediators of 5-HT turnover and have been shown to elicit a hyperthermic response that is similar to that of DDT. For example, rats dosed with 3,4-methylenedioxymethampetamine (MDMA) undergo marked 5-HT release in the CNS and succumb to hyperthermic death when housed at temperatures greater than ~22°C (Gordon et al., 1991c). Hyperthermic death following MDMA dosing is associated with vasoconstriction of blood flow to the tail. Moreover, when exposed to cold temperatures, rats treated with MDMA become hypothermic in spite of a normal metabolic rate. The neurotoxicity of MDMA and other amphetamines is clearly dependent on whether they induce hyperthermia (see Chapter 4).

Organochlorine insecticides such as dieldrin and chlordane, while producing neurotoxic effects similar to DDT, elicit a hypothermic response in the rat housed at standard room temperatures (Table 3.4). The hypothermic effects of chlordane were found to be associated with an increase in CNS release of norepinephrine (NE), a catecholamine that drives heat loss–hypothermic responses in the rat (Hrdina et al., 1974). Blocking the synthesis of NE by pre-treatment with methyl-*p*-tyrosine led to an attenuation in the hypothermic effects of α-chlordane.

## 3.6.2 Uncouplers of Oxidative Phosphorylation

2,4-dinitrophenol (DNP) is a classic uncoupling agent of oxidative phosphorylation that causes a marked stimulation in metabolic rate. It is interesting to find that there is so little known about the thermoregulatory mechanisms of DNP in spite of the long history of its use as a herbicide (Takehiro et al., 1979). Overall, DNP is very effective for raising body temperature in rodents and other species, including humans (see Chapter 6).

### *3.6.2.1 Thermogenesis*

At a relatively cool ambient temperature of 15°C, the metabolic rate of the rat was increased from 34 to 51 ml $O_2$/kg/min following repeated intravenous injections of DNP, with a total dose of 24 mg/kg. The increase in metabolic rate was associated with a 1.1°C increase in core temperature. However, making the rats hypoxic with a resulting 7 ml $O_2$/min/kg reduction in metabolic rate resulted in a 2°C decrease in core temperature. This experiment illustrates the nonlinearity in the relationship between heat production and core temperature that is dependent on the mechanism of the agent on thermoregulation. That is, with DNP a 50% elevation in heat production was associated with a 1.1°C hyperthermia, while a 20% decrease in heat production during hypoxia was associated with a 2°C hypothermia.

Intravenous infusion of DNP in the awake rat caused a near instantaneous elevation in metabolic rate, ventilatory frequency, and tidal volume concomitant with a steady rise in body temperature (Figure 3.12). Arterial blood pressure decreased transiently following DNP injection and then rose gradually along with an elevation in heart rate. The increase in metabolism and ventilation clearly preceded the change in body temperature, which means that DNP-induced hyperthermia is not driving the elevations in metabolism and ventilation (i.e., not a $Q_{10}$ effect; see Chapter 4). In view of the similarity in response of the cardiopulmonary and thermoregulatory responses to exercise and DNP exposure, it is possible that some aspects of the thermoregulatory response to DNP may be useful in studying the thermal physiology of exercise (Takehiro et al., 1979).

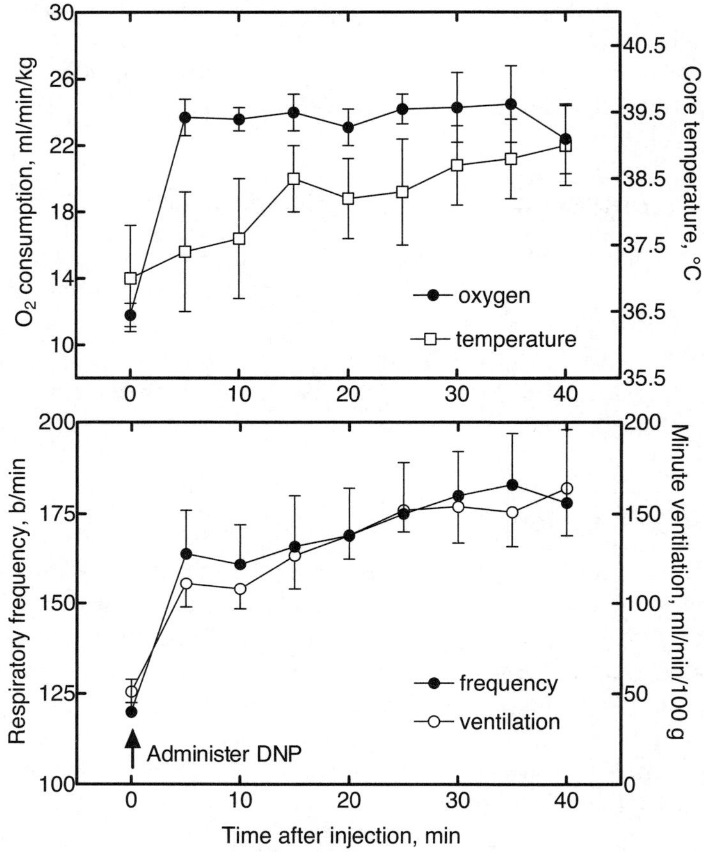


**Figure 3.12 Time-course of oxygen consumption, core temperature, respiratory frequency, and minute ventilation in rats following intravenous infusion of 9 mg/kg dinitrophenol (DNP). (Data taken from Takehiro, T.N., Shida, K., and Lin, Y.C. (1979). *J. Thermal Biol.* 4: 297–301.)**

### *3.6.2.2 Behavior*

Mice allowed to behaviorally thermoregulate in a gradient after intraperitoneal injection of 20 mg/kg DNP preferred cooler temperatures and maintained a normal core temperature provided they remained in the cool environment (Pertwee and Tavendale, 1979). However, if the DNP-treated mice were unable to leave a warm environment, their core temperature increased precipitously. The behavioral thermoregulatory response to prefer a cool ambient temperature to avoid hyperthermia indicates that DNP elicits a forced hyperthermic response. Additional information on the effects of DNP on behavioral thermoregulation and peripheral vasomotor tone would be useful to understand the mechanism of action of this class of agents on thermoregulation.

### 3.6.3 Pyrethroids

The pyrethroid-based insecticides are widely used in domestic and agricultural applications, but there is surprisingly little information on their effects on thermoregulation (Table 3.8). Pyrethroids are generally classified as Type I or Type II, depending on their neurophysiological and behavior effects (see McDaniel and Moser, 1993, for summary). Type I pyrethroids keep sodium channels open momentarily (i.e., milliseconds), whereas Type II agents keep channels in an open state for minutes. Unlike many of the toxicants discussed in this chapter, the toxicity of pyrethroids is generally exacerbated with decreasing temperature (Chapter 4). How the thermoregulatory response to either increase or decrease core temperature affects the recovery and survival to pyrethroids remains largely unknown in mammals but has been studied in insects (Chapter 8).

The few thermoregulatory studies in mammals suggest that Type I pyrethroids induce hyperthermia, while Type II agents induce hypothermia. Permethrin, a Type I pyrethroid, has been found to elicit a hyperthermic response in the rat when dosed orally. By 4 h after administration of 150 mg/kg permethrin, core temperature increased to nearly 40°C (Figure 3.13). Cypermethrin, a Type II pyrethroid, elicited a hypothermic response in the rat, and a high dose of 120 mg/kg led to a maximum hypothermic effect at 3 h after dosing. Interestingly, a significant hyperthermic effect of a lower dose of cypermethrin (20 mg/kg) was noted at 90 min after dosing. Gilbert et al. (1989) also found that cismethrin, a Type I pyrethroid, induced an elevation in core temperature, while fenvalerate and deltamethrin, Type II pyrethroids, led to a reduction in core temperature in the rat. While the Type I and II pyrethroids have apparent opposite thermoregulatory effects, it is interesting to note that both types elicit reductions in motor activity that coincide with their peak effects on body temperature. Overall, the effects of the pyrethroids on the autonomic and behavioral control of body temperature remain largely unknown.

## 3.7 PRE-NATAL AND POST-NATAL EFFECTS

The increased susceptibility of the animal exposed *in utero* or post-natally to toxicants is the crux of developmental neurotoxicology. Thermoregulatory endpoints are occasionally used to assess toxicity in animals exposed during gestation or following parturition. Moreover, during pregnancy

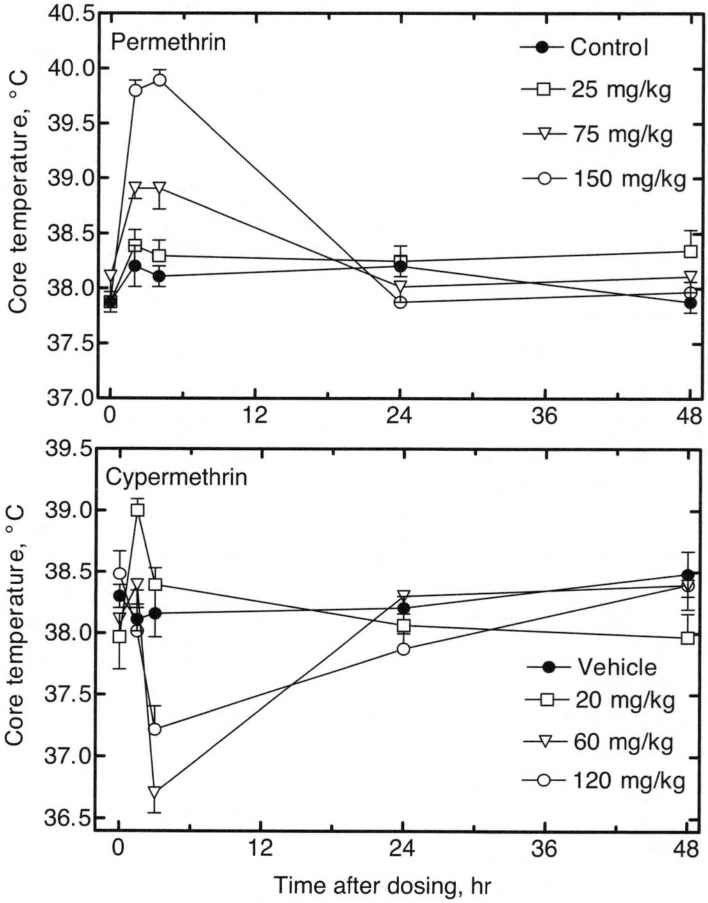


**Figure 3.13 Time-course of core temperature following oral dosing with a Type I (permethrin) and Type II (cypermethrin) pyrethroid in the rat. (Data modified from McDaniel, K.L. and Moser, V.C. (1993). *Neurotoxicol. Teratol.* 15: 71–83.)**

thermoregulatory reflexes are altered in rodents and other species, and such responses could alter the sensitivity to a toxicant. For example, the pregnant rat undergoes a gradual reduction in core temperature, a decrease in responsiveness to fever, and reduced capacity for cold-induced metabolic thermogenesis during the final week of pregnancy (for review, see Gordon, 1993). Following parturition, there is an explosive rise in metabolic rate and body temperature of the dam that persists until the pups are weaned. It is not known whether these changes in thermoregulation could affect the susceptibility of the dam, fetus, or newborn to a toxicant or drug.

Developmental toxicologists often study the responses of pre-weaned rats and mice and must consider the instability of the pups' core temperatures when tested at standard room temperature. Newborn rats, hamsters, and mice are essentially poikilothermic and do not show homeothermic capacity until an age of approximately 14 days (for review, see Gordon, 1990, 1993). Pups effectively use behavioral thermoregulation and huddle with each other and the warm dam and are able to maintain relatively stable internal temperatures of at least 30°C. However, when a pup is removed from the huddle for testing at standard room temperature, its internal temperature plummets because it has no insulation and little capacity for heat production. In a study on the toxicity of pyrethroids in neonatal mice, Pauluhn and Schmuck (2003) noted that mild changes in ambient temperature can have effects on other neurotoxic endpoints, including thermoregulation. In studies on fetal alcohol syndrome using rodent models, Zimmerberg et al. (1987) drew attention to the problem that behavioral teratologists face when conducting studies in rat pups. That is, the ambient temperature and the time during which the pup is isolated from the dam will have critical effects on body temperature.

### 3.7.1 Dioxin and PCBs

Dioxin (TCDD) is one of the most toxic xenobiotic agents, but its toxicity is expressed relatively slowly over days to weeks. Biliary excretion of thyroxine is enhanced by dioxin causing a type of hypothyroid state. For example, one week after intraperitoneal dosing with 45 μg/kg dioxin in adult rats, there was a 50% decrease in blood levels of $T_4$ but no change in $T_3$ along with a 0.6°C reduction in core temperature (Potter et al., 1983). The fetal and post-natal development of thermoregulation appears to be sensitive to very low levels of TCDD. Relatively high doses of TCDD have been found to target BAT function in adult rats. Injection of doses of TCDD that are eventually lethal (125 to 150 μg/kg) lead to functional deficits (reduced thermogenic response to norepinephrine) and histopathological alterations (fat and glycogen accumulation) in BAT (Rozman et al., 1986; Weber et al., 1987).

Development of thermoregulation in rodents appears to be sensitive to much lower doses of TCDD. Thermoregulatory responses were evaluated in adult rats and golden hamsters that had been exposed perinatally to TCDD. The adult offspring of the dams of rats exposed on gestational day 15 to 1.0 μg/kg TCDD were found to regulate their core temperature by 0.4 to 0.7°C below controls over an ambient temperature range of 10 to 24°C (Gordon et al., 1995). Thermoeffectors such as metabolism and evaporative water loss were unaffected, whereas thermal conductance was increased by TCDD treatment. Likewise, adult offspring from the dams of golden hamsters that were exposed on gestational day 12 to 2.0 μg/kg TCDD also showed marked thermoregulatory dysfunction. Hamsters treated with TCDD maintained their core temperature by approximately 1.0°C below controls over an ambient temperature range of 14 to 25°C (Gordon et al., 1996).

It is thought that disruptions in the hypothalamic–pituitary–thyroid axis during development may explain the permanent effects of TCDD on thermoregulation. When rats were made hypothyroid by chronic administration of propylthiouracil (PTU) in their drinking water, their core temperature was regulated at 0.5°C below controls over an ambient temperature range of 10 to 30°C (Yang and Gordon, 1997). Interestingly, PTU-treated rats housed in a temperature gradient preferred slightly warmer temperatures but maintained a hypothermic core temperature. The rats could have selected warmer temperatures in the gradient to reverse the hypothermic effects of PTU. Hence, toxicants that disrupt the hypothalamic–pituitary–thyroid axis may induce changes in the regulated core temperature.

PCBs (polychlorinated biphenyls) are also extremely toxic organochlorine compounds that bioaccumulate in the food chain. Many PCBs are thyrotoxic and result in thermoregulatory responses that are similar but not identical to hypothyroidism. Seo and Meserve (1995) fed rats a mixture of PCBs (termed Aroclor 1254) during pregnancy and lactation and measured core temperature and metabolic rate of the pups from day 4 to day 15. Aroclor 1254 treatment at doses of 125 and 250 ppm led to hypothermia in the pups, but it was not manifested until an age of 8 days, which corresponds with the time when control animals began to show improvements in thermoregulatory stability. PCB treatment also led to a profound reduction in circulating levels of $T_4$ and a slight reduction in metabolic rate.

### 3.7.2 Anticholinesterase Agents

The rat pup is generally considered to be more sensitive to organophosphate insecticides as based on parameters such as lethal dose and the efficacy to inhibit AChE activity (Pope and Chakraborti, 1992). Studying thermoregulation in the rat pup using conventional colonic probes is fraught with problems in addition to those discussed for adult rodents (see Chapter 7). Body temperature should be monitored in undisturbed pups that are housed with their dam and litter mates if one is interested in assessing the long-term effects of a toxicant under natural conditions. Radiotelemetry systems are being developed that are small enough to be implanted into rat pups at an early age (~12 days) and left in place such

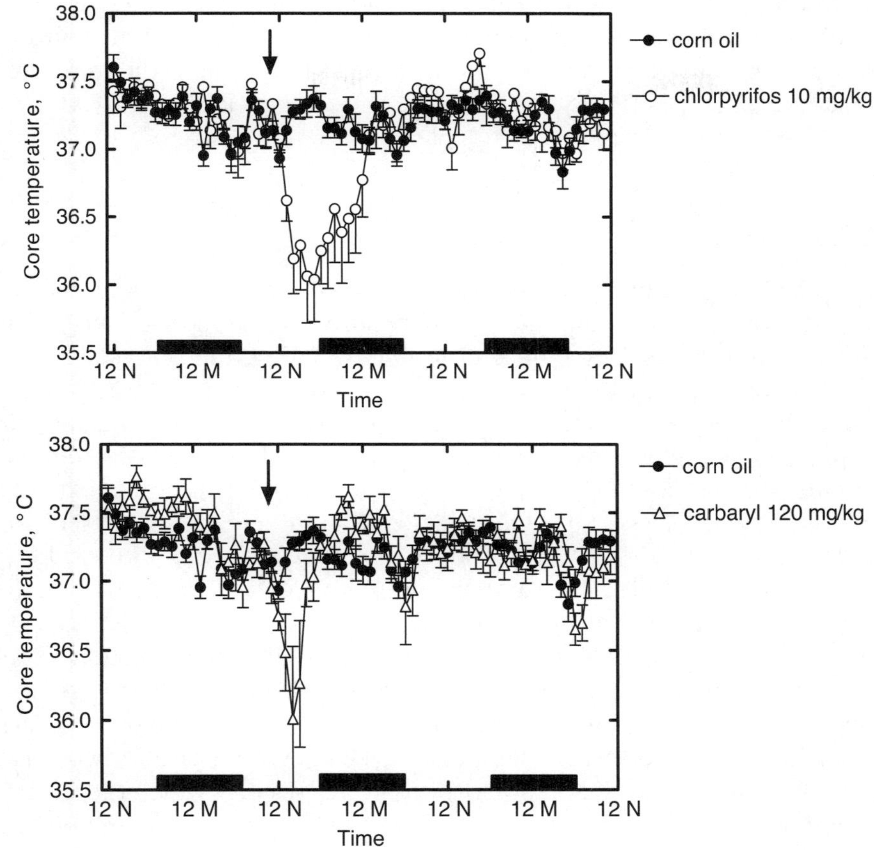


**Figure 3.14 Time-course of core temperature in male rat pups dosed by oral gavage with corn oil, 10 mg/kg chlorpyrifos, or 120 mg/kg carbaryl. Radiotelemetry units were implanted at 15 days of age and pups were dosed on day 17 while housed with dam and five litter mates at ambient temperature of 22°C. Compare with response of adults in Figures 3.1B and 6.2B. (From Mack and Gordon, unpublished observations.)**

that body temperature and other physiological parameters can be monitored in the undisturbed pup that is housed with dam and litter mates. Our laboratory measured core temperature by telemetry in pre-weaned rats dosed with an organophosphate and carbamate insecticide (Figure 3.14). Comparing these to the adult responses in Figures 3.1B and 6.2B, it can be seen that the hypothermic response of the 17-day-old rat pup to chlorpyrifos and carbaryl is qualitatively similar to that of adults. That is, in both age groups the hypothermic response to chlorpyrifos is much longer than that to carbaryl. However, the rat pup was found to be markedly sensitive to chlorpyrifos but relatively insensitive to carbaryl.

Interestingly, the rat pup regulates a stable core temperature of approximately 37.5°C but has no discernable circadian rhythm until around day 35. The lack of a circadian rhythm in the pup presents a different time course of recovery of body temperature, including lack of a delayed fever as is seen in the adult rat (see Chapter 6).

### 3.7.3 Alcohol

Fetal alcohol syndrome has prompted a variety of studies on the effects of maternal and post-natal alcohol exposure on thermoregulation. The studies with ethanol may be useful in understanding the potential effects of other toxicants on the development of thermoregulation. Zimmerberg and colleagues first noted alterations in behavioral thermoregulation in rat pups as young as 5 days that had been exposed pre-natally to ethanol (Zimmerberg et al., 1987, 1987a, 1993). The offspring from rats exposed from gestational days 6 to 20 to a diet containing ethanol as 35% of the total caloric intake or a pair-fed control diet were placed on the cold end of a simple thermocline (i.e., a temperature gradient). The ethanol-treated rats moved farther toward the warm end but had the same core temperature as control pups. Deficits in behavioral thermoregulation were detected when the rat pups were placed in the hot or cold end of a temperature gradient, thus forcing the pups to use their behavior or face a dangerous change in core temperature (Zimmerberg et al., 1993). For example, when control 10-day-old pups were placed in the cold end of the gradient with a temperature of 19°C or the hot end with a temperature of 42°C, they moved to a thermoneutral temperature of 31 to 33°C within an hour. Applying the same procedure with rat pups from alcohol-treated mothers revealed thermoregulatory deficits. Pups placed in the cold end selected a temperature of 25°C, while pups placed in the warm end selected 35°C over the same time period. Control and ethanol-treated pups placed in the thermoneutral zone of the gradient exhibited a similar behavioral temperature preference (Figure 3.15).

Thus, a deficit was detected only by challenging the behavioral response by forcing the pups to respond to being placed in a warm or cold area of the thermocline. It is interesting to note that ethanol treatment did not affect motor activity of the pups in the temperature gradient, suggesting that the effects on thermal preference were not a result of impaired movement within the gradient. Ethanol-treated 8- to 9-day-old pups also cool much faster than pair-fed controls when exposed to an ambient temperature of 22°C, suggesting that metabolic thermogenesis is impaired by ethanol treatment (Zimmerberg et al., 1987a). Overall, prenatal ethanol affects behavioral and autonomic thermoeffectors of the rat

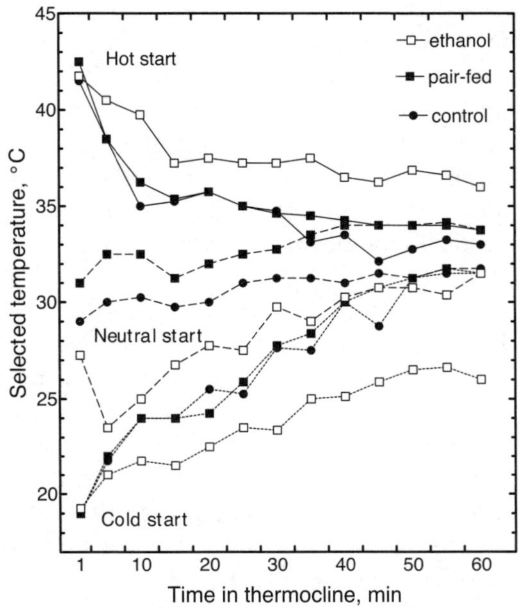


**Figure 3.15 Effect of pre-natal exposure to ethanol on behavioral thermoregulation of 10-day-old rat pups placed in the hot, cold, or thermoneutral location in a temperature gradient. Control indicates rats allowed to eat *ad libitum*; pair-fed indicates rat fed control diet calorically equivalent to that of rats fed ethanol diet. (Modified from Zimmerberg, B., Tomlinson, T.M., Glaser, J., and Beckstead, J.W. (1993). *Alcohol* 10: 403–408.)**

pup and leads to a widening of the zone for the regulation of core and selected temperature.

## 3.8 CHRONIC, SUBCHRONIC, AND REPEATED DOSING

The majority of toxicology studies on thermoregulation involve single, acute exposure to toxicants. Acute experiments are relatively simple and inexpensive to run compared to chronic and subchronic studies. There is a considerable data base on the thermoregulatory responses to repeated, subacute dosing with anti-ChE agents. However, there is hardly any information on how the thermoregulatory system is altered when an animal is exposed to relatively low levels of toxicants by either dietary, dermal, or inhalation routes.

Repeated administration of organophosphate and carbamate agents generally leads to tolerance to the toxicity of these agents (Costa et al., 1982). Tolerance means that the initial signs of anti-ChE poisoning, including behavioral and autonomic dysfunctions, abate or disappear with repeated administration of either the insecticide or a cholinergic agonist that mimics the primary acute effects of the insecticide. Studies of rodents have found that repeated dosing with subacute doses of organophosphates and carbamates results in the development of tolerance to the hypothermic effects of the agent.

Overstreet et al. (1973) conducted one of the first systematic studies on the CNS mechanisms of tolerance to DFP-induced hypothermia in the rat. Tolerance to the hypothermic effects of DFP (1.0 mg/kg followed by 0.5 mg/kg every third day) could be demonstrated by the absence of a hypothermic effect of the muscarinic agonist pilocarpine. Rats made tolerant to DFP also showed an attenuated hypothermic response to intrahypothalamic injection of carbachol, suggesting that tolerance involved a reduction in CNS sensitivity to muscarinic stimulation. The hypothermic effects of soman in the rat were abolished by the fourth injection when given every 2 to 3 days (Russell et al., 1986). The acute hypothermic effects of daily administration of chlorpyrifos in rat were attenuated with each daily oral administration (Rowsey and Gordon, 1997; Maurissen et al., 2000).

Telemetric monitoring of body temperature has recently been used to demonstrate the development of tolerance to the hypothermic and delayed hyperthermic effects of chlorpyrifos (Figure 3.16). Rats dosed in the morning for 4 consecutive days with chlorpyrifos at a dose of 10 mg/kg showed the greatest hypothermic response following the first dosing. The hypothermic response to the fourth dose was about 25% of that observed following the first dose. The following day, control rats given a larger dose of chlorpyrifos underwent a marked hypothermia and hyperthermia that persisted for 48 h, while tolerant rats had a smaller hypothermic response and a delayed hyperthermia that persisted for just 24 h. This study performed in female rats illustrates the greater sensitivity of female rats to chlorpyrifos (i.e., compared to responses of males in Figure 3.1).

Tolerance of the thermoregulatory system is especially important to understand since some of the toxic effects on the nervous system and other physiological processes may be directly dependent on changes in body temperature. Gupta and Dettbarn (1986) related the tolerance to the hypothermic effects of daily injections of DFP with recovery in protein metabolism in brain, skeletal muscle, and other regions. Acute hypothermia from DFP resulted in a suppression of protein synthesis, and it was not clear how recovery of protein synthesis and hypothermic tolerance to DFP would parallel each other. Rats were dosed on day 1 with 1.5

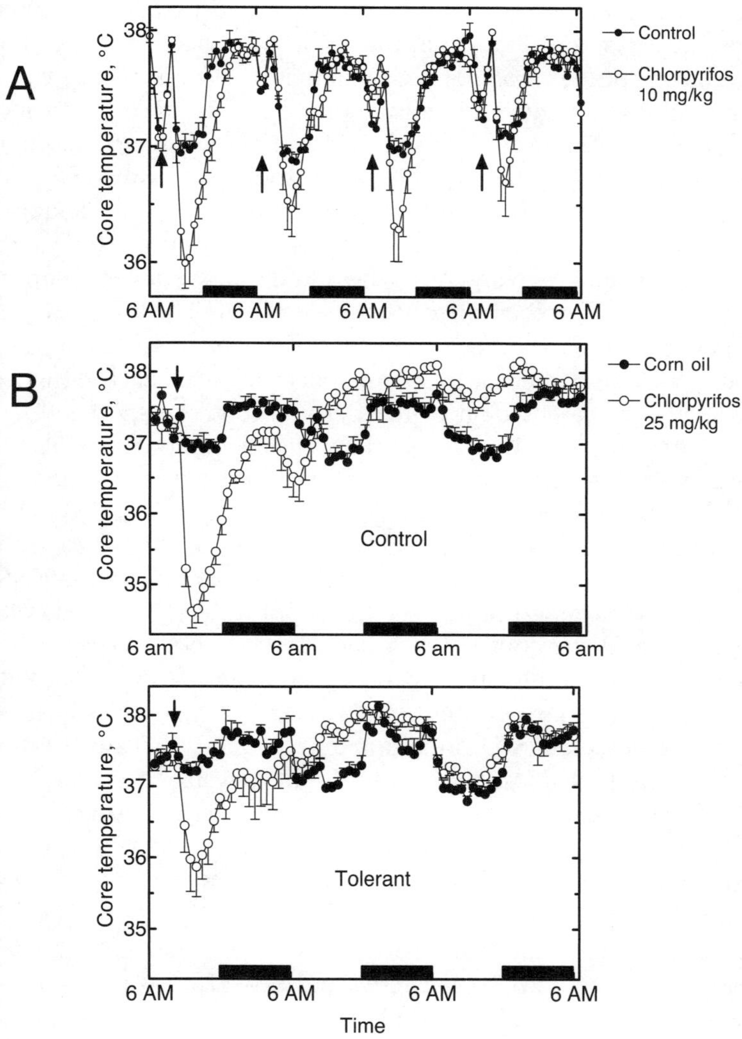


**Figure 3.16 (A) Development of tolerance to the organophosphate chlorpyrifos in female rats, given by oral gavage daily for 4 days. (B) The day after the fourth dose, rats were administered corn oil or a higher dose of chlorpyrifos. Tolerant rats exhibited smaller hypothermic and delayed hyperthermic response. (From Rowsey, P.J. and Gordon, C.J. (1997). *Toxicology* 121: 215–221.)**

mg/kg DFP and then daily for 14 days with 0.5 mg/kg DFP while incorporation of [$^{14}$C]valine was assessed at 7 and 14 days. The hypothermic effects of DFP peaked at day 5 and fully recovered by day 7. However, protein synthesis recovered by day 14 in skeletal muscle but not in discrete brain regions, liver, and kidney.

Not all thermoregulatory effects of the anti-ChEs are attenuated with repeated administration. Indeed, there is evidence of hypersensitivity depending on the thermoregulatory parameter and species. Smolen et al. (1986) found that repeated administration of DFP to mice led to increased sensitivity to DFP-induced hypothermia but tolerance to oxotremorine-induced hypothermia. The hypothermic response to soman in mice also did not improve with repeated dosing, whereas there was tolerance to the hypothermic effects of oxotremorine (Clement, 1991a). Repeated, daily dosing of the female rat with chlorpyrifos in the morning led to progressive attenuation in the hypothermic response with each dose for 4 days (Rowsey and Gordon, 1997). However, when the rats were dosed each day in the afternoon, their hypothermic response persisted with each administration and there was no attenuation in the delayed febrile response (Gordon & Rowsey, 1998; also see Chapter 6). Rats dosed intraperitoneally with the organophosphate malathion for 7 days displayed a 1°C elevation in their core temperature following the last dose of malathion (Haque et al., 1987). The toxic effects, and possibly the hyperthermic response to malathion, may be attributed to toxicant-induced lipid peroxidation in the CNS.

The studies discussed in this section report effects of repeated administration of low doses, but these doses are nonetheless very high when compared with most human exposures. It would be very important to understand the response of the thermoregulatory systems as well as other physiological processes when experimental animals are administered relatively low doses of toxicants in their food or water at levels that could be appropriately scaled to human exposure scenarios. This approach is also important in the risk assessment of wildlife that is exposed to pesticides and other toxicants in food and water (see Chapter 8). Such studies are rarely performed.

Mice administered the carbamate propoxur in their drinking water in increasing concentrations for 6 weeks developed tolerance as based on a higher $LD_{50}$ and a smaller hypothermic response to acute propoxur (Costa et al., 1981). In a recent study on the chronic effects of the organophosphate chlorpyrifos, rats were fed chlorpyrifos in their diet for a period of up to 6 months while core temperature was monitored by radiotelemetry (Gordon and Padnos, 2002). A 6-month exposure to a dose of 1 or 5 mg/kg/day chlorpyrifos led to a statistically significant increase in core temperature. However, the effect was inconsistent and hampered by an altered circadian rhythm as a result of the rats being placed on a restricted amount of food to control the daily intake of chlorpyrifos. When rats were allowed to feed *ad libitum* on a diet containing chlorpyrifos that resulted in a dosage of approximately 7 mg/kg/day, core temperature during the day increased by 0.2°C within a couple of days after the start of chlorpyrifos treatment, whereas nighttime core temperature was unaffected by the chlorpyrifos treatment. Chlorpyrifos treatment led to an 87% inhibition in serum ChE activity. The small albeit significant elevation in core temperature may represent a low-grade febrile response to chronic organophosphate exposure (see Chapter 6). This study illustrates how radiotelemetry is the only means to detect subtle changes in body temperature in a rodent treated chronically with a drug or toxicant.

# Chapter 4

# Temperature Effects on Chemical Toxicity

## 4.1 INTRODUCTION

A large part of this book is devoted to understanding an organism's integrative, thermoregulatory response to a toxicant. To briefly summarize, rodents generally respond to acute exposure to xenobiotic agents with a decrease in body temperature when tested at ambient temperatures that are equal to or less than their thermoneutral zone (e.g., ambient temperature <28°C; Chapter 3). However, when rodents are exposed to toxicants at environmental temperatures equal to or above thermoneutrality, a hyperthermic response is more likely to occur. In addition to the thermoregulatory responses of rodents and other mammals, many birds and lower vertebrates as well as invertebrates undergo marked changes in temperature when exposed to toxicants (Chapter 8). To this end, what are the consequences of the organism's thermoregulatory response? That is, does the change in body temperature affect the manifestation of toxicity?

As will be shown in this chapter, the thermoregulatory response to raise or lower body temperature will affect the toxicity of a chemical or drug. Even with a stable core temperature, the stress from ambient heat and cold stress can nonetheless influence toxicity. Humans and other relatively large mammals exposed to toxicants rarely show the hypothermic responses that are characteristic of rodents. A mild fever or hyperthermia is a more common thermoregulatory response in relatively large species (Chapter 6). It is essential to understand the role of environmental heat or cold stress in the modulation of toxicity, even when core temperature is unaffected. This topic is touched upon briefly in this chapter in Section 4.5 and in additional detail in Chapter 7.

## 4.2 SYSTEMIC, WHOLE-ANIMAL TOXICITY

Temperature affects the toxicity of essentially all biological end points, from the molecular to whole-animal level. In this section, the effects of temperature on toxicity at the level of the whole animal are discussed; the role of temperature on toxicity at the molecular and cellular level is discussed elsewhere in detail (Section 4.3.1). Studies on how the rate of a chemical reaction or physiological process is affected by temperature were underway as early as the first half of the 18th century (for review, see Cossins and Bowler, 1987, and Chapter 1). Because of the marked effects of temperature on the toxicity of xenobiotics, drugs, and other agents, studies in toxicology have improved our basic understanding of how temperature affects biological processes.

### *4.2.1 Temperature Coefficient and $Q_{10}$*

When studying the effects of temperature on a particular biological end point in the whole animal, organ, or tissue, thermal sensitivity is commonly described in terms of either the $Q_{10}$ or the temperature coefficient. $Q_{10}$ is defined as "the ratio of the rate of a physiological process at a particular temperature to the rate at a temperature 10°C lower, when the logarithm of the rate is an approximately linear function of temperature" and is calculated by the following equation (IUPS, 2001):

$$Q_{10} = (R_2/R_1)^{10/(T_2-T_1)} \tag{4.1}$$

where $R_1$ is the rate of activity at one temperature, $T_1$, and $R_2$ is the rate of activity at another temperature, $T_2$. Thus, a $Q_{10}$ of 2 means that the activity doubles with each 10°C elevation in temperature, a $Q_{10}$ of 3 is a tripling of activity, and so on. A $Q_{10}$ of 1 means there is no change in activity with temperature, and a $Q_{10}$ of less than 1 means that activity increases with a reduction in temperature. The $Q_{10}$ normally ranges between 2 and 3 for most physiological processes. Comparative physiologists and toxicologists studying the effects of toxicants and drugs on the lower vertebrates and invertebrates often rely on $Q_{10}$ to express the effects of temperature on physiological processes. Because $Q_{10}$ varies with temperature, usually decreasing in magnitude with a progressive rise in temperature, the use of this parameter is limited unless one reports $Q_{10}$

values with the temperature range of their determination (Cossins and Bowler, 1987).

The *temperature coefficient* is also a commonly used parameter to describe thermal sensitivity and is defined as "the ratio between the change in any temperature dependant activity and the defined temperature range within which this change occurs" (IUPS, 2001). In other words, the temperature coefficient is an expression of the change in rate or activity of a process for a 1.0°C change in temperature. Unlike the $Q_{10}$ parameter, the temperature coefficient is either positive or negative, depending on how temperature affects the physiological process. Temperature coefficient is probably used more often than the $Q_{10}$ in studies on the effect of temperature on toxicological and pharmacological processes in birds and mammals as well as in other species. Like $Q_{10}$, temperature coefficient will also be dependent on the temperature range used in its calculation.

When body temperature changes, the temperature coefficients of the components of a physiological system will determine, in large part, the efficacy and toxicity of a chemical or drug. In a typical diagram used to model pharmacological and toxicokinetic properties (Figure 4.1), one can better visualize the complexities of how a change in temperature affects intake, metabolism, and excretion of a toxicant. The intake of the toxicant into the body through the gastrointestinal tract, lungs, and skin, the metabolic activation and deactivation in the liver, and excretion in the kidneys and other sites are essentially dependent on temperature. Specifically, it is the temperature coefficients of the enzymes that govern the absorption, metabolism, and excretion of the toxicant.

The temperature coefficient can be applied to all levels in a biological system in which temperature can change, such as activity of an enzyme, rate of axonal transport, heart rate, and essentially any physiological process of interest (see Table 1.1). The toxicant can also directly change the temperature coefficient of an enzyme. To this end, one can see how body temperature and the characteristics of the toxicant govern the pharmacokinetics of the toxicant and its metabolites. The time versus concentration of the toxicant and its metabolites in a given site or compartment in the body such as outlined in Figure 4.1 is dependent primarily on temperature and the temperature coefficient.

### 4.2.2 Magnitude Versus Duration

The response or sensitivity to a toxicant is characterized by both the magnitude and length of time that the toxicant persists in the body (Doull, 1972). In most cases, hypothermia is going to prolong the duration but reduce the magnitude of toxicity. That is, when body temperature is lowered, the mechanism(s) responsible for metabolizing and excreting the

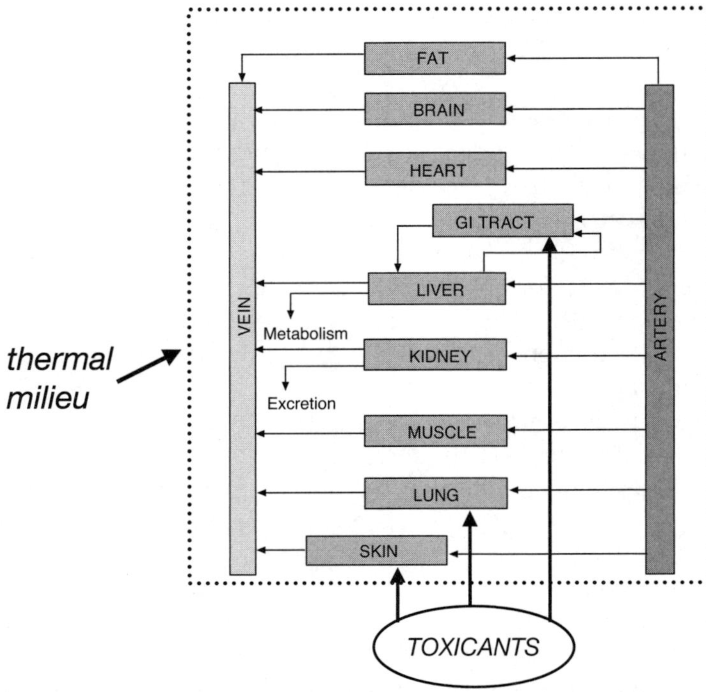


**Figure 4.1 General mammalian model for physiologically based pharmacokinetic modeling. Temperature governs the uptake, metabolism, and excretion of a toxicant and its metabolites. (Model adapted from Mordenti, J. (1986). *J. Pharmacol. Sci.* 75: 1028–1040.)**

toxicants are going to be depressed because the temperature coefficients for these processes are positively affected by temperature. Doull (1972) developed a general prediction of the response of biological systems to toxic levels of drugs that can also be applied to xenobiotic agents: "Temperature is directly correlated with the magnitude and inversely correlated with duration of drug response in biological systems." In other words, while the concentration of a toxicant will persist longer during hypothermia, the toxicity of the agent is reduced. This tenet of Doull's was stated over 30 years ago and remains a critical issue in toxicology and pharmacology. The magnitude and duration of toxicity have a bearing on experimental and clinical toxicological studies. First, rodents are the primary species for toxicological studies, and their thermoregulatory system is extremely sensitive to toxic insults. Their integrated thermoregulatory responses will have a marked impact on the pharmacokinetics and overall toxicity of the chemical or drug. Second, if hypothermia affords protection to toxicants, then this should have important implications in the way in which humans and domestic species are treated following episodes of poisoning (see Chapter 5).

### 4.2.3 Lethality

The direct effect of body temperature on the toxicity of drugs and toxic chemicals was recognized well over 100 years ago (for review, see Fuhrman, 1946). Prior to the development of modern analytical methods, the frog was commonly used as a species for bioassays of drugs and other chemicals. Because the body temperature of frogs and other poikilotherms is directly dependent on ambient temperature, it was quickly determined in these early studies that body temperature has a marked effect on the pharmacokinetics of drugs and toxicants. For example, studies in the 1890s on the toxicity of cholchicine showed that the frog was 400 to 500 times as sensitive when dosed at an ambient temperature of 32°C as compared to 20°C. Toxicity studies of amphibians maintained at different temperatures led to a general conclusion for cholchicine that "the temperature of the body is the chief factor influencing toxicity" (Fuhrman, 1946).

The lethal dose (e.g., $LD_{50}$) has conventionally been used in rodent studies as a benchmark of toxicity. While tests using death as an end point are rarely used in current research studies, these older studies are very useful for illustrating how overt toxicity is affected by body and ambient temperature. Keplinger et al. (1959) were among of the first to perform a systematic analysis of the lethal dose of a variety of drugs and chemicals in the rat. The approximate lethal dose of 58 compounds was determined in rats housed for 45 min before and 72 h after dosing at environmental temperatures of 8, 26, and 36°C (Figure 4.2). When compared to a temperature of 26°C (i.e., slightly below the rats' thermoneutral zone) the lethal dose was consistently lowered for all compounds administered to rats at an ambient temperature of 36°C. In the selected examples in Figure 4.2, the lethal dose of DDT, toluene, methacholine, and chlorpromazine was less than 50% of those observed at 26°C. There were inconsistent effects of exposure to an ambient temperature of 8°C on the lethal dose. The toxicity of some compounds was increased by cold exposure, while the toxicity of others was reduced or unchanged when compared to the lethal dose assessed at an ambient temperature of 26°C.

It is interesting to note the consistent effect of higher environmental temperatures on chemical lethality in spite of marked differences in the chemicals' mechanisms of action. For example, DDT causes tremor and hyperthermia and would be expected to be more toxic at higher temperatures by inducing hyperthermic death. However, many other agents that normally induce hypothermia such as ethanol, toluene, and chlorpromazine at 26°C are nonetheless more toxic in the heat.

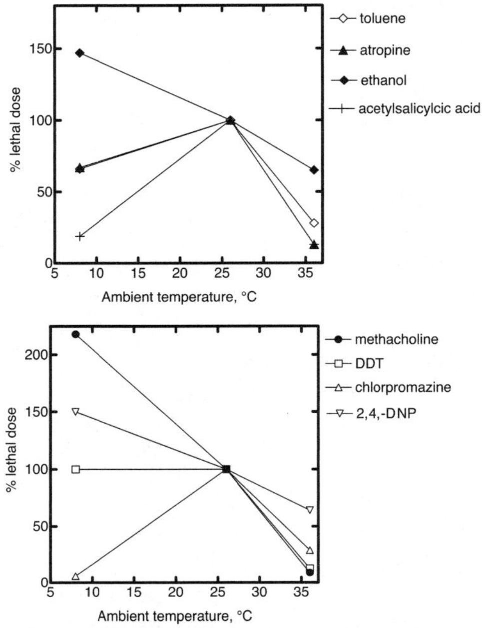


**Figure 4.2 Effect of ambient temperature on the approximate lethal dose of a variety of toxicants and drugs injected intraperitoneally in rats. All lethal doses are reported as the percent of the lethal dose measured at an ambient temperature of 26°C. Note difference in scales of ordinate. (Modified from Keplinger, M.L., Lanier, G.E., Deichmann, W.B. (1959). *Toxicol. Appl. Pharmacol.* 1: 156–161.)**

The resiliency of the mouse to withstand relatively large reductions in body temperature is remarkable, and it is not surprising to find that acute exposure to toxicants led to profound reductions in body temperature. Similar to the findings in rats, the lethal effects of xenobiotics in mice are directly related to ambient and body temperature (Figure 4.3). In these examples, administering mice approximate $LD_{50}$s for ethanol, chlordimeform, sulfolane, and lead acetate resulted in a high percentage of lethality when the mice were housed at a temperature of 35°C, whereas lethality was markedly reduced when housed at cooler temperatures of 20 to 30°C. An ambient temperature of 35°C is approximately 3°C above the upper critical temperature of the thermoneutral zone, and the ability of the mouse or rat to become hypothermic when housed at this temperature is attenuated or completely blocked. At this warm temperature, mice and rats have to rely primarily on evaporative water loss to dissipate heat, and it is likely that a toxicant would impair this critical thermoeffector. The inability to become hypothermic following the acute toxic insult is likely responsible for the increased lethality in a warm environment.

Ethanol is one of the most abused drugs throughout the world and is also used as a solvent and as a prototypic toxicant in a variety of studies.

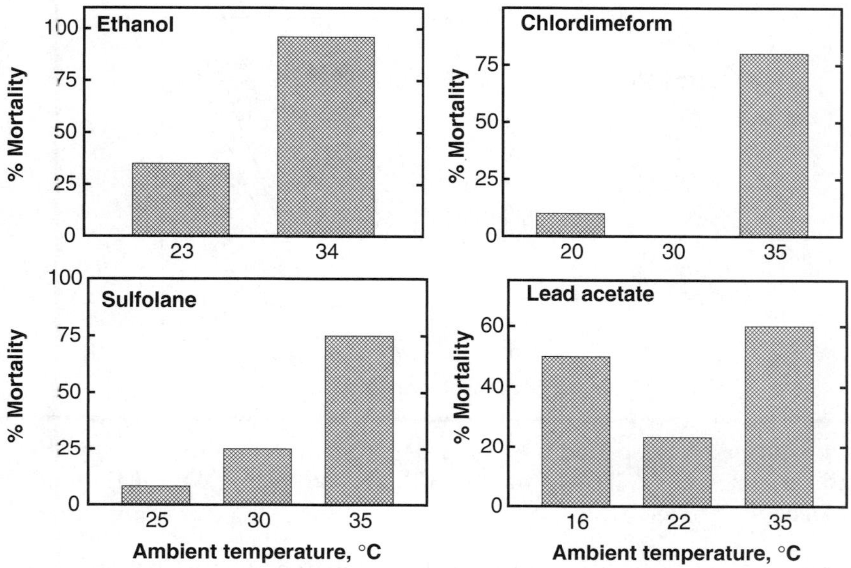


**Figure 4.3 Examples of how ambient temperature affects lethal effects of a variety of chemical toxicants administered intraperitoneally to mice; ethanol, 7.0 g/kg (Finn et al., 1989); chlordimeform, 90 mg/kg (Gordon et al., 1985a); sulfolane, 1270 mg/kg (Gordon et al., 1986); and lead acetate, 125 mg/kg (Baetjer et al., 1960). Note difference in scales of ordinate.**

Acute ethanol intoxication in humans is very common under a wide range of ambient temperatures, and there is a considerable data base on the impact of body temperature on the toxic effects of ethanol (Kalant and Le, 1983). The lethal dose of ethanol as a function of ambient temperature has been well characterized in mice (Malcolm and Alkana, 1983). For example, by lowering the ambient temperature for housing ethanol-exposed mice from 35 to 30°C, there was a reduction in core temperature from 37.7 to 34.1°C and a resulting 42% increase in ethanol's $LD_{50}$. In rats, lowering ambient temperature from 36 to 26°C resulted in a 35% increase in ethanol's lethal dose (Figure 4.2).

### *4.2.4 Patterns of Toxicity as a Function of Temperature*

The toxicity of drugs as a function of ambient temperature, as first developed by Fuhrman and Fuhrman (1961), can also be applied to chemical toxicants (Figure 4.4). The toxic potency can describe lethal and nonlethal end points. In general, there are three basic functions that describe the toxicity of drugs with changes in ambient temperature: (Type A) a V- or U-shaped function with minimum toxicity (i.e., high $LD_{50}$) at

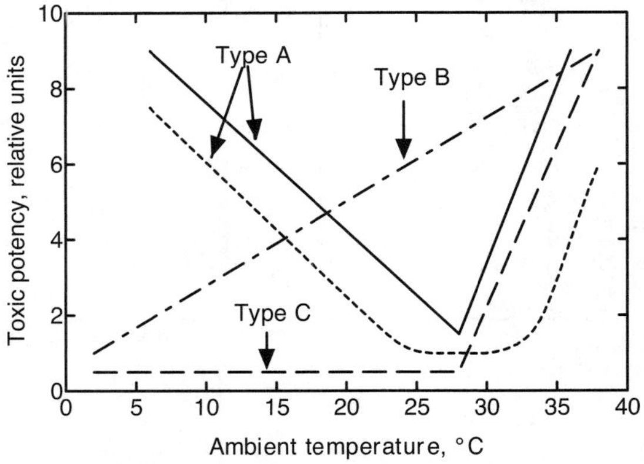


**Figure 4.4 General patterns of drug and chemical toxicity as a function of ambient temperature. (From Fuhrman, G.J. and Fuhrman, F.A. (1961).** ***Annu. Rev. Pharmacol.*** **1: 65–78.)**

a narrow range of temperatures usually below the metabolic thermoneutral zone and increased toxicity as temperature increases or decreases from the nadir of the curve, (Type B) a linear function with toxicity increasing with temperature, and (Type C) a function with a constant toxicity over a wide range of cool ambient temperatures and then a sudden increase in toxicity at a threshold temperature around or above the upper portion of the thermoneutral zone. The Type C response is questionable in the sense that if temperature is reduced to sufficient levels, a toxicant would eventually respond in a Type A response.

Like the efficacy of many drugs, the toxicity of xenobiotics can follow one of the three responses depicted in Figure 4.4. In general, the *in vivo* toxicity of mice and rats as a function of temperature exhibits the Type A response (V-shaped). Examples of a Type A response include chlorpromazine and toluene. Toxicants such as ethanol and dinitrophenol display a Type B response, while DDT shows a Type C response (Figure 4.2). Whether a toxicant fits the Type A, B, or C function is dependent in part on the mechanism of action (Weihe, 1973). For example, agents that are sympathomimetics and stimulate metabolic thermogenesis would be expected to take on a Type B response. That is, the effectiveness of a compound to cause a dangerous elevation in core temperature, such as oxidative phosphorylation uncouplers, would increase with ambient temperature.

The Type A response is most likely the most relevant function for many of the toxicants discussed in this book. The V- or U-shaped function is thought to be a result of agents that interfere with the CNS control of body temperature. Although Weihe (1973) suggested that Type A drugs would not be involved with a resetting of the thermoregulatory set-point, the evidence from many toxicants would suggest otherwise. This function is a result of the combined effects of the toxicant and ambient temperature on thermoregulatory control. That is, many toxic chemicals induce a regulated hypothermic response in rats and mice. There is an ideal range of ambient temperatures where the treated rodent can maintain a moderate hypothermic core temperature (Chapter 5). If ambient temperature changes above or below optimal levels, the thermoregulatory effectors are overwhelmed and are unable to maintain core temperature, possibly resulting in death from thermoregulatory failure.

### 4.2.5 Nonlethal End Points

Other than death as an end point, there are many other pathological and physiological end points of toxicity that are affected by body and environmental temperature. Indeed, toxicological and pharmacological researchers must deal with the potential impact of a drug or toxicant on the body temperature of their rodent model and how it may affect a particular end point. The hypothermic response to a toxicant will most likely affect a variety of physiological and pathological processes. Blocking the toxicant-induced hypothermia by raising ambient temperature is bound to alter the expression of a toxicological end point. One can peruse the literature and find a multitude of toxicological studies where the rodent's core temperature was blocked such that it would not be factored into the toxicity of the chemical. In fact, blocking the hypothermic effect of a toxicant is often found to worsen the toxicological outcome. In some instances, excessive ambient warming raises body temperature above normal, further exacerbating the toxicity of the chemical.

### 4.2.6 Nervous System

The temperature sensitivity of the nervous and cardiovascular systems and their responses to toxicants have been well studied. These systems have high aerobic demands and are expected to be more sensitive to temperature and toxic insults. Neurotoxicologists have devoted a considerable effort to understanding how hypothermia and hyperthermia affect a variety of electrophysiological end points (for review, see Janssen, 1992). The thermoregulatory effects of a suspected neurotoxicant should be assessed to determine if the toxic effect on the nervous system is attributable to the indirect change in body temperature or is due to a direct effect of the toxicant on CNS function.

Exposing the rat to the formamidine insecticide chlordimeform results in a variety of neurotoxic effects, including reduced amplitude, increased latency of visual (light flash) evoked response (VER), and slowing of axonal transport in the optic nerve (Boyes et al., 1985). These effects were observed along with a 2 to 3°C reduction in core temperature when the rat was exposed at an ambient temperature of 22°C. When chlordimeform was administered at an ambient temperature of 30°C, which blocked the hypothermia, the effects on the axonal transport and the latency and amplitude of the VER were nearly eliminated; however, other deficits in visual function persisted, including peak-to-peak amplitudes and increased pattern-reversal latencies.

In another example, rats treated with the organotin compound triethyltin underwent marked dysfunction in their VER and developed hypothermia (Figure 4.5). Under standard room temperature conditions, the rats developed a transient hypothermia and the VER recovered within 24 h after triethyltin (TET) exposure. If the hypothermic response was attenuated by housing the rats at an ambient temperature of 30°C for just 7 h following TET treatment, the recovery of the VER was markedly delayed (Figure 4.5). In fact, VER in rats subjected to the warm temperature did not recover for at least 300 h after TET exposure (Dyer and Howell, 1982).

### 4.2.7 Cardiovascular System

Thermoregulation is tightly integrated with the function of the cardiovascular system. As ambient temperature is varied, the sympathetic modulation of peripheral vasomotor tone to either increase or reduce heat loss places constraints on the regulation of cardiac output and blood pressure. Heart rate and cardiac output increase concomitantly with metabolic rate in rodents subjected to ambient temperatures below their thermoneutral zone (see Figure 2.2). The interplay between cardiovascular demand and thermoregulation during exercise is also well studied (for review, see Simon, 1999).

Toxicants that affect heart rate and blood pressure will most likely affect the control of body temperature. Moreover, the sensitivity of the cardiovascular system to the toxicant will in turn be affected by body temperature. For example, chlordimeform is cardiotoxic and was shown to induce marked reductions in heart rate, blood pressure, and hypothermia; however, the cardiotoxic effects persisted even when the core temperature of the anesthetized rat was maintained at 37°C (Watkinson et al., 1989). The chlorpyrifos-induced reductions in heart rate and blood pressure were minimized if core temperature was reduced to 35°C; the cardiovascular effects were augmented when core temperature was maintained at 33 or 37°C. The lethality of chlordimeform was also minimal at a core temperature of 35°C. Hence, a moderate 2°C decrease in core temperature, which happens to be the same core temperature that the

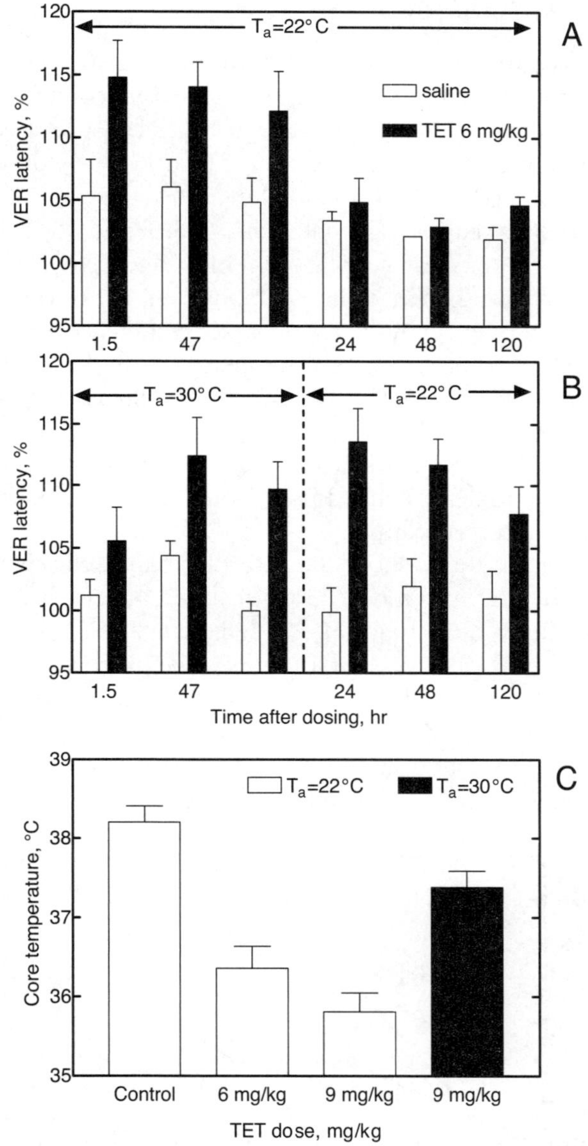


**Figure 4.5 Time-course of the visual evoked response (VER) in rats dosed with triethyltin (TET) and maintained continuously at 22°C (A) or maintained at 30°C to attenuate the hypothermic effects of TET (B). (C) Dose–response effect of TET on core temperature at 22 and 30°C. (Modified from Dyer, R.S. and Howell, W.E. (1982). *Neurobehav. Toxicol. Teratol.* 4: 267–271.)**

awake, unrestrained rat maintains when allowed to behaviorally thermoregulate, is protective (Chapter 3). Temperatures above or below this level exacerbate the cardiotoxicity of chlordimeform.

## 4.2.8 Liver and Kidney

The metabolism and excretion of toxicants in the liver and kidney will be affected by body temperature. A xenobiotic agent may be incorrectly identified as a hepatoxicant or renal toxicant because the ensuing hypothermia leads indirectly to a reduction in function. For example, paraquat was suspected as a hepatoxicant in the mouse because it reduced the biliary clearance of sulfobromophthalein (BSP) and indocyanine green (ICG), agents used to assess hepatic function (Cagen et al., 1976). Paraquat induced a profound hypothermic response in the mouse, with body temperature remaining below normal for several days after a single exposure (Table 3.8). When the body temperature of the paraquat-treated mouse was prevented from decreasing, the biliary clearance of ICG returned to normal, whereas the clearance of BSP was still impaired. Thus, some facets of liver metabolism appear to be directly dependent on temperature, and toxicants that lower body temperature will affect the activity of this excretory pathway.

Body temperature influences the rate of ethanol elimination, presumably through the thermal modulation of alcohol dehydrogenase activity in the liver (Romm and Collins, 1987; Bejanian et al., 1990). Modulating the hypothermic response of ethanol by housing mice at different ambient temperatures has a marked impact on the clearance of ethanol from the circulation (Figure 4.6). Mice dosed with 3.6 g/kg ethanol and allowed to become severely hypothermic with a core temperature of 26°C had an ethanol elimination rate of 0.01g/ml/min as compared to 0.016 mg/ml/min

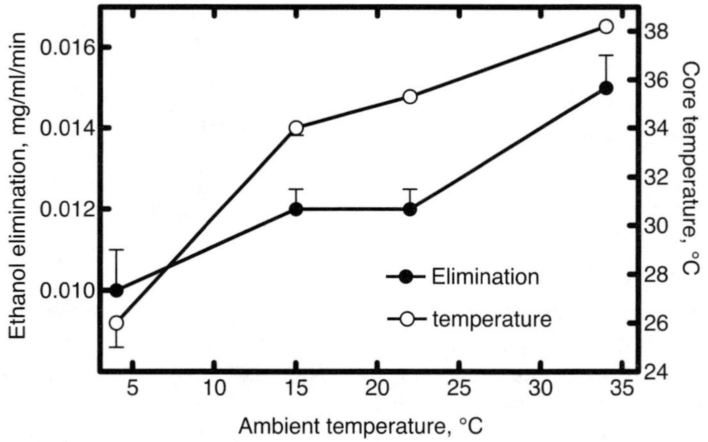


**Figure 4.6 Effect of ambient temperature on body temperature and rate of ethanol clearance in mice. (Modified from Bejanian, M., Finn, D.A., Syapin, P.J., and Alkana, R.L. (1990). *Alcohol* 7: 331–337.)**

in normothermic animals. Of course, the lower elimination rate and hypothermia are more favorable for surviving the ethanol exposure (Figures 4.2 and 4.3).

Plasma levels of enzymes that are indicators of liver function have been found to be affected by ambient and body temperature (Watanabe et al., 1990a). Mice were dosed with sodium selenite and maintained at ambient temperatures of 9, 22, or 33°C for just 4.5 h (1.5 h before and 3 h after injection) and were then housed at a temperature of 22°C. Plasma levels of lactate dehydrogenase (LDH) were found to be markedly affected by ambient temperature by 3 days after selenite exposure (Figure 4.7). Blocking the hypothermic effects of selenite by exposing the mice to an ambient temperature of 33°C led to a near fourfold increase in LDH when compared to mice maintained at a cooler temperature that allowed for a hypothermic response to develop. Plasma levels of creatine kinase and glutamic-oxaloacetic transaminase were similarly affected by the temperature and selenite treatment.

It is also interesting to note the correlation between liver function and loss in body weight. Selenium exposure led to loss in body weight that reached a nadir approximately 10 days after exposure. Weight loss was exacerbated by exposure to the warm temperature and nearly eliminated by brief exposure to a cold environment. Cold exposure was associated

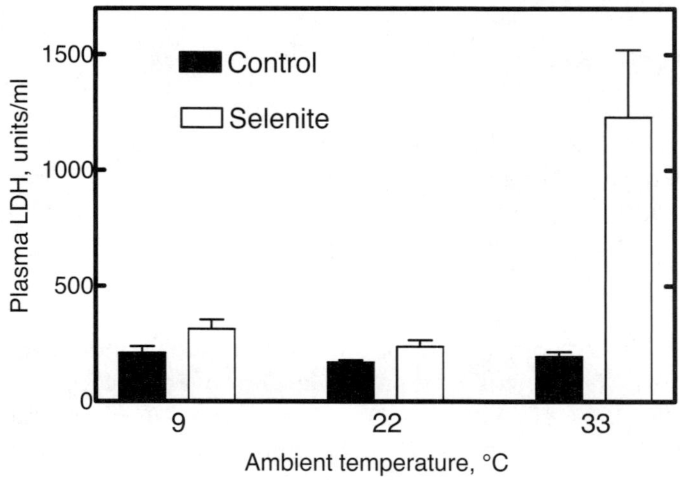


**Figure 4.7 Effect of 45 μmol/kg sodium selenite and simultaneous exposure to ambient temperatures of 9, 22, or 33°C for 3 h on plasma levels of lactate dehydrogenase (LDH) in the mouse when measured 3 days after selenite exposure. Data plotted as mean + S.E. (Data taken from Watanabe, C., Weiss, B., Cox, C., and Ziriax, J. (1990). *Fund. Appl. Toxicol.* 14: 578–588.)**

with greater levels of selenite in the liver, while heat exposure resulted in greater levels in the kidney when measured 3 h after exposure (Watanabe et al., 1990a). The induction of metallothionein, a protein involved in the detoxification of heavy metals, was apparently not affected by blocking the hypothermic response to selenium (Iwai et al., 1988). Nonetheless, a relatively brief exposure to an ambient temperature to either augment or attenuate selenite's hypothermic effect had a remarkable effect on hepatic function and weight loss.

## 4.3 CELLULAR AND MOLECULAR MECHANISMS OF TOXICITY

Assessing how temperature affects the toxicity of drugs and other chemical agents using *in vitro* cellular preparations has been an invaluable approach in pharmacology and toxicology. *In vitro* preparations provide ideal opportunities to manipulate and study intracellular processes while maintaining a precisely controlled thermal environment. In addition, temperature can be easily changed with *in vitro* preparations, allowing researchers to better understand mechanisms that would be extremely difficult to acquire in a whole-animal study.

### *4.3.1 Temperature and Cell Death*

It seems intuitive that the effects of a toxicant on cellular and intracellular processes such as membrane permeability, membrane transport, receptor binding, enzyme inhibition, and other processes would be directly related to temperature ($T$) in a first-order, exponential function:

$$v = a\mathrm{T}^b \qquad (4.2)$$

where $v$ is the velocity or rate of the reaction and $a$ and $b$ are constants. A plot of log T versus log $v$ is a straight line with a slope of $b$ (Cossins and Bowler, 1987). Equation (4.2) is an extreme oversimplification, especially for biological systems. The Arrhenius equation takes into account the critical incremental energy of activation (μ) to describe the effects of temperature on a chemical reaction and has been found to be useful for quantifying the behavior of biological systems subjected to a change in temperature:

$$\ln (k_1/k_2) = \mu/\mathrm{R}(1/\mathrm{T}_1 - 1/\mathrm{T}_2) \qquad (4.3)$$

where $k_1$ and $k_2$ are velocity constants at absolute temperatures $T_1$ and $T_2$ and R is the universal gas constant (8.3 J °K$^{-1}$ mol$^{-1}$). A plot of ln $k$ versus 1/T yields a straight line with a slope of μ/R. Arrhenius plots are extremely useful in biophysical and biochemical reactions but become more limited as the complexity of the preparation increases (see Cossins and Bowler, 1987). *In vitro* cultures of cells are certainly not simple systems, and one must always be cautious in using such simplified first-order kinetic reactions to study these complex systems.

The kinetics of cell death as a function of temperature and toxicant exposure is often studied using an Arrhenius plot. When the relationship between temperature and cell killing is linear, then it could be assumed that the toxicant follows an Arrhenius type of temperature dependence (Figure 4.8). However, a sudden departure from linearity would indicate a nonadditive cytotoxic effect of the chemical with temperature. For example, Arrhenius plots for *in vitro* cell inactivation by the genotoxicants bleomycin and paraquat show a distinct, nonlinear break, while other anti-cancer agents show a linear temperature-dependent cytotoxicity (Takayoshi et al., 1987). A nonlinear Arrhenius plot such as that depicted in Figure 4.8 could result from simultaneous activation of the toxicant of two or more processes that have different activation energies.

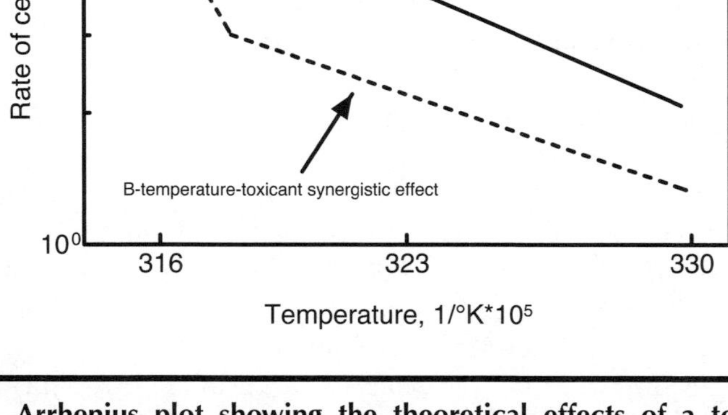


**Figure 4.8 Arrhenius plot showing the theoretical effects of a toxicant that exhibits (A) no synergistic interaction with temperature on cell death, meaning there is a linear increase in toxicant-induced cell death with rising temperature or (B) a synergistic or nonadditive interaction leading to a nonlinear break in the Arrhenius plot. See Takayoshi et al. (1987).**

## 4.3.2 Chemotherapy

The study of chemotherapy has brought forth a large data base on the interaction between hyperthermia and the toxicity of chemicals and solvents that are considered as candidates for use in chemotherapy. This data base is a useful resource in the study of *in vitro* and *in vivo* models of toxicology. Researchers in oncology and chemotherapy have attempted to capitalize on the nonadditive interaction between hyperthermia and cytotoxicity to enhance the effectiveness of chemotherapeutic agents (Hahn, 1979; Falk and Issels, 2001). *In vitro* preparations have frequently been utilized in these endeavors. The viability of Chinese hamster cells was essentially unaffected from temperatures of 37 to 41°C and then dropped steadily as temperature increased from 41 to 44°C. The thermal sensitivity of cultured cells in terms of killing efficacy was exacerbated by a variety of chemotherapeutic agents when the cells were incubated at hyperthermic temperatures (i.e., 41 to 45°C). In addition, commonly used solvents in temperature–toxicity drug studies are also more cytotoxic with increased temperature (Li et al., 1977). For example, solvents such as dimethyl sulfoxide (DMSO) and ethanol, which have no effect on cell viability at normal body temperature, were found to lower the lethal temperature by approximately 1°C.

Ideally, the use of localized hyperthermia with relatively low concentrations of chemotherapeutic agents would allow for more effective therapeutic treatment of cancerous tissues without harming healthy tissue. However, many toxicants as well as the solvents impart thermal tolerance, making the cells more resistant to hyperthermic death. Chemical-induced thermotolerance is likely to counter the efficacy of the chemotherapeutic agent. The phenomenon of tolerance has clearly hampered advances in the field of hyperthermia and chemotherapy. For example, Chinese hamster ovary cells that are exposed briefly to ethanol, other alcohols, and anesthetics undergo a marked increase in their heat tolerance (Hahn et al., 1985). Induction of heat shock proteins by these chemicals is most likely responsible for their increased heat tolerance (see Chapter 9). Another critical phenomenon in this field is the differential thermal sensitivity of tumor cell lines between rodents and humans (Hahn et al., 1989). Human cell lines appear to be more heat tolerant than rodent cells. The threshold for an abrupt increase in thermal sensitivity is 42 to 43°C for rodent cell lines and 44°C for human cell lines. Human cell lines appear to develop thermal tolerance more easily than rodent cells. These *in vitro* species differences present a challenge in the extrapolation of data from rodent to human and could have significant implications in predicting the efficacy of chemotherapeutic agents (Hahn et al., 1989).

## *4.3.3 Membrane Fluidity and Toxicity*

Cell membrane fluidity is a crucial aspect of cellular function. The membrane lipids can undergo a change in their physical state over a physiological range of temperatures (i.e., bulk phase transition). As temperature decreases from optimal conditions, the membrane physical state transcends from a liquid-crystalline to gel state. An Arrhenius plot of the reciprocal of temperature versus fluidity of the membrane generally shows a discontinuity that reflects the temperature of liquid-crystalline to gel transition (Cossins and Bowler, 1987). The fluid characteristics of the lipid bilayer affect membrane-bound enzyme activity by maintaining an optimal conformation of protein structure. Moreover, the physical characteristics of the lipid bilayer are influenced by temperature and the chemical composition of the phospholipids that make up the lipid bilayer; the proportion of saturated versus unsaturated fatty acids in the bilayer is especially critical. In general, more unsaturated fatty acids in the lipid membrane result in increased fluidity and a decrease in the gel-to-liquid transition temperature. In addition to the proportion of unsaturated fatty acids, other components of the phospholipids influence membrane fluidity (for review, see Hazel, 1995). It follows that toxicants that affect the fluidity of the cellular membrane could alter the function of membrane-bound enzymes, and this process will show a marked interaction with temperature.

Ethanol is a toxicant that has been shown to interact with the lipid bilayer and alter membrane fluidity. Membrane-bound $Na^+$-$K^+$ ATPase and acetylcholinesterase are examples of enzymes that are affected by ethanol via changes in fluidity of the lipid bilayer (Guerri and Grisolia, 1983; Collins et al., 1984). Short-term exposure to ethanol causes expansion of the lipid bilayer and a resulting increase in membrane fluidity (Rubin and Rottenberg, 1983). However, with chronic exposure to ethanol, there are biochemical compensations to restore membrane fluidity, but these are accompanied by changes in the gel-to-fluid transition temperature (Guerri and Grisolia, 1983). For example, an Arrhenius plot of $Na^+$-$K^+$ ATPase activity of synaptosomes from rats subjected to chronic ethanol treatment for 4 weeks showed a discontinuity (i.e., gel-to-liquid transition) at 33.7°C as compared to 28.5°C for control animals (Figure 4.9). Organophosphate and pyrethroid insecticides have also been shown to interact with the membrane phospholipids and alter membrane fluidity. Parathion was shown to reduce membrane fluidity, while fenvalerate increased fluidity (Sarkar et al., 1993; Antunes-Madeira et al., 1994). These *in vitro* studies provide alternatives to explain mechanisms of toxicity and should lead one to pursue how membrane fluidity and temperature interact to affect pesticide toxicity.

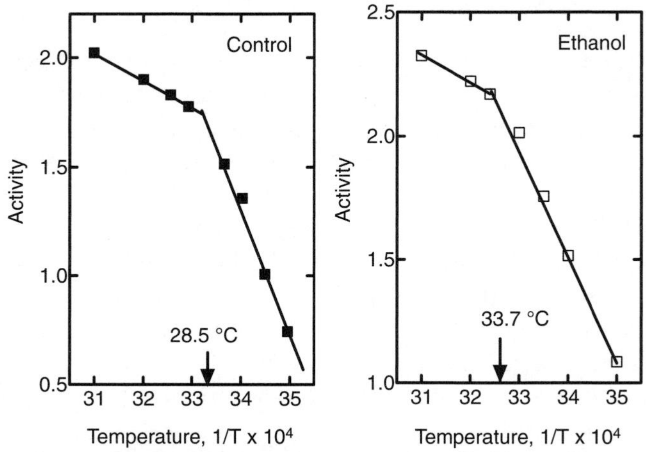


**Figure 4.9 Arrhenius plot of $Na^+$-$K^+$ ATPase activity of synaptosomal membranes from rats treated chronically with ethanol (36% of total calories) or control liquid diet for 4 weeks. Ethanol treatment causes increase in the temperature threshold for liquid-crystalline to gel state. Numbers above arrows on abscissa represent threshold temperatures converted to Celsius. (Data from Guerri, C. and Grisolia, S. (1983). *Pharm. Biochem. Behav.* 18: 45–50.)**

The interaction between temperature, membrane fluidity, and the efficacy of toxicants should be especially notable in poikilothermic species. Organisms that live at low temperatures have higher percentages of unsaturated fatty acids in their cellular membranes when compared to the same or similar species living at warmer temperatures. These molecular adaptations assure a stable membrane function in the face of seasonal changes in air and water temperature (Prosser, 1986). In view of the daily and annual changes in the internal temperature of poikilothermic organisms, it would seem that changes in membrane fluidity should be a critical variable to consider in the study of the mechanisms of toxicity. This would seem to be a very fruitful area of study for toxicants that mimic the actions of hormones and neurotransmitters (i.e., endocrine disputers) but also toxicants that can directly alter membrane fluidity, such as organic solvents.

## 4.3.4 Toxic Mechanisms Attenuated by Hypothermia

With the exception of the pyrethroids and DDT, it is clear that hypothermia affords protection to a variety of toxicants. Systemic toxicological studies show that mild hypothermia improves the recovery of organ function and chances of survival following exposure to a variety of toxicants, but there is relatively little known about the mechanisms of action (Section 4.2). Based on the Arrhenius relationship [Equation (4.2)], it is reasonable to assume that hypothermia most likely protects tissues by slowing the rate at which the toxicant exerts its damaging effects at the cellular and intracellular level. This is a rather simplistic explanation and there is a need to have a better understanding of how toxicant damage is ameliorated with hypothermia.

Recently, there have been numerous studies designed to show the mechanisms of hypothermic protection to the CNS of rodents treated with substituted amphetamines (O'Callaghan and Miller, 2002). Methamphetamines are drugs of abuse and are also well-known neurotoxicants that cause depletions in striatal levels of dopamine and serotonin. These drugs also affect thermoregulatory control in rodents, inducing hypothermia at relatively cool temperatures below thermoneutrality and marked hyperthermia that can be lethal at relatively warm ambient temperatures (Gordon et al., 1991c; Bowyer et al., 1992; Miller and O'Callaghan, 1994). The data base for the methamphetamines is considerable, and a review of the mechanism of action of the amphetamines and their interaction with body temperature is relevant to understanding the mechanisms of hypothermic protection exhibited by many of the other xenobiotics covered in this book (see Chapter 3).

Allowing rodents to become hypothermic imparts remarkable protection on the neurotransmitter-depleting actions of amphetamines. This observation has spurred numerous investigations on the mechanisms of amphetamine-induced striatal dopamine depletion and the protection afforded by hypothermia (e.g., Malberg and Seiden, 1998). For example, after treating rats with the amphetamine analog methylenedioxymethamphetamine (MDMA), there is a marked depletion of serotonin in the striatum, frontal cortex, and other CNS sites when measured 14 days after treatment (Malberg et al., 1996). Pre-injection with ketanserin (KET) or α-methyl-p-tyrosine (AMPT) with MDMA causes a profound hypothermia and no depletion in serotonin levels. Blocking the hypothermia in MDMA-treated animals given KET or AMPT led to marked serotonin depletion, suggesting that the reduction in core temperature afforded neuroprotection by preventing loss of neurotransmitters. However, fluoxetine given with MDMA also blocks serotonin depletion, but this combination of drugs has no effect on body temperature. Overall, these experiments serve as just one example to illustrate the difficulty in separating the protective actions of a drug or toxicant that is directly dependent on a hypothermic response or completely independent of body temperature.

Researchers studying the methamphetamines have been frustrated by the difficulty to separate a true neuroprotective mechanism of a chemical treatment from the general protective effects of hypothermia. This frustration was best summed up in a review of the mechanisms of amphetamine neurotoxicity by O'Callaghan and Miller (2002): "Although painful to consider, what these data suggest is that the results of all 'mechanistic' studies of amphetamine neuropharmacology or neurotoxicity are compromised unless temperature can be ruled out as a contributing factor."

There has also been a resurgence of attempts to understand how hypothermia affords protection to the CNS following stroke as well as other types of traumatic brain injuries (Dietrich, 1992). Like the amphetamine studies, there has been tremendous success in connecting a specific cellular or molecular end point that is manifested with hypothermia, but it has been difficult to identify whether the mechanism is a direct result of hypothermia (Table 4.1). Nonetheless, many of the protective effects of hypothermia can likely be applied to toxicological studies because many toxicants cause the same sequelae as those seen from ischemia, including lipid peroxidation, free radical formation, $Ca^{+2}$ imbalance, and many others (Gultekin et al., 2000; Stohs and Bagchi, 1995). The stroke and ischemia literature may well be helpful to the endeavors of toxicologists and pharmacologists to understand the protective role of hypothermia.

#### *4.3.4.1 Reactive Oxygen Species*

Formation of reactive oxygen species (ROS) and the protective role of hypothermia have emerged as a key mechanism to understanding the mechanisms of toxicant-induced degeneration of the CNS and other tissues and organs (Halliwell, 1992). ROS cause damage to proteins, lipids, and nucleic acids, leading to cell damage and death. Slikker et al. (2001) found that the percentage of surviving Chinese hamster ovary cells subjected to oxidative stress by exposure to hydrogen peroxide was inversely related to the temperature of incubation. This is not surprising in view of the earlier discussion on how hypothermia attenuates lipid peroxidation and free radical formation.

Furthermore, it was shown that reducing the incubation temperature led to greater expression of *bcl-2*, an anti-apoptotic protein that affords marked protection against oxidative stress. This experiment shows how hypothermia protects cells by more than just a simple Arrhenius effect (Figure 4.10). That is, lower temperatures induce and higher temperatures suppress the expression of a protein that is critical for cell survival. In this scheme, hypothermia suppresses the activity of damaging processes that have a positive temperature coefficient (e.g., lipid peroxidation, ROS formation) and activates expression of *bcl-2* that further protects cells from oxidative stress. Overall, hypothermic protection may play a pivotal role in understanding the mechanisms of how toxicants and toxins lead to the manifestation of neurodegenerative diseases. Even though body temperature of humans is usually stable in the face of exposure to toxicants and other insults, hypothermia is nonetheless a critical parameter to study in *in vitro* and small-animal studies. Incorporating temperature as an exper-

**Table 4.1 A Summary of Studies with Plausible Mechanisms to Explain How Hypothermia Protects Tissues in the CNS from Pathological Insults[a]**

| *Insult* | *Hypothermia Mechanism* | *References* |
|---|---|---|
| Amphetamine toxicity | Attenuated 5-HT and DA release from striatum | Bowyer et al., 1992; Malberg et al., 1996 |
| Amphetamine toxicity[b] | Attenuated DA release and GFAP accumulation in striatum | Miller & O'Callaghan, 1994 |
| Amphetamine toxicity | Prevented $Ca^{+2}$ accumulation in nigrostriatum | Corbett et al., 1990[b] |
| CNS ischemia | Reduced free radical formation | Globus et al., 1995 |
| CNS ischemia | Blocked translocation of PKC in striatum | Cardell et al., 1991 |
| CNS ischemia | Reduced ATP depletion in hippocampus | Zeevalk and Nicklas, 1993 |
| CNS ischemia | Reduced glutamate and glycine release in hippocampus | Illievich et al., 1994 |
| Trimethyl tin | Reduced accumulation of GFAP in hippocampus | Gordon & O'Callaghan, 1995 |

[a]Key: 5-HT, serotonin; DA, dopamine; PKC, phosphocreatine kinase C; GFAP, glial fibrillary acidic protein

[b]Various substituted amphetamines studied

imental variable is undoubtedly leading to advances in understanding the mechanisms of toxicity.

## *4.3.5 Toxicant Mechanisms Exacerbated by Hypothermia*

In general, one finds a positive temperature coefficient for most classes of toxicants, meaning that the efficacy of the chemical to alter cellular function and induce toxic sequelae is exacerbated with increasing temperature. However, two principal types of insecticides, DDT and pyrethroids, manifest their toxicity on the nervous system with a negative temperature coefficient (or a $Q_{10} < 1.0$; Narahashi, 2000). The sodium channel is the primary target for these pesticides.

Pyrethroids interact with sodium channels and affect the mechanisms of nerve membrane depolarization and repolarization. Pyrethroids cause sodium channel gates to be stuck in an open position resulting in prolonged depolarization and repetitive firing activity. These processes are magnified by lowering temperature (Narahashi, 2000). For example, the potency of tetramethrin to depolarize the crayfish giant axon was increased

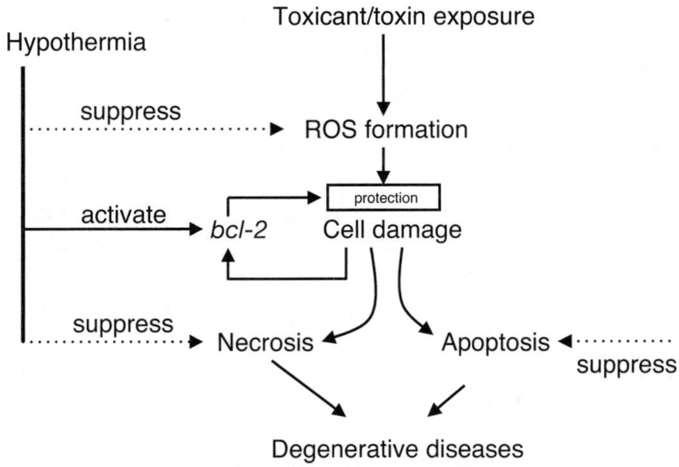


**Figure 4.10 A general mechanism of how hypothermia can modulate the toxic-induced cell damage and the manifestation of neurodegenerative disease. Damaging processes with a positive temperature coefficient, or $Q_{10} > 1.0$, including formation of reactive oxygen species (ROS), are exacerbated with an increase in temperature. Expression of proteins that protect cells from ROS formation such as *bcl-2* is a process that has a negative temperature coefficient and is activated with hypothermia. Hypothermia suppresses cell necrosis and apoptosis (see Slikker et al., 2001).**

fourfold by lowering the temperature from 21 to 10°C (Figure 4.11). Similar results were seen with the pyrethroid fenvalerate.

The mechanism of pyrethroid neurotoxicity in mammals was elucidated by Song and Narahashi (1996). They showed that tetramethrin's effects on sodium channels were a result of prolongation of the tail current in rat cerebellar purkinje neurons. A prolonged tail current augments the depolarization after-potential, with a resultant increase in propensity for repetitive activity. This leads to repetitive nerve firing and increased release of transmitters at the neuromuscular junction, an effect that is exacerbated by lowering temperature (Salgado et al., 1983; Vijverberg and van den Bercken, 1990). Because of the difference in body temperature and other species-specific differences between mammals and insects, pyrethroids are fortunately much more toxic in insects than in mammals (Table 4.2). That is, assuming a body temperature of 37°C for mammals and 25°C for insects (a temperature that could actually be higher during the day due to behavioral regulation; see Chapters 2 and 8), the toxicity of pyrethroids would be fivefold more toxic in insects. The higher body temperature also leads to a greater rate of detoxification in mammals (threefold difference). Finally, intrinsic factors in the nerve sensitivity of mammals

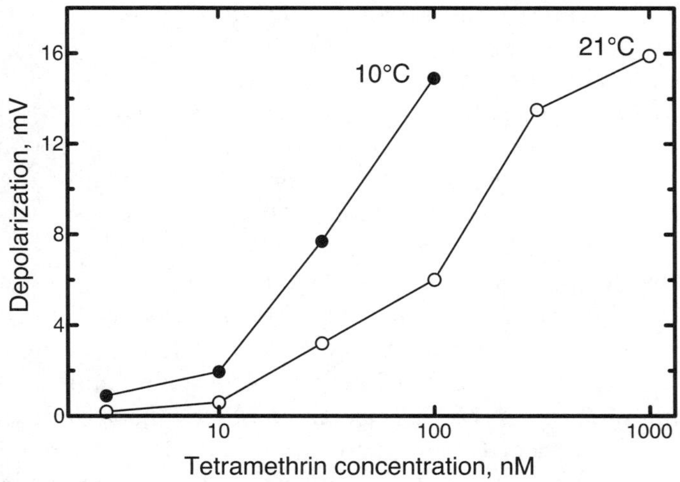


**Figure 4.11** Effect of increasing dose of the pyrethroid tetramethrin on depolarization in crayfish giant axons. Lowering the temperature from 21 to 10°C exacerbates depolarizing effects of the pyrethroid insecticide. (Modified from Salgado, V.L., Herman, M.D., and Narahashi, T. (1989). *Neurotoxicology* 10: 1–14.)

**Table 4.2 Summary of Factors Contributing to Selective Toxicity of Pyrethroids in Insects and Mammals**

| *Selectivity Factor* | *Mammals* | *Insects* | *Difference* |
|---|---|---|---|
| **Potency on nerve membrane** | | | |
| Temperature dependence[a] | Low | High | 5 |
| Intrinsic sensitivity | Low | High | 10 |
| Recovery | Fast | Slow | 5 |
| **Detoxification rate** | | | |
| Enzymatic action (temperature-dependent) | High | Low | 3 |
| Due to body size | High | Low | 3 |
| **Overall difference in sensitivity to pyrethroids = 2,500** | | | |

[a]Assuming that body temperature of insects is 25°C and of mammals is 37°C.

*Source*: Song, J.H. and Narahashi, T. (1996). *J. Pharm. Exp. Ther.* 277: 445–453.

and the smaller size of the insects further exacerbate their sensitivity to pyrethroids, culminating in an overall 2500-fold difference in toxicity between insects and mammals (Table 4.2).

DDT behaves similarly to pyrethroids by prolonging the opening of sodium channels at lower temperatures and inducing repetitive nerve activity. The lateral-line organ of the clawed frog (*Xenopus laevis*) was a useful model in these types of studies. This preparation showed marked spontaneous activity when exposed to DDT and subjected to decreasing temperature (van den Bercken and Akkermans, 1971). The $Q_{10}$ for DDT on nerve activity is approximately 0.2 (see Narahashi, 2000 for review) and explains why its lethal effects in poikilothermic vertebrates and invertebrates are accentuated at lower ambient temperatures (Chapter 8). Botulinium A toxin is another example of an agent with a negative temperature coefficient (Dreyer and Schmitt, 1983). Its effects (i.e., neurotransmitter release) on nerve-stimulated end plate potentials in the mouse hemidiaphragm preparation are enhanced when temperature is decreased from 37 to 20°C.

## 4.4 PHYSIOLOGICALLY BASED PHARMACOKINETIC MODELS

### 4.4.1 Pulmonary Uptake

Physiologically based pharmocokinetic (PBPK) models quantify the disposition of a toxicant or drug based on an organism's physiological and biochemical processes. PBPK models take into account tissue solubility, binding kinetics of the toxicant, metabolism, and many other factors (Gearhart et al., 1993, 1994). Body temperature is also recognized as a crucial aspect of PBPK models, especially in rodents and species in which body temperature can readily be altered by exposure to a toxicant. The effects of body temperature on PBPK models have focused on the pulmonary uptake of airborne contaminants as well as the hepatic metabolism of circulating toxicants and their metabolites.

Body temperature indirectly influences the intake and absorption of airborne pollutants (Mautz, 2003). That is, when body temperature and metabolic rate are depressed as a result of exposure to an airborne contaminant, one would expect a reduced rate of ventilation and thus, a lowered rate of intake of the toxicant. Such an effect would be manifested in relatively small mammals that show a marked hypothermia in response to a toxicant exposure. As discussed in Chapter 3, ozone elicits a rapid reduction in body temperature of the mouse and rat when they are exposed at ambient temperatures below thermoneutrality. Mautz and

Bufalino (1989) monitored oxygen consumption, respiratory frequency, tidal volume, and body temperature in rats exposed to various concentrations of ozone (Figure 4.12). Although metabolic rate decreased during ozone exposure, respiratory rate actually increased by 27%. However, as respiratory rate increased, there was a 35% decrease in tidal volume, resulting in an overall 20% reduction in minute ventilation (i.e., liters/minute) and a concomitant decrease in ozone uptake. Moreover, using an isolated lung preparation in which the rate of perfusion and temperature could be carefully controlled, a direct relationship between the temperature of the lung tissue and the rate of absorption of ozone was demonstrated (Postlethwait et al., 1994). Thus, lowering body temperature during ozone exposure imparts protection through a variety of mechanisms, including the suppression in minute ventilation and reduction in absorption of ozone by the respiratory epithelium.

That a lower body temperature protects an animal from ozone has also been demonstrated using a genetic model (Slade et al., 1997). Mice of the C57Bl/6J strain are more sensitive than C3H/HEJ mice when they are exposed to equal concentrations of ozone. The latter strain showed a greater hypothermic response and, hence, reduced intake of ozone when compared to the former strain. However, when rats were exposed

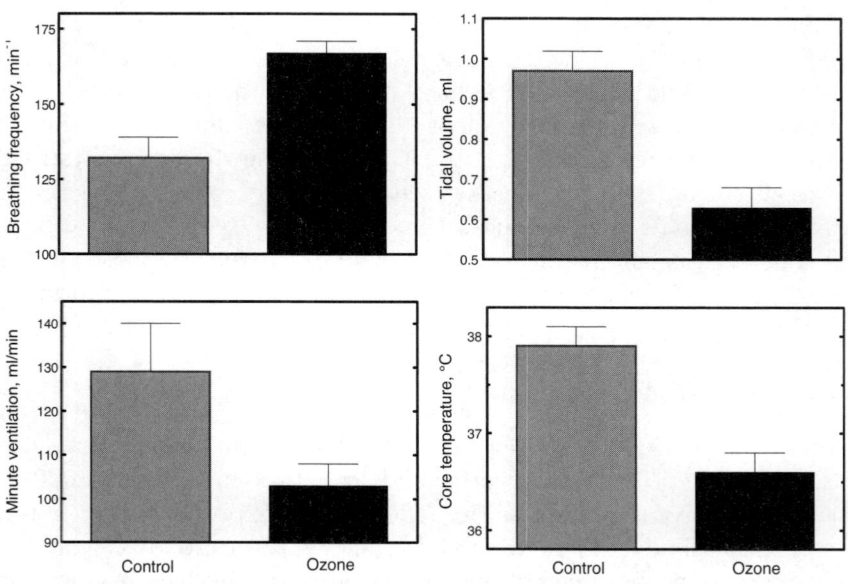


**Figure 4.12 Relationship between respiratory and thermoregulatory response of the rat exposed for 3 h to 0.8 ppm ozone. Hypothermia during ozone exposure leads to increased ventilatory frequency but reduced tidal volume, resulting in decreased minute ventilation and pulmonary intake of ozone. (Data from Mautz, W.J. and Bufalino, C. (1989). *Respir. Physiol.* 76: 69–78.)**

to ozone for five days while maintained continuously at ambient temperatures of 10, 22, or 34°C, animals exposed to the warmer temperatures sustained less damage from ozone (Wiester et al., 1996). This paradoxical protective effect of warmer temperature from ozone is likely due to the reduced minute ventilation and lower intake of ozone in rats housed at 34°C as compared to housing at the cooler ambient temperatures. Although rats housed at 10°C become more hypothermic in response to ozone as compared to animals at warmer temperatures, the cold-exposed animals also have greater metabolic requirements to thermoregulate. A higher metabolic rate in the cold results in an overall greater dosage of ozone. Overall, developing a mild hypothermia at ambient temperature of ~22°C is protective because it leads to a net reduction in ozone uptake. But with severe cold exposure, any protective effect of hypothermia is likely to be overwhelmed by the thermoregulatory demand for an increased metabolic rate.

Oxygen-enriched environments are routinely used in clinical settings for a variety of treatments and therapies. Hyperoxia, while beneficial in some situations, is also very toxic, resulting in damage to the lungs that is similar to that seen with ozone and other airborne toxicants. Allowing mice to breathe 95% oxygen for 36 h leads to pulmonary vascular injury, neutrophil infiltration, increase in lavage proteins, and pulmonary edema (Hasday et al., 2003). Raising the temperature of the mice to a febrile range (39 to 40°C) by increasing ambient temperature to 34°C led to a marked exacerbation in the aforementioned biomarkers of lung damage. Exposure to the warm environment also shortened the survival time to hyperoxia by approximately 2 days. It is surprising that there is so little known about the thermoregulatory effects of hyperoxia considering its common use in clinical settings. Hyperoxia may be a useful tool for studying the toxicology of some airborne pollutants because of their similarity in pulomonary pathology and effects on thermoregulation.

### 4.4.2 Hepatic Metabolism

Gearhart et al. (1993) recognized that understanding the temperature–activity relationship of P-450 in liver would lead to a more accurate PBPK model of chloroform metabolism. The mouse exposed to chloroform quickly developed a marked hypothermia, a response that would be expected to slow the metabolism of chloroform in the liver. Cytochrome P-450 activity for the metabolism of chloroform exhibited a $Q_{10}$ of 2.2 over a temperature range of 24 to 37°C. It was shown that exposure to 2,000 ppm chloroform for 3.5 h led to a ~3°C decrease in core temperature. This reduction in liver temperature translates to a 20% decrease in P-450 activity and metabolism of chloroform. Incorporating the temperature sensitivity data for P-450 into a PBPK model significantly improved the predictability for the metabolism of chloroform (Gearhart et al., 1993).

Most pharmacokinetic studies have been performed in rodents acclimated to standard room temperature (e.g., 22°C). This type of approach does little to mimic the potential exposure scenarios of species that can be acclimatized to a wide range of environmental conditions. Prolonged exposure to a warm or cold environment results in marked alteration in function of the liver and kidneys that can impact on the pharmacokinetics of toxicants. The weight of the liver and kidneys of mice and rats generally varies inversely as a function of the ambient temperature of rearing (Yamauchi et al., 1981, 1983). For example, the weight of the liver (normalized to body weight) in the rat acclimated to an ambient temperature of 35°C for 10 weeks is 26% smaller than that of animals acclimated to 22°C (Ray et al., 1968). By modifying the growth and development of the liver and kidney, acclimation temperature has been shown to alter the oxidation and metabolic deactivation of drugs. Clearance of antipyrene has served as a model for hepatic drug metabolism. Rats acclimated to an environmental temperature of 35°C for 32 days show a 57% decrease in antipyrene clearance (Zvi and Kaplanski, 1980). There is little known about how warm acclimation affects the activation and deactivation of xenobiotics; however, the studies done with antipyrene would lead one to expect a depressed ability to metabolize toxicants in warm-acclimated animals.

## 4.5 TEMPERATURE ACCLIMATION

Most of this chapter has centered on the toxicological response of rodents and other species' thermoregulatory systems when exposed to heat or cold stress for relatively short durations. When thermal stress persists for longer periods (ca. >24 h), homeotherms undergo a variety of autonomic and behavioral adaptations that augment their ability to survive in adversely warm or cold environments. These adaptations are bound to influence the thermoregulatory as well as other biological responses to toxic agents. In spite of the fact that humans and domestic species can be exposed for long durations to thermal environments that are stressful, there is relatively little known about how acclimation to thermal stress affects the response to toxic agents. In this section, attention will be drawn primarily to the experimental animal studies that show how temperature acclimation affects toxicological response and possible mechanisms of action. The interaction between environmental stress and response of humans to toxicants is discussed in Chapter 7.

## 4.5.1 Terminology

The terms *acclimation, acclimatization,* and *adaptation* are often used interchangeably to describe the physiological, behavioral, and morphological changes of an organism subjected to thermal stress. These terms are distinct in the field of environmental physiology and characterize certain types of environmental conditions (IUPS, 2001). *Acclimation* describes the adaptive changes that occur within the lifetime of an organism in response to experimentally induced changes in particular climate factors such as ambient temperature in a controlled environment. *Acclimatization* is used to describe those changes occurring within the lifetime of an organism as a result of inhabiting a particular natural climate. *Adaptation*, specifically, *genotypic adaptation*, refers to the physiological and behavioral characteristics fixed genetically that favor a species' survival in a particular environment. Most toxicological studies have focused on the effects of temperature acclimation. There is very little known about temperature adaptation and toxicological response other than some very old studies on the responses of hibernating animals to toxic doses of drugs (see Fuhrman, 1946).

## 4.5.2 Lethality

There has been an interest in understanding how temperature acclimation affects the lethality of various types of toxicants in rodents. Prolonged housing (ca. >7 days) in a cold or warm environment leads to physiological and morphological changes to enhance the efficiency and capacity of the thermoregulatory system (for review, see Gordon, 1993). In general, cold acclimation increases the capacity for nonshivering thermogenesis. Rodents are more sensitive to sympathomimetic agents such as norepinephrine and are able to sustain high rates of heat production for long periods during acute cold exposure. With heat acclimation, there is a development of more sensitive reflexes to increase skin blood flow and dissipate heat by evaporation. There is also a gradual decrease in the basal metabolic rate, a critical response to lessen the total heat load of the rodent maintained in a hot environment.

In nearly all temperature acclimation/toxicology studies, the animals are housed for several weeks in a cold, thermoneutral, or warm temperature, and the lethal dose of the toxicant is determined while the animals are maintained at the acclimation temperature. These tests can be misleading because the animal acclimated to a cold ambient temperature and exposed to a toxicant may still succumb to thermoregulatory failure. That is, one cannot be sure if the lethal effects should be attributed to a direct toxic effect or were indirect results of the disabling of thermoeffector mechanisms and subsequent thermoregulatory failure in the cold. There is little known about how acclimation at a warm or cold temperature affects the toxic response when the animal is tested at a neutral temperature. Such studies could provide a better understanding of how the processes of acclimation can alter the toxicity of the agent.

Nomiyama et al. (1980) assessed the effects of acclimation to temperatures of 8, 22, and 38°C for 2 weeks on the lethal dose in mice exposed to a variety of toxic agents, including solvents, heavy metals, and agricultural chemicals (Figure 4.13). Cold acclimation consistently led to a marked lowering of the lethal dose for the solvents, pesticides, and all but one heavy metal (mercuric chloride). Heat acclimation led to an elevation in the lethal dose for mercuric chloride and fratol but had little effect on the lethal dose of dieldrin. The effects of temperature acclimation on the lethal dose of methylmercury are similar to the other metals in mice (Nomiyama et al., 1980a). Mice acclimated to an environmental temperature of 8°C succumbed faster than mice at 22°C, which died faster than mice maintained at 38°C. However, if 8°C-acclimated mice were dosed with methyl mercury and then housed at 38°C to block the toxicant-induced hypothermia, their survival improved dramatically. It is important to note in these mouse lethality studies that an ambient temperature of 38°C is well above the thermoneutral zone and extremely stressful. This is likely to be the highest acclimation temperature for mice ever reported (see Gordon, 1993).

### 4.5.3 Renal Toxicity and Temperature Acclimation

Considering the likely importance of diuresis as a factor in the renal toxicity of metals and other toxicants, it is surprising that there is so little known about the effects of brief and prolonged heat and cold stress on renal toxicity. Heat-acclimated rodents drink more water but excrete less urine because of the increased demand for evaporative heat loss. The glomerular filtration rate is reduced by 67% in rats following 3 weeks of acclimation to 35°C (Chayoth et al., 1984). During cold acclimation, rodents initially consume more food, but their ratio of food to water consumption is reduced. Clearly, the study of renal toxicity requires an environmental physiological approach to understand how ambient and body temperature modulate diuresis and the potential toxicity.

There is relatively little known about how temperature acclimation affects renal toxicity. Brief periods of cold exposure have been found to protect rats from the renal toxicity of mercuric chloride (Burgat-Sacaze et al., 1982). Exposing rats to an ambient temperature of 1.0°C for 24 h after exposure to low doses of mercuric chloride attenuated the damage to the nephron's proximal brush border. The diuretic effects of cold exposure may account for the reduced nephrotoxicity of mercury.

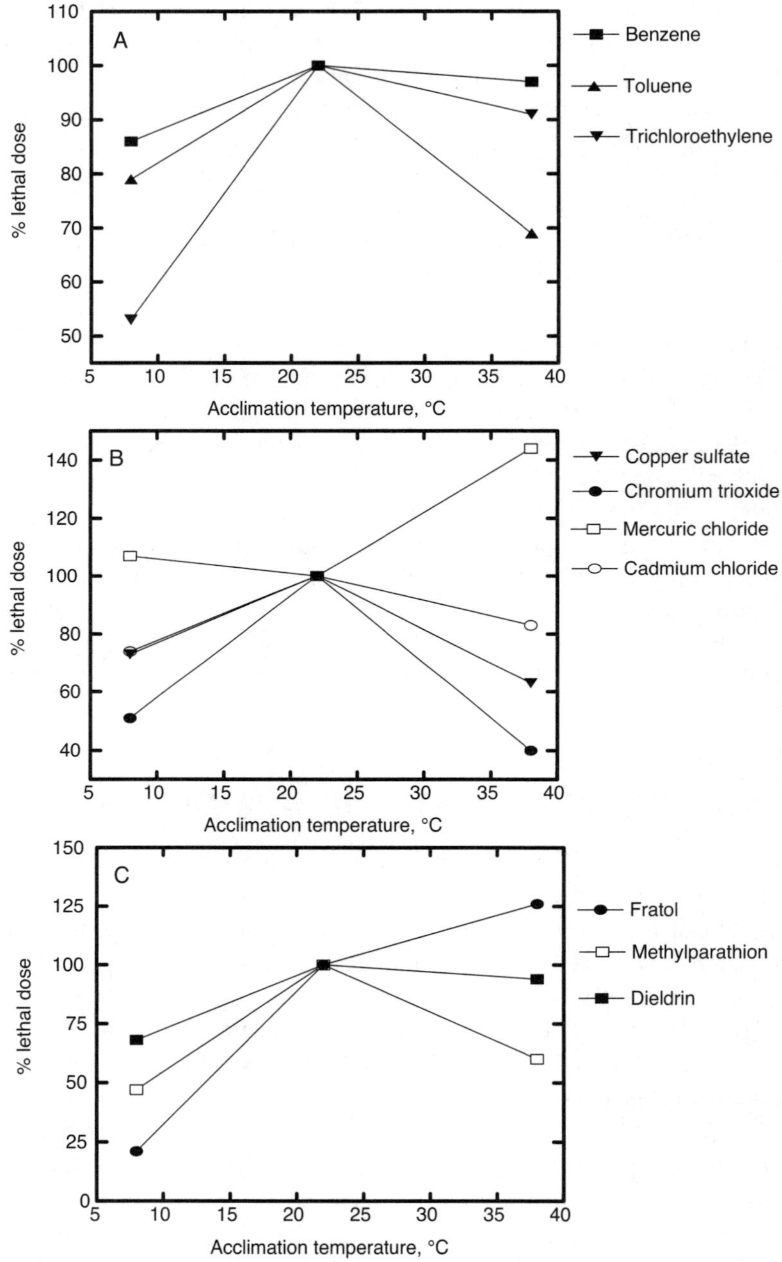


**Figure 4.13 Effect of 4 weeks of acclimation to ambient temperatures of 8, 22, or 38°C on the lethal dose of solvents (A), heavy metals (B), and pesticides (C) in mice. All solvents and metals administered intraperitoneally. Pesticides administered orally. Note difference in scales of ordinate. (Data modified from Nomiyama, K., Matsui, K., and Nomiyama, H. (1980). *Toxicol. Lett.* 6: 67–70.)**

### 4.5.4 Anticholinesterase Agents

Organophosphates were some of the first toxicants to be evaluated for their effects on rodents acclimated to cold and warm environments. This is becoming an increasingly important area of study because of the potential exposure of military personnel and civilians to organophosphate nerve gas agents in hot, arid environments (see Chapter 7). That is, in view of the possible exposure of humans to organophosphates while acclimatized to hot desert conditions, it would be important to know if the physiological changes from heat acclimatization improve or exacerbate the toxicological effects of nerve gas agents. One of the first studies in this area reported that the survival time to 16.5 mg/kg parathion was 17.0, 11.1, and 2.3 h in mice acclimated for 3 days to ambient temperatures of 15.6, 22.8, and 35.6°C, respectively (Baetjer and Smith, 1956). Unfortunately, these toxicity studies were only performed at the acclimation temperature. Testing for parathion sensitivity in heat-acclimated animals at a thermoneutral test temperature would allow one to determine if heat acclimation improves or exacerbates the toxic response of the toxicant.

Such an approach has been performed in the cold-acclimated rat (Ryhanen et al., 1988). When rats were acclimated to a temperature of 5°C for 14 days and then administered diisopropyl fluorophosphate (DFP) while housed at 20°C, the $LD_{50}$ for DFP was 32% higher when compared to rats acclimated to 20°C. Not surprisingly, cold acclimation was also shown to afford greater protection from DFP when rats were treated with the toxicant and exposed to cold temperatures. Cold-acclimated rats showed a doubling in their resistance to DFP, as assessed by the $LD_{50}$, when tested at an ambient temperature of 5°C as compared to 20°C. Cold acclimation most likely improves thermoregulatory control in a cold environment following exposure to a toxicant that limits thermoeffector function.

Overall, cold acclimation increased the sensitivity of mice to a variety of toxicants; however, this was most likely due to the fact that the animals were maintained in the cold environment while exposed to the toxicant, with the cause of death resulting from hypothermia and not the toxicant per se. It is likely that cold acclimation affords protection to the rodent given the toxicant and subjected to cold exposure, but this was not possible to determine because of limitations in the experimental design. That is, if animals in these studies that were acclimated to warmer temperatures had been exposed to the toxicant and placed in the cold environment, they most certainly would have succumbed to hypothermia and death faster than the cold-acclimated animals.

### 4.5.5 Lead Poisoning

The systematic studies by A.M. Baetjer and colleagues on the impact of heat and cold stress on lead toxicity were some of the first to document a link between environmental temperature and chemical toxicity (Baetjer et al., 1960). These studies were spurred by observations in humans on the seasonal effects of lead toxicity, which suggested that warm temperatures were associated with increased susceptibility to lead toxicity. In addition to demonstrating that the toxicity of lead in mice and rats was exacerbated at a warm temperature of 35°C, they found that pre-exposure of mice for 3 days to the warm temperature lowered their susceptibility to lead poisoning when compared to mice that experienced no heat acclimation (Baetjer et al., 1960). For example, the survival time of mice exposed to 35°C for 3 days, dosed intraperitoneally with 100 mg/kg lead acetate, and then exposed again to 35°C was 105 h; when mice were exposed to an ambient temperature of 22°C, dosed with the same dose of lead acetate, and placed in a 35°C environment, the survival time was only 77 h. Body temperature was not measured in these studies, but it is possible that the 3 days of heat acclimation improved the ability of mice exposed to lead to maintain thermal homeostasis during heat exposure compared to the nonacclimated mice.

Since the incidence of lead poisoning in humans was seasonally related with increased exposures during the summer months, there was an interest in understanding if heat acclimation alters toxicity during chronic lead exposure. Wright and Lessler (1979) fed rats a diet containing lead while acclimating to temperatures of 23 or 33°C and found that heat acclimation exacerbated the effects of dietary lead on growth rate. Mice exposed for 8 h per day, 5 days per week for 6 weeks while dosed each test day with lead acetate accumulated lead in the liver, kidney, and brain at different rates depending on the acclimation temperature (Martinez-Garcia et al., 1995). Interestingly, brain lead levels were initially higher in animals maintained at 22°C after 3 weeks of exposure. However, by 6 weeks of exposure, the mice exposed to 35°C displayed significantly greater increases in lead levels in the brain (Figure 4.14). During the first 3 weeks of exposure to a warm temperature, it is expected that mice excrete the lead faster because they salivate and drink more water. However, as warm acclimation progresses, there are adaptations to the heat stress, primarily a reduction in metabolic rate, which lowers the requirements for heat dissipation by evaporation. Thus, less water is consumed, which may explain the greater retention of lead in the tissues.

Rats exposed repeatedly to methyl mercury show differences in accumulation of mercury in the brain and signs of toxicity depending on whether they are exposed to ambient temperatures of 11, 22, or 33°C

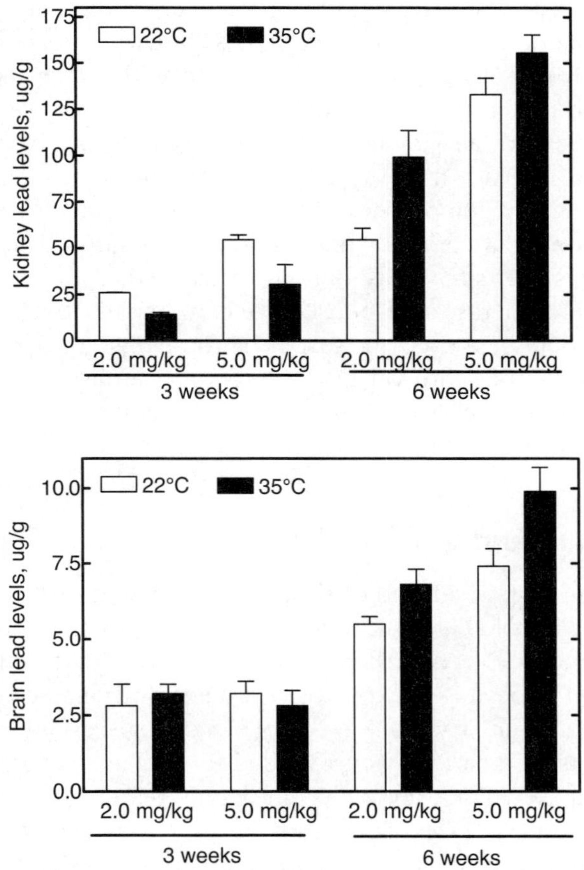


**Figure 4.14 Effect of warm acclimation on the accumulation of lead in the kidneys and brains of mice exposed repeatedly to lead acetate for 3 and 6 weeks. (Data modified from Martinez-Garcia, F., Martinez-Ruiz, F., Vicente, I., Penafiel, R., and Cremades, A. (1995).** ***Eur. J. Pharmacol.*** **293: 271–275.)**

(Yamaguchi et al., 1984). Average days to onset of hind leg crossings, a behavioral sign of methyl mercury poisoning, was found to be 31 days in rats exposed to 33°C, 36.7 days at 22°C, and 47 days in rats exposed to 11°C. Methyl mercury accumulation in brain and blood followed the same pattern as the manifestation of behavioral symptoms. It is interesting to note that sweating in humans results in excretion of significant quantities of some metals such as zinc, cadmium, and nickel (see Chapter 7).

The effects of environmental temperature on the recovery of mice suffering from acute lead poisoning were also investigated (Baetjer et al., 1960). Mice with chronic lead poisoning began to die sooner following cessation of lead exposure when they were housed at a warm temperature.

When dosed repeatedly with lead acetate over 26 days at 22°C such that the mice sustained significant weight loss, housing the recovering mice at temperatures of 16, 22, or 35°C was associated with a mortality of 29, 56, and 77%, respectively. Severe dehydration also exacerbated the toxicity of lead in mice and rats (Baetjer et al., 1960; Baetjer and Horiguchi, 1964). The excretion of lead in the feces was reduced when rats were housed at a warm ambient temperature; fecal content of lead in rats housed at 21°C was more than double that found in rats housed at 31.5°C. This means there is a greater body burden of lead when animals are housed in warm environments, a response that may explain the greater toxicity of lead in rodents exposed in warm environments (Baetjer et al., 1960). The effect of temperature on lead excretion is unusual since one finds that the excretion of toxicants such as ethanol is increased when rodents are exposed to warm environments (see Figures 4.2 and 4.3).

### 4.5.6 Ethanol and Cold Acclimation

In view of the high incidence of death from hypothermia in humans who abuse alcohol in cold climates, there has been an interest in understanding how warm and cold acclimation affects the sensitivity to ethanol (Kalant and Le, 1983). One might expect that a subject acclimated to a thermoneutral or warm environment would be more susceptible to the hypothermic effects of ethanol when exposed to cold as compared to a cold-acclimated subject. For example, guinea pigs reared for 6 to 10 months at a temperature of 17 to 18°C showed a remarkable improvement in survival to extreme cold exposure and treatment with ethanol as compared to guinea pigs reared at 22 to 23°C (Huttunen and Hirvonen, 1977). Guinea pigs reared in the cooler environment showed, in some instances, a doubling of their survival time when they were dosed with ethanol and subjected to acute cold exposure of −20°C. It is interesting to note that the preferred thermal environment of the guinea is close to 30°C (Gordon, 1993). Hence, the dramatic effects of rearing at 17 versus 22°C would likely be even more marked had a comparison been made between animals reared in a thermoneutral versus cool environment.

Cold acclimation also increases the resistance to ethanol-induced hypothermia in rats. When rats were acclimated to an ambient temperature of 4°C as compared to 20°C, they showed increased resistance to the hypothermic effects of ethanol when administered the ethanol solutions and exposed to 4°C (Lomax and Lee, 1982). This study is one of the few to document the time-course of improved resistance to ethanol and cold exposure. The investigators found an exponential increase in resistance to ethanol from day 1 to day 8 of cold acclimation, with little change in resistance from day 8 to day 20 of acclimation. The ethanol studies suggest that cold acclimation may improve the ability of animals to survive cold environments when exposed to other types of toxicants.

### 4.5.7 Chemical Carcinogens

Rodents have been the primary model used to evaluate the carcinogenicity of xenobiotics. The temperature of acclimation is rarely taken into account in spite of the fact that temperature modifies the spontaneous formation of tumors as well as the carcinogenic potential of many chemical toxicants. Indeed, there was considerable interest in the 1940s in evaluating the role of environmental temperature on spontaneous and chemical-induced carcinogensis (e.g., see Mills, 1945). With the ongoing breakthroughs in genomics and molecular biology, it is easy to overlook the importance of environmental temperature. The potential impact of environmental temperature on oncological research was perhaps best summed up by Mills (1945): "Whatever the mechanism by which the differences are produced, it is obvious that temperature control should be exercised in all tumor studies."

Skin tumorigenesis resulting from topical exposure to several doses of benzo(a)pyrene in mice was markedly affected by the temperature of acclimation (Weiss et al., 1981). Mice acclimated to an ambient temperature of 16°C developed tumors 4 weeks prior to mice acclimated to 23°C. Warm acclimation (32°C) further delayed the incidence of tumor formation by 2.5 weeks compared to mice acclimated to 23°C (Figure 4.15). These temperature treatments had little effect on body weight or body temperature. However, skin temperature was directly affected by ambient temperature with cold-acclimated and warm-acclimated mice exhibiting skin temperatures that were 1 to 2°C below and above those of mice acclimated to 23°C. Such temperature treatment and accompanying changes in blood flow may affect the efficacy of topically applied carcinogens. However, that mice with cooler skin temperature develop tumors faster than mice with warmer skin temperatures was an unexpected finding. The physiological response to cold and warm acclimation, possibly the adrenal stress response, may override any protective effect of a lower skin temperature on tumorigenesis.

Another interesting facet of the interaction between local temperature and carcinogenesis is the novel use of scalp cooling to prevent alopecia (i.e., hair loss) in patients taking chemotherapeutic agents (Dean et al., 1979). Hair loss can be a devastating side effect in cancer patients on chemotherapy. Alopecia in patients taking the therapeutic agent doxorubicin was greatly suppressed when ice packs were applied to the scalp for 5 min before and 30 min after drug treatment. Scalp temperature was reduced from about 37 to 23°C by the ice treatment. This cooling appar-

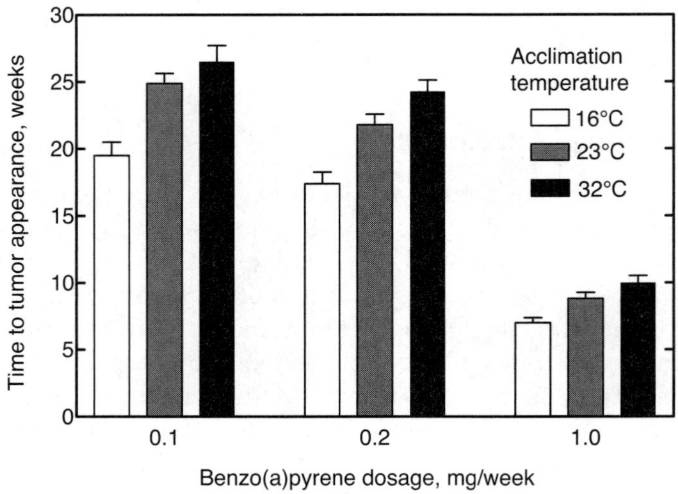


**Figure 4.15 Effect of acclimation to 16, 23, and 32°C on the time for appearance of tumors in mice exposed to a topical application (2 doses/week) of the carcinogen benzo(a)pyrene. Data plotted as mean + S.E. (Data taken from Weiss, H.S., Pitt, J.F., Kerr, K.M., Hakaim, A.G., Weisbrode, S.E., and Daniel, F.B. (1981).** ***Proc. Soc. Exp. Biol. Med.*** **167: 122–128.)**

ently reduced the toxicity of doxorubicin by (a) peripheral vasoconstriction in the scalp, which probably lessened the amount of drug that reached the hair follicles, and (b) slowing uptake of the chemotherapeutic agent into the hair follicles (Dean et al., 1979). Localized change in skin temperature has also been shown to modulate the dermal toxicity of doxorubicin in the mouse. Skin ulceration was accentuated with topical heating that raised the dermal temperature from 32 to 38.5°C, while ulceration was suppressed by cooling the skin to 17°C (Dorr et al., 1985). The protection afforded by tissue cooling in these studies is remarkable, but the skin temperatures are extreme and would elicit marked discomfort. It would seem that, based on the fact that protection is seen with *in vitro* preparations with temperature reductions of just a few degrees below normal, protection from chemotherapeutics at the whole-animal level could possibly be achieved with more mild cooling.

Another example illustrates the effects of temperature acclimation on tumorigenesis but with thermal effects opposite to the rodent studies described above. The incidence of lung tumors induced by 7,12-dimethylbenz(a)anthracene (DMBA) injected subcutaneously was markedly reduced when mice were cold acclimated by repeated exposure to 0°C for 2 h every other day for 3 months (Yamamoto et al., 1995). When compared to mice acclimated to an environment of 23°C, the incidence of lung tumors from DMBA was reduced anywhere from 30 to 50%, depending on gender. These types of studies exemplify the impact of environmental temperature on the efficacy of chemical carcinogens in rodent models. Clearly, assessing how the efficacy of chemical carcinogens can be modulated by temperature acclimation will lead to a better understanding of the mechanism of action of chemical carcinogens.

# Chapter 5

# Regulated Hypothermia: An Adaptive Response to Toxic Insult

## 5.1 INTRODUCTION

Some of the key aspects discussed in the previous two chapters are that (*a*) rodents administered or exposed to a variety of xenobiotic agents are likely to become hypothermic and (*b*) the ensuing hypothermic response attenuates the pathological damage and increases the likelihood of surviving exposure to the toxicant. It follows that the hypothermic response may represent a physiological strategy that is activated to protect the animal from toxic insults. This chapter presents the toxic-induced hypothermia of rodents as a physiological adaptation that may in fact be linked to a general pathophysiological response that has evolved as a protective mechanism to internal and environmental insults.

## 5.2 FEVER VERSUS HYPOTHERMIA AS ADAPTATIONS

The importance of the development of the thermoregulatory system is unquestionable as a key reason for the successful evolution of birds and mammals. The thermal homeostasis of the internal milieu is the quintessence of an organism's life and survival in a wide range of environments. In addition, the ability of the thermoregulatory system to mount a fever has also evolved as a mechanism that improves the survival from infection. This being the case, one can also contend that other thermoregulatory responses, such as hypothermia, may also have evolved as a protective response.

Fever most likely evolved hundreds of millions of years ago as a mechanism to enhance the ability of the host to resist infection (see also Chapter 6). The elevation in body temperature during fever represents a large metabolic cost that would seem to be counterproductive to survival. However, the febrile response to infection has been shown to be beneficial to the host organism, allowing it to recover more quickly and survive the infection. Briefly, it has been shown that febrile temperatures increase mobility and function of white blood cells, stimulate cytokine production and activation of T lymphocytes, and impair the growth of bacteria in a hypoferremic environment. That is, plasma iron concentrations decrease during fever, and the growth of some bacteria in a low iron media is markedly hampered with a febrile rise in temperature (for review, see Kluger, 1986). Fever is thus adaptive and has been conserved in the evolution of vertebrates. Likewise, the hypothermic response to toxicants may also represent an adaptive response. However, unlike fever, the hypothermic response to xenobiotics is not omnipresent across species but is likely restricted to those with a relatively small mass. To this end, the following sections elaborate on the critical nature of body mass and the behavioral and autonomic regulation of a hypothermic response to xenobiotic agents and other insults.

## 5.3 BEHAVIORAL THERMOREGULATION: A TOOL TO STUDY REGULATED VERSUS FORCED HYPOTHERMIA

Providing animals with an environment that allows them to display thermoregulatory behavior provides investigators with a valuable tool to determine how a toxicant affects the thermoregulatory state. By measuring behavioral thermoregulation along with core temperature in a temperature gradient or similar device, one can better assess if a drug or toxicant induces a forced or regulated change in core temperature (see Chapter 2). In rodents, shivering and nonshivering thermogenesis produce additional heat during cold stress, and grooming saliva on the fur is an effective means of dissipating excess heat during heat stress, but these autonomic thermoeffectors are metabolically costly. Rodents and other homeotherms strive to minimize energy expenditure to thermoregulate. Behavioral thermoregulation, such as moving about in a temperature gradient, requires little metabolic energy. To this end, it is not surprising to find that in studies where animals are subjected to a scenario where behavioral and autonomic thermoeffectors can be utilized, it has been shown that behavior generally takes precedence over autonomic systems (Gordon, 1993; Satinoff, 1974). Hence, monitoring behavioral thermal preference in undisturbed animals is likely to provide one with the most sensitive indication of drug- or toxic-induced changes in thermoregulatory status.

When rodents are housed in a temperature gradient and allowed to select from a range of ambient temperatures they will, with few exceptions, select cooler temperatures and undergo a moderate reduction in core temperature following exposure to a variety of toxicants (Chapter 3). Many toxicants appear to induce a regulated hypothermia, characterized by behavioral and autonomic thermoeffectors working together to maintain core temperature at a level that is from 1 to 6°C below normal. This is essentially opposite to a fever, where core temperature is maintained at 1 to 2°C above normal and the elevated temperature is maintained against changes in environmental temperature (see Chapter 6).

### 5.3.1 Defining the Limits of Normothermia in Toxicant-Exposed Subjects

The ambient temperature limits of normothermia of rodents and other homeotherms are quite broad (see Chapter 2). For example, rats maintain a stable core temperature, varying by less than 0.2°C, over an ambient temperature range of 15 to 30°C (see Figure 2.2). Thus, when housed in a temperature gradient, the core temperature of mice and rats is relatively stable in spite of the fact that the animals will move about from the cold to warm end of the gradient.

Rodents subjected to an acute toxic insult develop a hypothermic core temperature over a moderate range of ambient temperatures but have an impaired ability to regulate core temperature against extreme changes in ambient temperature. Even though the toxicant-treated rodent housed in a temperature gradient is hypothermic, it can nonetheless be viewed in a physiological state where it retains a modicum of thermoregulatory control and has a limited range of normothermia (Figure 5.1). In this scenario, the toxicant alters the profile between environmental temperature and core temperature and there is a limited range of ambient temperatures where a moderate 2 to 3°C reduction in core temperature is maintained. Assuming that this moderate decrease in core temperature represents the new set-point for temperature regulation, then the limits of normothermia can now be defined as the ambient temperatures above and below which core temperature increases or decreases from the moderate hypothermic state. The limits of normothermia are restricted to a relatively narrow range in the toxicant-exposed animal because the thermoeffectors for heat

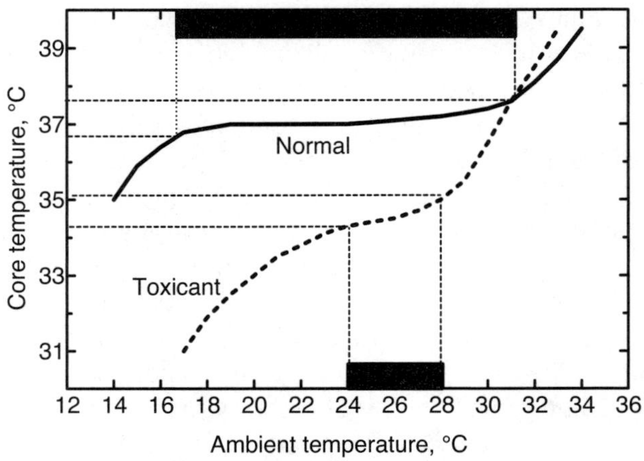


**Figure 5.1 Idealized plot of ambient temperature versus core temperature in normal and toxicant-exposed rodents that are allowed to behaviorally thermoregulate in a temperature gradient. The range of normothermia (indicated by upper black bar) is relatively wide in normal animals. With exposure to a toxicant, there is a regulated decrease in core temperature, and thermoeffectors for heat production and heat loss are impaired, resulting in a limited range of normothermia (indicated by lower black bar).**

gain and heat loss are usually impaired, making the animal more susceptible to hyperthermia or hypothermia. Note that the range of core temperature within the limits of normothermia, depicted by the horizontal dashed lines in Figure 5.1, is similar for both the normal and toxicant-treated animal. Because autonomic thermoeffectors are most likely impaired by a toxicant, the rodent's thermoregulatory behavior in the gradient will now have a marked impact on core temperature that would not be observed in normal animals. By selecting the appropriate range of ambient temperatures, the rodent maintains a moderate degree of hypothermia that is likely to enhance recovery and survival from the toxicant. Of course, the same relationship could be developed in Figure 5.1 by measuring core temperature in a toxicant-treated rodent that is maintained at different ambient temperatures for fixed times.

The response in the temperature gradient shows the critical nature of behavioral temperature regulation and also illustrates the potential flaws in toxicology studies where environmental and core temperature are not viewed with scrutiny. That is, when allowed to behaviorally thermoregulate, mice and rats select relatively warm temperatures of 28 to 31°C (Chapter 2). Under the ideal conditions of a temperature gradient, if exposure to a toxicant leads to a preference for a 3 or 4°C lower temperature, this new thermal environment is still somewhat warmer than the typical laboratory (e.g., 20 to 24°C). Indeed, mice and rats are often exposed to toxicants and tested while housed in wire-screen cages or other environments with very little insulation. Under these conditions, a rodent has little option for behavioral thermoregulation and must rely primarily on autonomic thermoeffectors to regulate body temperature. Such an environment is likely to be below the range of normothermia in the toxicant-exposed rodent (i.e., lower black bar of Figure 5.1), meaning that thermal homeostasis is severely compromised.

Researchers in toxicology and pharmacology have on occasion recognized how a change in body temperature during an experiment will affect their particular biological end point of interest. In some instances, precautions are taken to block the effects of the toxicant on body temperature. A system to warm the animals during and following exposure to the toxicant by utilizing a heating pad or infrared lamp can be effective to maintain normothermia (see Chapter 4). However, such an approach can create additional problems because application of external heat prevents the rodent from using its behavioral and autonomic thermoeffectors to regulate a hypothermic state and thereby ameliorate the symptoms of toxicity (Watkinson and Gordon, 1993). Overall, clamping the core temperature at 37°C is likely to be deleterious to the rodent exposed to a variety of toxicants.

## 5.4 THERMOREGULATORY RESPONSE TO TOXICANTS: RELATIONSHIP TO OTHER PATHOLOGICAL INSULTS

In addition to the response to xenobiotics, a regulated hypothermic response appears to be a common physiological response in rodents and some other species when subjected to a variety of pathological insults. Hypoxia, hypoglycemia, uremia, hemorrhage, endotoxemia, vascular ligation, and hypergravity have all been found to induce marked reductions in the body temperature of rodents (Table 5.1). The behavioral and autonomic responses to these insults point toward a regulated hypothermia. Behavioral thermoregulatory responses have not been assessed for all of the insults listed in Table 5.1, but the data generally show a reduction in selected ambient temperature. Measurements of peripheral heat loss made indirectly by measuring skin temperature also suggest that the CNS thermoregulatory centers respond with a regulated decrease in core temperature. Most importantly, the hypothermic response to these pathological insults appears to improve the chances of recovering from the insult. As alluded to in Chapter 1, there is a coordinated, integrated thermoregulatory response to lower core temperature with the apparent goal of improving the chances of survival of the toxic insult.

**Table 5.1 Summary of Effects of Pathological Insults on the Behavioral and Autonomic Thermoregulatory Responses of Rodents and Rabbits[a]**

| *Insult* | $T_c$ | $T_{sk}$ | $T_{sel}$ | *HP* | *Hypothermia Beneficial?* |
|---|---|---|---|---|---|
| Hypoglycemia | ↓ | ? | ? | ↓ | Yes |
| Hypoxia | ↓ | ↑ | ↓ | ↓ | Yes |
| Uremia | ↓ | ↑ | ↓ | ↓ | Probably |
| Skin burn | ↓ | ? | ? | ↓ | Yes |
| Vascular ligation | ↓ | ? | ? | ↓ | Yes |
| Endotoxemia | ↓ | ↑ | ↓ | ↓ | Yes |
| Hemorrhage | ↓ | ↓ | ↓ | ↓ | Yes |
| Hypergravity | ↓ | ↑ | ? | ↓ | Yes |

[a] $T_c$, core temperature; $T_{sk}$, skin temperature; $T_{sel}$, selected ambient temperature; HP, metabolic heat production; uremis studies for rodents and rabbit. All other studies performed in rodents.

*Source:* Most data modified from Gordon (1993, 1996); endotoxemia (Romanovsky et al., 1996); hemorrhage (Henderson et al., 2000; Brown et al., 2001).

Toxicologists must then be aware that rodents have evolved a response to lower core temperature in response to pathological insults that they may encounter in nature. The thermoregulatory responses to hypoxia and endotoxemia are discussed in more detail in Sections 5.4.1 and 5.4.2 because there is a considerable data base of their effects on behavioral and autonomic thermoeffectors. Both insults are relevant to this book because the mode of action for many types of toxicants appears to be similar to that of hypoxia and endotoxemia. The evolutionary development of thermoregulatory responses to these insults may provide researchers with a means to understand the mechanisms of some classes of xenobiotic agents.

## 5.4.1 Hypoxia

Hypoxia is a particularly well-studied insult that has been shown to elicit a regulated hypothermic response in a variety of vertebrate and invertebrate species (Wood, 1991). Indeed, even single-celled paramecia have been shown to prefer cooler temperatures in a micro temperature gradient when subjected to hypoxia (see Chapter 8). The thermoregulatory response of rodents to hypoxia has been under intensive study because of the relevancy to the development of anti-ischemic drugs for treating stroke and myocardial infarction. Rodents react to hypoxia with a vigorous heat dissipatory response. For example, rats adapted to a temperature gradient for several days and then exposed to an acute hypoxic environment (6.9% oxygen) underwent a rapid preference for cooler ambient temperatures followed by a gradual reduction in core temperature (Figure 5.2). Core temperature decreased from 37 to about 34.5°C and remained

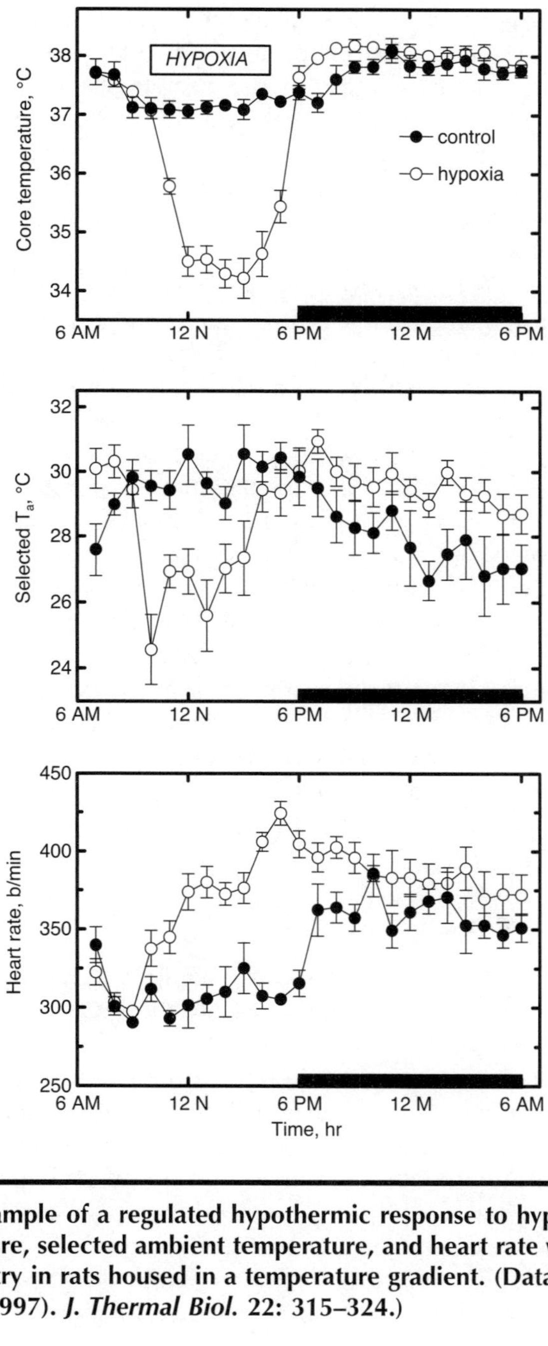


**Figure 5.2 Example of a regulated hypothermic response to hypoxia in the rat. Core temperature, selected ambient temperature, and heart rate were monitored by radiotelemetry in rats housed in a temperature gradient. (Data modified from Gordon, C.J. (1997). *J. Thermal Biol.* 22: 315–324.)**

relatively stable while hypoxia exposure continued for over 6 h. As hypoxia progressed, selected temperature increased but remained below normal as the rat appeared to use its behavior to regulate core temperature at a hypothermic level. The hypothermic response was accompanied by a 50 to 75 beat/min elevation in heart rate, reflecting the attempt of the cardiovascular system to deliver more blood to hypoxic tissues. Hypoxia also causes a transient elevation in dry heat loss and reduction in $CO_2$ production, reflecting peripheral vasodilation and a reduced metabolic rate (Gordon, 1997a; Tattersall and Milsom, 2003). Replacing hypoxic air with normoxic air resulted in an abrupt increase in selected ambient temperature, transient overshoot, and then return of core temperature to normothermic levels.

The hypothermic response to hypoxia has been shown to be beneficial to the animal's survival. That is, a moderate reduction in core temperature has been proven to be neuroprotective during exposure to hypoxia and CNS ischemia (Wood, 1991; Katz et al. 2001). Hypothermia lowers oxygen requirements for aerobic tissues in the CNS and also increases the affinity of hemoglobin for oxygen, assuring greater delivery of oxygen to hypoxic tissues. It is well known that rodents maintained in a normothermic state in the face of hypoxia show a greater incidence of mortality as compared to animals allowed to become hypothermic. For example, Minard and Grant (1982) used several methods to raise or lower the body temperature of mice and then measured their survival time to acute hypoxia (4.5% oxygen; Figure 5.3). There was a very predictable relationship between body temperature and survival during hypoxia. Above a threshold core temperature of 31.6°C, there was a logarithmic decrease in survival time as core temperature increased to over 41.5°C. Hence, the thermoregulatory response to hypoxia and resulting improvement in survival is similar to the acute response of many toxicants as discussed in Chapter 4.

## 5.4.2 Endotoxemia

Small doses of bacterial endotoxins such as lipopolysaccharide (LPS; e.g., ~10 μg/kg, IV; 50 μg/kg, IP in rats) activate the fever pathway leading to a regulated elevation in core temperature in the rat (see Chapter 6). However, relatively large doses of endotoxin elicit an acute hypothermic response that appears to be regulated (Romanovsky et al., 1996). Administering large doses of endotoxin is a useful model to study septicemia. Rats housed in a temperature gradient and dosed intravenously with 0.5 mg/kg LPS (i.e., 50 times the dose needed for a fever) exhibited an abrupt preference for cooler ambient temperatures and a subsequent reduction in core temperature (Figure 5.4). Endotoxemia was accompanied by a ~2°C reduction in the threshold hypothalamic temperature for metabolic

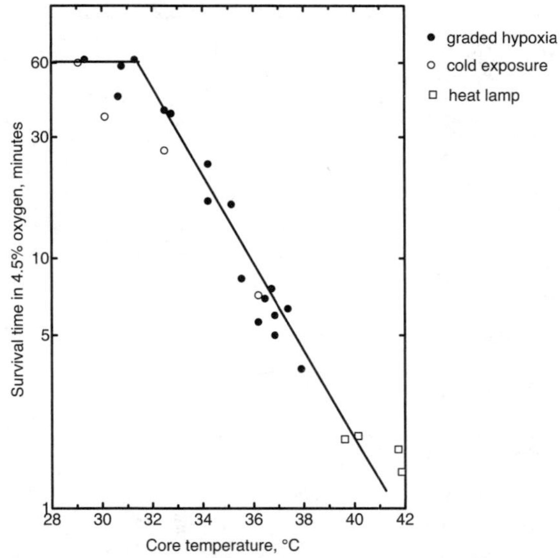


**Figure 5.3 Relationship between core temperature and survival of acute hypoxia in mice. Core temperature was modulated as indicated and the mice were then exposed to 4.5% oxygen. (Modified from Minard, F.N. and Grant, D.S. (1982). *Biochem. Pharmacol.* 31: 1197–1203.)**

heat production. This widening of the interthreshold zone resulted in a reduction in thermogenesis during septic shock (Romanovsky et al., 1996). Loading the rat with endotoxin elicited behavioral and autonomic responses resulting in a profound heat loss response that apparently overwhelms the activated febrile pathway, the net result being a hypothermic response.

Septic shock is a debilitating condition and significant human health issue. A better understanding of the alterations in temperature regulation such as described in the aforementioned studies is essential to improve methods for treatment. Hypothermia is indeed a typical thermoregulatory response in patients suffering from septic shock. Some evidence suggests that survival to endotoxic shock in rodents is improved with a moderate degree of hypothermia (Romanovksy et al., 1997). However, the protective hypothermia mechanism has to be viewed with caution. For example, sheep infused with acute doses of endotoxin were more likely to survive when their body temperature increased. Nonsurviving sheep were unable to mount a fever and became mildly hypothermic (Pittet et al., 2000). In another study, severe hypothermia was a primary indicator of mortality in mice subjected to enterotoxic shock, but surviving mice nonetheless developed a mild hypothermia (Ylach et al., 2000). In rodents, a moderate hypothermic response may be beneficial to survival, but a severe hypo-

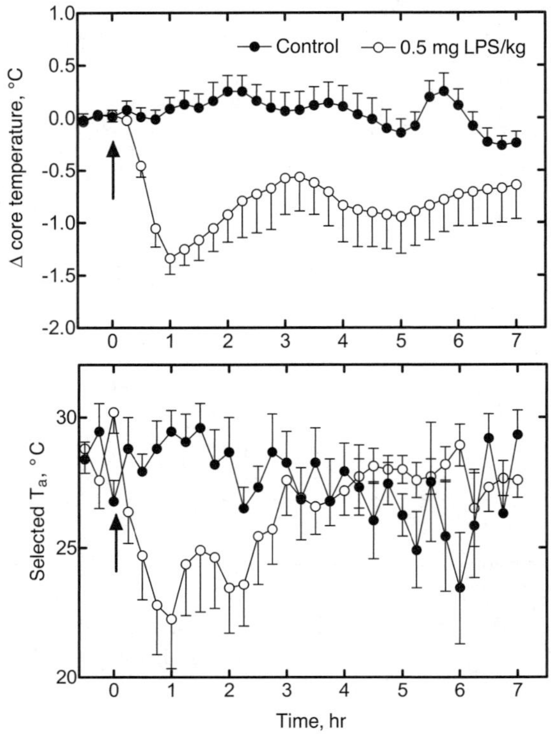


**Figure 5.4 Time-course of core temperature and selected ambient temperature of rats subjected to endotoxic shock. Animals were injected intravenously with 0.5 mg/kg LPS endotoxin while housed in a temperature gradient that allowed the rats to behaviorally thermoregulate. (Data from Romanovsky, A.A., Shido, O., Sakurada, S., Sugimoto, N., and Nagasaka, T. (1996). *Am. J. Physiol.* 270: R693–R703.)**

thermic response is detrimental. The severe hypothermic response may represent a failure in metabolic thermogenesis leading to thermoregulatory dysfunction and death. The role of hypothermia in large mammals with such a debilitating insult as septic shock is less well understood and its role in survival merits further work.

Overall, the behavioral thermoregulatory response of rodents to endotoxemia appears to have some similarities to hypoxia as well as to that following exposure to various toxicants. In spite of the differences between hypoxia and endotoxemia, it is interesting to note that they both elicit a regulated hypothermic response in the rat. Does this imply that there is a common physiological response that can be activated by a variety of insults? If true, then one can see how a xenobiotic agent could interact with a common pathway, leading to regulated hypothermia that improves the survival from the insult.

## 5.5 EXTRAPOLATION FROM RODENT TO HUMAN

A fundamental premise in assessing the potential risk to humans following exposure to chemical toxicants, drugs, and other agents is the assumption of a basic similarity in biological response between the animal models and humans. When there are dissimilarities between the experimental species and humans, appropriate scaling and correction factors have to be developed in order to normalize the differences between species. To this end, it behooves one to assess how the thermoregulatory responses of rodents and other test species can be extrapolated such that a response in humans can be predicted with reasonable accuracy. It is clear from the data presented in Chapters 3 and 6 that humans rarely show the degree of hypothermic response to toxic insult that is commonly found in rodents. In fact, a fever or hyperthermic response is commonly seen in humans subjected to anticholinesterase insecticides, metal fumes, and a variety of other toxicants (Chapter 6). The minimal or complete absence of a hypothermic effect in humans necessitates questioning the importance of understanding the thermoregulatory responses of rodents to toxicants. That is, if a benchmark of toxicity (i.e., hypothermia) is rarely seen in humans, then why bother to study the phenomenon in experimental animals? Indeed, it is the dissimilarity in response to toxicants between rodents and humans that calls for a better understanding of the thermoregulatory responses of rodents exposed to toxicants.

### 5.5.1 Principles of Allometric Scaling

Differences in body mass as well as other species-specific physiological differences account for the marked differences in thermoregulatory response to toxicants between rodents and humans. That is, assuming that a toxicant affects the thermoregulatory centers equally across species and induces a decrease in the set-point, the magnitude of the hypothermic response will then depend on the efficacy of thermoeffectors to dissipate heat and the subjects' physical characteristics, including insulation and thermal inertia. Humans possess effective mechanisms to dissipate heat by evaporation and peripheral vasodilation. Thermal inertia and possibly species-specific differences appear to be the key factors that limit the hypothermic response in humans and other large species.

Allometric analysis, or the scaling of physiological processes to body mass, is a mainstay of comparative physiology (Schmidt-Nielsen, 1975; Calder, 1981). The standard allometric equation often found in physiological publications is

$$Y = aM^b \tag{5.1}$$

This equation expresses the function of a physiological or morphological variable (*Y*) as affected by body mass (*M*) raised to a fractional power (*b*), and *a* is constant for a given set of phylogenetic and environmental conditions (Calder, 1981). In the logarithmic form, the slope is a straight line:

$$\log Y = \log a + b \log M \qquad (5.2)$$

The allometric equation is an ideal empirical relationship to explain how body mass affects a variety of physiological processes, including the effects of toxicants on body temperature.

An animal's weight, mass, and volume are proportional to the cube of its radial dimensions, whereas its surface area is proportional to the square of its radial dimension (Schmidt-Nielsen, 1985). The surface area is crucial because it represents the site where the most heat is exchanged between an organism and its environment. Not all of the surface area is exposed to the environment, so the effective radiating surface area would be a more appropriate term, but the total and effective areas are proportional (IUPS, 2001). The effects of body mass on surface area can be described using the allometric equation, with surface area increasing by approximately the 0.67 power (*b*) with body mass (*M*). However, the ratio of surface area to body mass decreases with increasing size; the logarithm of surface area to mass versus body mass in mammals has a slope of –0.33 (Figure 5.5). For example, the surface area to body mass ratio of a 30-g mouse is 0.35 $m^2/kg$,

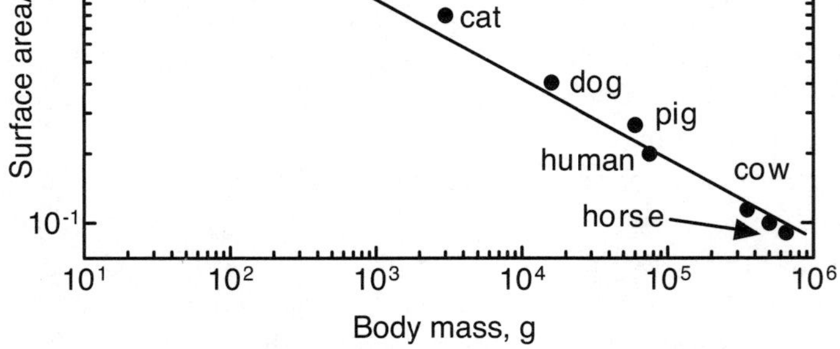


**Figure 5.5 The relationship between body mass and the surface area to body mass ratio of various species of mammals. (Data from Van Miert, A.S. (1989). *Arch. Exp. Veterinarmed.* 43: 481–488.)**

while that of a 70-kg human is more than 10 times less, 0.025 $m^2$/kg. While there are other factors to consider, the ratio of surface area to body mass has a marked impact on many physiological processes.

Heat loss and metabolic rate are key factors to consider in allometric scaling. Comparing a particular group of homeotherms such as eutherian mammals, body temperature among all species varies by a few degrees (see Chapter 2). If body temperature is essentially constant over a wide range in body mass, then heat loss (and metabolic rate) per unit body mass must increase with decreasing body mass (Figure 5.6). That is, a 0.3-kg rat and a 70-kg human regulate core temperature at 37°C, but the rat must maintain a metabolic rate (normalized to body weight) that is 5 times greater than that of the human. Overall, the logarithm of metabolic rate of eutherian mammals is inversely related to the logarithm of body mass with a linear slope of –0.24. Each particular group of homeothermic organisms, such as marsupials, birds, and others, displays a similar relationship depicted in Figure 5.6 with specific slopes and intercepts that are affected by ambient temperature and other environmental conditions (Schmidt-Nielsen, 1985). Because metabolic rate decreases logarithmically with increasing body mass, other physiological processes such as heart rate, respiratory rate, cardiac output, minute ventilation, and many other processes will likewise be affected in an allometric fashion with body mass (Hayssen and Lacy, 1985).

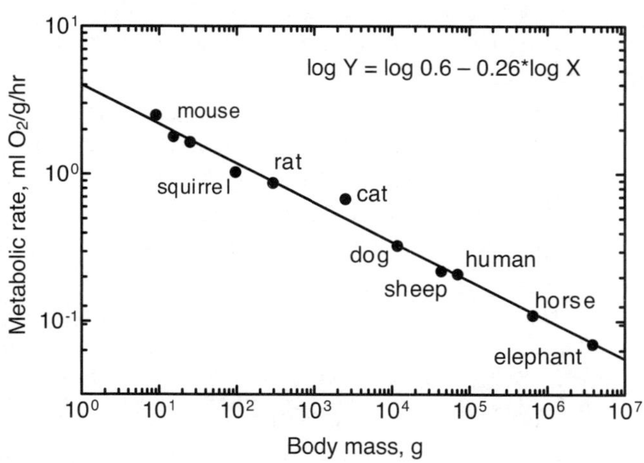


**Figure 5.6 Relationship between body mass and metabolic rate of various species of mammals. All measurements were made on resting subjects under thermoneutral conditions. (Data from Schmidt-Nielsen, K. (1975). *Animal Physiology: Adaptation and Environment.* London: Cambridge University Press.)**

The allometric scaling of metabolism has long been considered a crucial facet in the extrapolation of toxicological and pharmacological data from experimental animals to humans (Davidson et al., 1986; Vocci and Farber, 1988; Van Miert, 1989). The metabolic activation, deactivation, and clearance of drugs and toxicants are inescapably linked to a species' basal metabolic processes, with the rate of clearance of a toxicant from the blood generally found to be inversely dependent on body mass. As will be discussed in the next section, body mass also limits the hypothermic response to a toxicant. The surface area to body mass ratio thus impacts on two key processes: the metabolic clearance and the thermal response to the toxicant. Both are affected by body mass, with larger species impacted in a negative manner. An increase in body size means a lower clearance and smaller hypothermic response, leading one to predict greater toxicological sensitivity.

### 5.5.2 Thermal Conductance and Toxicant-Induced Hypothermia

The principles of allometric scaling used to study metabolism can also be applied to understand how body mass affects the thermoregulatory response to a toxicant. The dependence of thermal conductance on body mass is likely a key factor that limits the toxicant-induced hypothermia. Thermal conductance is defined as the rate at which heat is transferred between the unit area of two parallel surfaces in a medium when unit temperature difference is maintained between them (IUPS, 2001). It is generally expressed in dimensions of $W/m^2/°C$. Thermal physiologists specializing in rodent responses generally normalize metabolic responses to body weight rather than surface area and use a modified version of the thermal conductance term that can be termed whole-body thermal conductance. This is calculated from a modified version of the heat balance equation (see Chapter 2):

$$C = M/(T_c - T_a) \tag{5.3}$$

where metabolic rate ($M$) is in dimensions of ml $O_2$/min/kg, core temperature ($T_c$) is measured at ambient temperatures ($T_a$) below the thermoneutral zone, and thermal conductance ($C$) is expressed in dimensions of ml $O_2$/min/kg/°C; or if oxygen consumption is converted to heat production, $C$ can be expressed in units of W/kg/°C (Aschoff, 1981; Gordon, 1993). Whole-body thermal conductance can be viewed as a single coefficient that is a summary of the complex ways in which heat can enter or leave the body through convection, radiation, conduction, and evaporation. It represents the minimal rate of heat transfer at ambient temperatures below thermoneutrality and assumes a constant rate of heat loss by evaporation (~5 to 15% of the total heat loss) (Lovegrove et al., 1991). To this end, the value of *C* is also termed *minimum wet thermal conductance.*

There is an inverse allometric relationship between thermal conductance and body weight when it is calculated as metabolic rate normalized to body weight (McNab, 1980; Aschoff, 1981). The relationship between body size and thermal conductance is affected by factors such as activity level. Overall, calculation of thermal conductance in a variety of mammals during rest with a body mass range of 10 to 4000 g indicates a slope of –0.52 (Figure 5.7). That is, the thermal conductance of a 30-g mouse is approximately 3.4-fold greater than that of a 300-g rat. This relationship essentially shows that small homeotherms cool faster and have greater metabolic cost to thermoregulate when compared to larger species when all other environmental factors are equal. This principle should have a bearing on how a toxicant affects body temperature, especially when the mechanism of action of the toxicant is an inhibition in metabolic thermogenesis.

## 5.6 HUMAN VERSUS RODENT

Based on the aforementioned allometric relationships, one would expect that a decrease in body mass will have a greater impact on thermal homeostasis in the face of impaired thermoeffector function. Moreover, the ability of a homeotherm to develop regulated hypothermia will also be magnified as body mass is reduced.

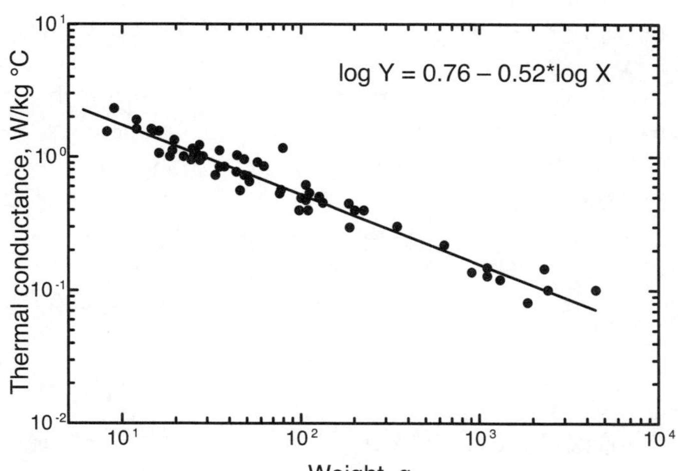


**Figure 5.7 Effect of body mass on whole-body thermal conductance in various species of mammals during rest. (Data from Aschoff, J. (1981).** ***Comp. Biochem. Physiol.*** **69A: 611–619.)**

It is thus not surprising to find a relatively small hypothermic response in humans exposed to toxicants that induce marked hypothermia in rodents. Many of the observations of body temperature in humans exposed to toxicological agents are based on emergency room observations. It is difficult to analyze these studies because of the tremendous variability in age, dose of toxicant, time of measurement, environmental conditions, etc. However, there are a few clinical studies with controlled exposures to ethanol and other agents that shed additional light on the impact of body mass on thermoregulatory response to a toxicant. Freinkel et al. (1972) determined the thermal response of humans and mice when made hypoglycemic via the administration of 2-deoxy-D-glucose (2-DG). Human subjects were administered 50 mg/kg 2-DG (IV) over a 30-min period and underwent a 1.1°C decrease in rectal temperature at 2 h after dosing. Mice dosed with ~700 mg/kg 2-DG (IV) sustained a 5°C decrease in core temperature by 1 h after injection. Peak plasma levels of 2-DG were lower in humans (16 mg/100 ml) than in mice (31 mg/100ml). Nonetheless, this represents one of the few studies that compare the hypothermic responses of a human to mouse to the same drug under controlled conditions. The results confirm the aforementioned allometric analysis that the hypothermic response to a metabolic blocking agent is limited with increasing body mass.

Many of the toxicological studies discussed in Chapter 3 also point to a reduced hypothermic response with increasing body mass. Mice consistently show greater reductions in core temperature compared to rats when administered similar doses of metal salts, sulfolane, and anti-ChE agents (see Chapter 3). The data base on ethanol toxicity in humans provides an ideal means of assessing the effects of body mass on the hypothermic efficacy of a toxicant. In a survey of the literature reporting on the hypothermic effects of ethanol under standard ambient temperature conditions of 20 to 24°C, there is an inverse relationship between body mass and magnitude of hypothermia (Figure 5.8). Of course, when drawing upon studies from many different laboratories separated in time by many years, one has to consider the potential variability that would affect this relationship (i.e., dose of ethanol, time to peak effect, environmental conditions, etc.). Nonetheless, the allometric equation (5.2) can be used to show an inverse relationship between body mass and the hypothermic efficacy of a toxicant.

## 5.7 RELEVANCE OF REGULATED HYPOTHERMIA IN TOXICOLOGY

It is clear that a hypothermic response to a toxicant that has been well documented in mice, rats, and some other test species is either markedly

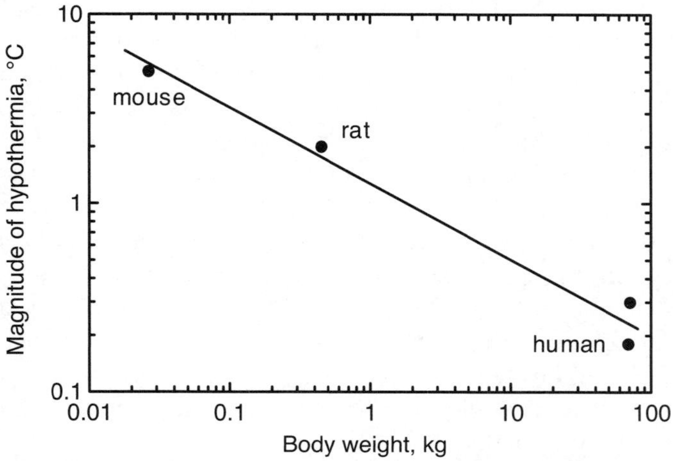


**Figure 5.8 Relationship between body mass and maximal hypothermic effect of ethanol administered to the mouse (Gordon and Stead, 1986a), rat (Myers, 1981), and adult humans (Fellows et al., 1984; Danel et al., 2001).**

attenuated or absent in adult humans. This being the case, why should one be so concerned about a thermoregulatory response in rodents that is nearly benign in humans? As explained in more detail in the following sections, a better understanding of the thermoregulatory response in rodents and other species may improve the procedures for assessing the health and safety of toxicant exposure in humans. This information could also be useful in the development of therapeutic treatments for poisoning. Finally, these studies may shed light on how birds, mammals, and other species have evolved mechanisms to ameliorate the cellular damage from dietary and environmental toxicants as well as other types of insults.

## *5.7.1 Assessment of Risk*

Risk assessment of chemical and drug safety relies heavily on data collected from laboratory rodents. The exaggerated hypothermic response of rodents to a toxic insult may in fact lead the risk assessor to underestimate the toxicity of drugs and chemicals when extrapolating to humans (Gordon, 1991, 1996; Watkinson and Gordon, 1993). That is, a moderate amount of hypothermia is protective to toxic insult, and the hypothermic response is attenuated with increasing body mass. Hence, a given dose of a toxicant that elicits a hypothermic response in a rodent would theoretically be more toxic in larger species that are unable to lower body temperature.

The development of neuroprotective drugs to treat stroke and other ischemic diseases is inexorably linked to rodent thermoregulation. The fact that rodent body temperature readily decreases when subjected to pathological insults has led to a huge controversy over the supposed neurotoxicity of toxicants and the neuroprotective nature of certain drugs (also see Chapter 4). If body temperature was as stable in experimental rodents as it is in humans when administered these toxic agents, there would probably be little interest in understanding the role of hypothermia as a neuroprotectant. For example, dizocilipine, an N-methyl-D-aspartate channel antagonist, has been thoroughly studied as a neuroprotectant using the gerbil as a principal experimental model (Buchan and Pulsinelli, 1990). The neuroprotective actions of dizocilipine were attributed to its hypothermic efficacy rather than to an anti-ischemic action per se. Since dizocilipine would have little if any hypothermic effect in an adult human, it is unlikely that it would be an effective neuroprotectant to a stroke when the patient is maintained in a normothermic state.

## 5.7.2 Hypothermia as Therapy in Poisonings?

In view of the neuroprotective role of hypothermia in stroke and ischemia, there has been an explosive development of both experimental and clinical studies on the potential use of hypothermia as a therapy (see Chapter 4). Since hypothermia also affords protection to many types of toxicants in experimental animals, one should question if hypothermia could be useful as a therapeutic tool in acute poisonings of humans as well as pets and agricultural species. To the best of my knowledge, this has been rarely studied. In fact, keeping a victim of a poisoning warm would most likely be a standard protocol in most emergency room settings. Hypothermia is assumed to be detrimental to recovery from most traumatic injuries (Luna et al., 1987). An exception would be the treatment of poisonings from oxidative uncouplers, which can cause lethal hyperthermia (see Chapter 6).

The current drive to develop and approve drugs to safely lower body temperature in stroke victims could well lead to a similar application of using hypothermia in the treatment of certain types of poisonings. One major obstacle in this endeavor is developing drugs that induce a regulated reduction in core temperature. Only recently has it been recognized that developing a drug that could induce a decrease in the set-point would be an ideal means of minimizing physiological stress while inducing a regulated decrease in core temperature (e.g., Gordon, 2001; Katz et al., 2001). Attempting to forcibly lower core temperature with ice packs, cooling pads, etc. will result in myriad of thermoeffector responses to counter the decrease in core temperature. This diminishes the efficacy of the method used to induce hypothermia and imparts undue physiological and psychological stress, effects that may counter the benefits of the hypothermic treatment (Gordon, 2001). Inducing a regulated hypothermic state could be an ideal means of applying hypothermia as a therapy in some poisoning victims.

Compared to rodents, humans and other large mammals are considered to be more sensitive to the pathological effects of hypothermia. A decrease in core temperature of just a few degrees centigrade can be adverse by increasing the duration of action of some drugs, impairing clotting, aggravating wound healing, and other sequelae (for review, see Sessler, 1994). To this end, the feasibility of using hypothermia as a therapy in acute poisoning would be fraught with complexities. One would also have to deal with how hypothermia would affect the metabolism and excretion of the toxicant.

In spite of the potential drawbacks and uncertainties, a relatively old case report on the treatment of a victim of carbon monoxide poisoning illustrates the potential use of hypothermia as therapy (Craig et al., 1959). Five hours after admission to the hospital, an 18-year-old victim of carbon monoxide poisoning had a core temperature of 39.8°C, a bilateral Babinski reflex (but no other reflexes could be elicited), severe hypotension, and dilated pupils that were unreactive to light. With the patient having little chance of survival, the physicians decided to induce hypothermia, and his core temperature was lowered over 2 h to 32°C (chlorpromazine was given to control shivering). Twelve hours after hypothermia was started, the patient responded to painful stimuli and attempted to talk. After 32 h of continuous hypothermia the patient showed remarkable recovery, could move his extremities, and spoke normally. Hypothermia was discontinued and the patient was slowly warmed for the next 12 h. The patient was discharged two weeks later with no signs of neurological deficits. It was concluded that "the knowledge that hypothermia may reverse many of the deleterious effects of carbon monoxide poisoning on the central nervous system makes hypothermia a logical form of therapy" (Craig et al., 1959).

#### *5.7.2.1 Changing the Set-Point to Treat Poisonings*

Just as hypothermia is being extensively studied to treat victims of stroke and myocardial ischemia, one should wonder if hypothermia could also be used to treat some victims of poisoning. Inducing a reduction in the set-point temperature would seem to be the least stressful way of lowering body temperature (Gordon, 2001). The fundamental neural mechanisms of temperature regulation discussed in Chapter 2 are a useful starting point to show how the set-point could be reduced to treat a poisoning victim. A simple neural network based on the illustrations in Figure 2.4 can be used to show how the set-point temperature could be reduced for relatively long periods of time (Figure 5.9). The set-point temperature is

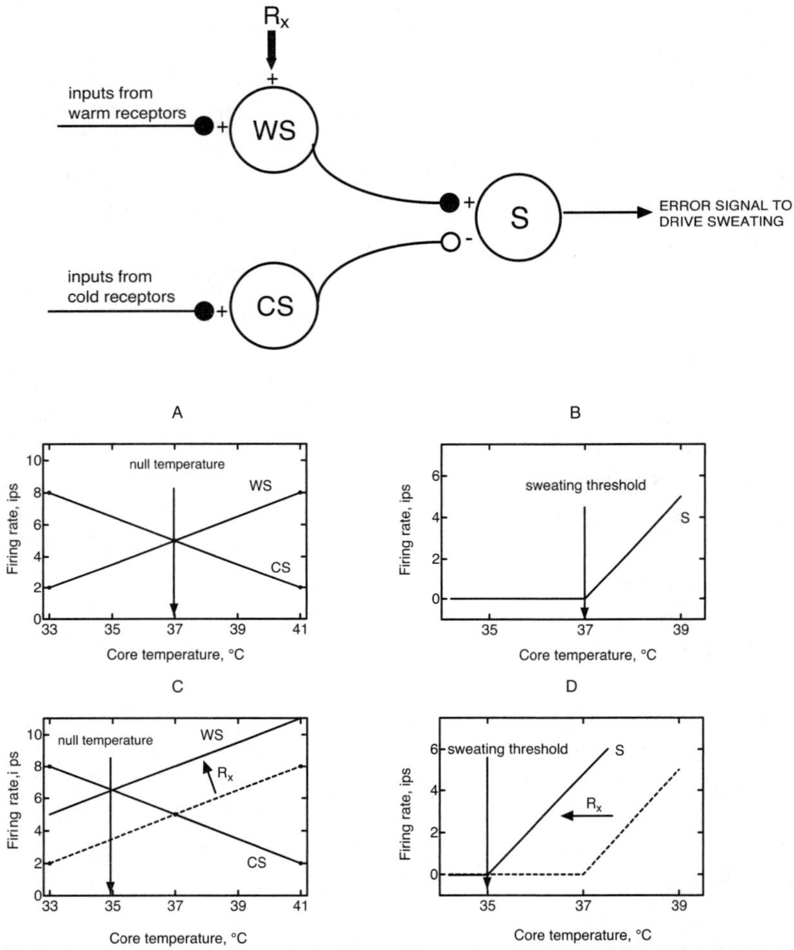


**Figure 5.9 Caption on next page.**

considered to be a result of the interaction in activity between warm-, cold-, and thermal-insensitive neurons in the preoptic area and anterior hypothalamus. An agent that shifts the slope and intercept of the thermal-sensitive neurons will result in a shift in the set-point temperature. In this example of a neural network to control sweating (Figure 5.9), an agent ($R_x$) has increased the intercept of the warm-sensitive neurons. This causes the null temperature of equivalent activity for the warm- and cold-sensitive neurons to decrease by 2°C. The threshold temperature to increase sweating is thus shifted to a lower temperature. In this new state, actual core temperature is 37°C, but sweating is activated as if the core temperature were 39°C. That is, the subject reacts as if he or she were hyperthermic. Evaporative cooling is elicited and body temperature is decreased until the null temperature is reached.

**Figure 5.9 Conceptual neural network showing how a set-point temperature can be generated and modulated to control sweating and mediate a regulated decrease in core temperature. In this scheme, a warm-sensitive (WS) neuron facilitates and cold-sensitive (CS) neuron inhibits the activity of an integrating (S) neuron in a 1 to 1 ratio. The activity from neuron S is an error signal exerting proportional control over the rate of sweating. In panel A, the intersection in activity of the WS and CS neurons results in no activity of the S neuron and no sweating. A temperature below the null point results in no activity from the S neuron. As temperature increases above the null point, the activity of the WS neuron exceeds that of the CS neuron, and activity of the S neuron also increases, leading to a proportional increase in sweating with rising core temperature (panel B). An agent that increases the intercept of the firing rate–temperature relationship of the WS neuron results in a 2°C lowering of the null point where activity of the WS and CS neurons are equal (panel C). This lowers the threshold temperature for sweating by 2°C (panel D), and the thermoregulatory system behaves as if body temperature is 2°C higher than the actual body temperature. Hence, profuse sweating is elicited at a core temperature of 37°C and would theoretically continue until the core temperature is reduced to 35°C. (Reproduced with permission from Gordon, C.J. (2001).** ***Emerg. Med. J.*** **18: 81–89.)**

The same type of neural network also drives thermoeffectors for heat production. Depending upon the ambient temperature, a neural network could theoretically be driven to reduce the set-point by either increasing heat loss or reducing heat production. This simple approach to the set-point temperature may represent the mechanism by which many toxicants induce a regulated decrease in core temperature. The activation of sweating in Figure 5.9 could be substituted with one or more other thermoeffector systems, depending upon the species.

The challenge for thermal physiologists is to ascertain the agent(s) that could evoke the regulated hypothermic response. Considering that a form of regulated hypothermia is rapidly evoked whenever an antipyretic agent is administered to a febrile subject, it seems possible that regulated hypothermic responses could also be elicited in afebrile humans. A myriad of agents could be used to lower the set-point, including neurotransmitters, hormones, electrolytes, and other agents (see Chapter 2). A recent study from this laboratory reported how an analog of neurotensin can be injected peripherally and penetrate the CNS thermoregulatory centers to activate a profound regulated hypothermic response in the rat (Gordon et al., 2003). It is not known how long a reduction in set-point temperature could be maintained by a chemical agent without causing harm to the subject. A therapeutic benefit of hypothermia may require that the hypothermic state be maintained for several days. Studies with neurotransmitters, hormones, or other chemicals to modulate the set-point temperature independently of other physiological processes would be invaluable to basic and applied biomedical research.

### 5.7.3 Evolution of Homeothermy and Resistance to Toxicants

Is it possible that the phylogenetic development of mechanisms to respond to toxicants and toxins has co-evolved with the development of homeothermy? By understanding the hypothermic responses to these insults, one may be better able to assess how and why xenobiotics induce similar types of thermoregulatory responses. That is, is it possible that xenobiotics impart their effects on thermoregulation through a common pathway that responds to other pathological insults? Wood and Malvin (1991) postulated that regulated hypothermia in reptiles and amphibians may be an evolutionary adaptation to a variety of stressors including toxicants and toxins, hypoxia, anemia, hypercapnia, dehydration, and starvation (Figure 5.10). In general, reptiles and amphibians will become hypothermic when subjected to these insults (see Chapter 8). Hypothermia improves survival to hypoxia because of a leftward shift in the oxyhemoglobin dissociation curve resulting in improved oxygen loading in the lungs. Hypothermia also reduces the activity of energetically costly processes such as ventilation and cardiac output. Hypothermia protects these species from hypercapnia by reducing ventilation and from dehydration by lowering the rate of water loss. The reduced metabolic rate during hypothermia also protects these species from starvation. This is essentially a generalized response that may have evolved in amphibians and reptiles because their body

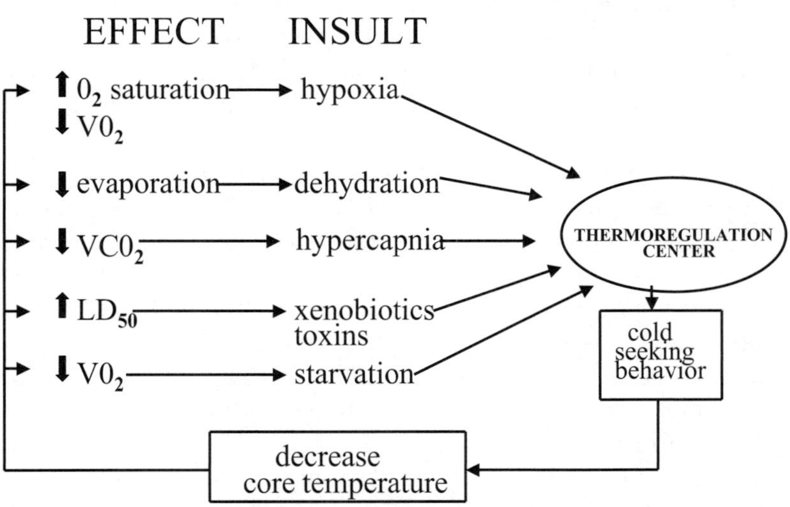


**Figure 5.10 Generalized hypothermic response to pathological insults postulated to occur in amphibians and reptiles. The hypothermic response to the insults is thought to lead to a common pathway, activating thermoregulatory processes to lower body temperature with resulting protection to the insults. (Modified from Wood, S.C. and Malvin, G.M. (1991).** ***J. Exp. Biol.*** **159: 203–215.)**

temperature is controlled primarily by behavior. Moreover, with exception of dehydration, the same scheme can probably be applied to fish. In the evolution of homeothermy in birds and mammals, the adaptive hypothermic response to these and other insults was not abandoned, but the response is muted due to the limitations placed by body mass.

In the process of mammalian evolution and the development of species with larger body mass, thermal stability was improved but with a diminished capacity to lower core temperature in response to an insult (cf. Figure 5.8). Did this co-evolution of size and homeothermy impact the way toxicants and toxins are metabolized and excreted? The ability of the mouse to become hypothermic in response to a toxicant such as ethanol means that there is a certain degree of protection provided by the thermoregulatory response. However, the inability of the adult human to become hypothermic means that the potential for toxic damage is greater and mechanisms of metabolic deactivation and excretions of the toxicant would be expected to be different in the species whose body temperature is unchanged. Large homeothermic species can be viewed as being essentially trapped in their *thermal milieu*. Their body mass and thermal inertia are a major impediment for these species to quickly lower core temperature in response to an insult. Moreover, if large mammals do become hypothermic, the recovery to normothermia is slow and they would face an additional risk of remaining hypothermic for an extended period of time. It is postulated that the inability of relatively large species to lower core temperature must have been a driving force in the evolution of mechanisms to protect the animal from toxicants and other pathological insults when compared to smaller, thermally labile species (Gordon, 1996).

# Chapter 6

# Fever and Hyperthermia

## 6.1 INTRODUCTION

Hypothermia is the most common thermoregulatory response in laboratory rodents and other small mammals subjected to acute exposure to xenobiotic toxicants as well as natural toxins (see Chapters 3 and 10). The labile nature of core temperature regulation of rodents when they are exposed to a toxicant at relatively cool ambient temperatures (i.e., below thermoneutrality) has led to the notion that hypothermia is a predominant thermoregulatory response. However, hyperthermia is frequently observed in humans following acute exposure to a variety of toxicants. In fact, the divergence of thermoregulatory responses between rodents and humans has often impeded the extrapolation of toxicological effects from experimental animal to human. Additionally, the febrile effects of toxicants should be assessed in view of the role of the immune system in the mediation of fever and the fact that many toxicants have inflammatory properties.

Fever is defined as "a state of elevated core temperature which is often, but not necessarily, part of the defensive responses of organisms (host) to the invasion by live (microorganisms) or inanimate matter recognized as pathogenic or alien to the host" (IUPS, 2001). Hence, the definition of fever covers the potential hyperthermic responses that may be elicited by a toxicological agent. The febrile responses to microorganisms and their components are certainly better understood than the response to inanimate matter. Only recently, with the development of radiotelemetry, have the hyperthermic responses of rodents to toxicological agents been assessed. To this end, this chapter addresses the hyperthermic and febrile effects of toxicants in experimental animals and humans, including anticholinesterase agents, oxidative uncouplers, alcohol, and metal fume fever.

## 6.2 MECHANISM OF FEVER

It is difficult to imagine where our understanding of thermoregulation would be without the study of fever. Investigations into the mechanism of fever during the past century are responsible for much of our knowledge of the mechanisms of thermoregulation in humans and laboratory mammals. Fever is one of the most dynamic fields in thermoregulation and is abundant with new discoveries on its neural, physiological, and immune mechanisms of action (for review, see Kluger et al., 1995; Kozak et al., 2000b; Leon, 2002). The development and recovery from a fever in mammals and other phyla involve an integration of the thermoregulatory and immune systems (Figure 6.1). The fever pathway is initiated by exposure to an exogenous pyrogen that is defined as any substance that causes a fever when it enters the body. The most potent pyrogens are infectious agents or their components (e.g., bacterial endotoxins such as lipopolysaccharide; LPS). These pyrogens activate circulating lymphocytes and fixed macrophages (e.g., Kupffer cells) to produce anti-inflammatory and pro-inflammatory cytokines that enter the circulation. Interleukin (IL)-1 and IL-6 are key pro-inflammatory cytokines that play a major role in the development of an elevated body temperature. Anti-inflammatory cytokines such as IL-10 and TNFα act as antipyretics (or cryogens) and

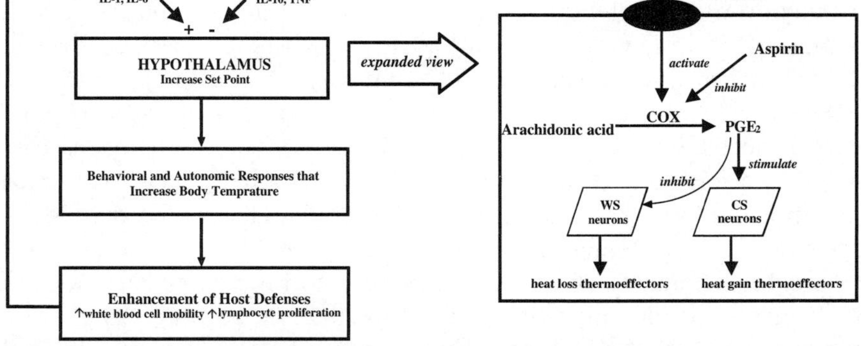


**Figure 6.1 Summary of the fever pathway. Diagram is a compilation of reviews from Kluger et al. (1995), Kozak et al. (2000a), and Leon (2002). IL, interleukin; TNF, tumor necrosis factor-α; OVLT, organovasculosum lamina terminalis; COX, cyclo-oxygenase; WS, warm-sensitive neuron; CS, cold-sensitive neuron.**

serve, among many functions, to limit the magnitude of the fever. Circulating cytokines are thought to enter the hypothalamus and preoptic area via fenestrated (i.e., leaky) areas of the organovasculosum lamina terminalis or by active transport mechanisms (Figure 6.1; expanded view). Constitutive and inducible cyclo-oxygenase (COX-1 and COX-2, respectively) are activated by IL-1 and IL-6, leading to the conversion of arachidonic acid to prostaglandins and other products.

Prostaglandin $E_2$ ($PGE_2$) is considered a key mediator in fever. When released into the preoptic area of the anterior hypothalamus (POAH), $PGE_2$ selectively stimulates the activity of cold-sensitive neurons and suppresses the activity of warm-sensitive neurons, leading to suppression of heat-dissipating thermoeffectors and stimulation of heat gain/conserving thermoeffectors (see Chapter 2). This pattern of activity of the thermoregulatory neural networks leads to an elevation in the set-point and a regulated elevation in core temperature. The fever, in conjunction with an increase in white blood cell motility and lymphocyte proliferation, enhances the host's defense against the infectious agent. The high body temperature of the host also suppresses growth of some pathogens. The fever pathway can also be activated by other traumatic insults that cause cellular injury such as a severe sun burn, physical injury to the CNS, and stroke.

### 6.2.1 Rodents as a Model for Fever and Toxicology Studies

The advent of radiotelemetry has revolutionized whole-animal pharmacological and toxicological studies because subtle physiological effects of a drug or toxicant can be assessed in undisturbed (i.e., unstressed) and conscious animals. Telemetry has opened the study of fever with the rodent becoming the key experimental model. As will be shown below, the telemetry-monitored rodent has been a successful approach to studying the febrile and delayed hyperthermic responses to anticholinesterase–based insecticides and other toxicants.

The laboratory rabbit has historically been the species of choice for experimental studies of fever because it adapts well to restraint and is amenable for intravenous injection of pyrogens through the marginal ear veins. Rabbits typically respond to pyrogens with a vigorous and well-defined fever. Until the advent of radiotelemetry, rodents were not the species of choice to study fever because the repeated handling and measurement of core temperature with conventional probes obviated the effect of the pyrogen on body temperature. The daytime core temperature of the rat is approximately 37°C; however, when core temperature is measured at intervals of 1 or 2 h with a colonic probe, the stress of handling and measurement leads to a sustained increase of core temperature to 38°C or higher (Chapter 7). This stress-induced temperature elevation equals or surpasses the core temperature of animals with fever, making it nearly impossible to detect and study the febrile response to a suspect pyrogen. Indeed, it was thought that mice and rats exhibited hypothermic responses to pyrogens because very large doses of agents such as LPS were given in an attempt to document a fever in these species.

## 6.3 FEVER AND CHOLINESTERASE-INHIBITING INSECTICIDES

My laboratory first used radiotelemetry to study the long-term thermoregulatory effects of organophosphate agents. In a study on the thermoregulatory responses to a single dose of diisopropyl fluorophosphate (DFP), an organophosphate that irreversibly blocks AChE activity, a sustained elevation in core temperature was observed following recovery from the acute period of hypothermia (Gordon, 1993a). The magnitude of the hyperthermia was relatively small when compared to the hypothermic effect. However, while the hypothermic effect of DFP ended 5 h after exposure, the hyperthermic response persisted for 3 days after dosing. Acute oral dosing with organophosphate insecticides such as chlorpyrifos and diazinon (Gordon, 1997; Gordon et al., 1997; Gordon and Mack, 2003) and carbaryl, a carbamate-based insecticide (Gordon and Mack, 2001), elicited a similar pattern of hypothermia followed by a delayed elevation in core temperature that persisted for at least one day (Figure 6.2).

While the hyperthermic response is likely to be a relevant manifestation of the pathology of organophosphate insecticides, it has largely gone undetected in most thermoregulatory studies because of a limitation in technology. Similar to the study of infectious fever, it was thought that rodents simply became hypothermic in response to acute exposure to anti-ChE insecticides. The hypothermic response can be marked and is easy to detect with conventional probes since core temperature can decrease by as much as 7 or 8°C and remains depressed for several hours after exposure to a toxicant (see Chapter 3). However, a hyperthermic or febrile response is relatively small, generally less than 1.0°C. Since a fever appears to be a predominant thermoregulatory response of humans exposed acutely to organophosphate and carbamate insecticides, it behooves one to understand the mechanism of the delayed fever in the rodent model. The delayed fever following exposure to anti-ChE agents in rodents may represent an important biological end point and mechanism of neurotoxicity that can be extrapolated from experimental animals to humans.

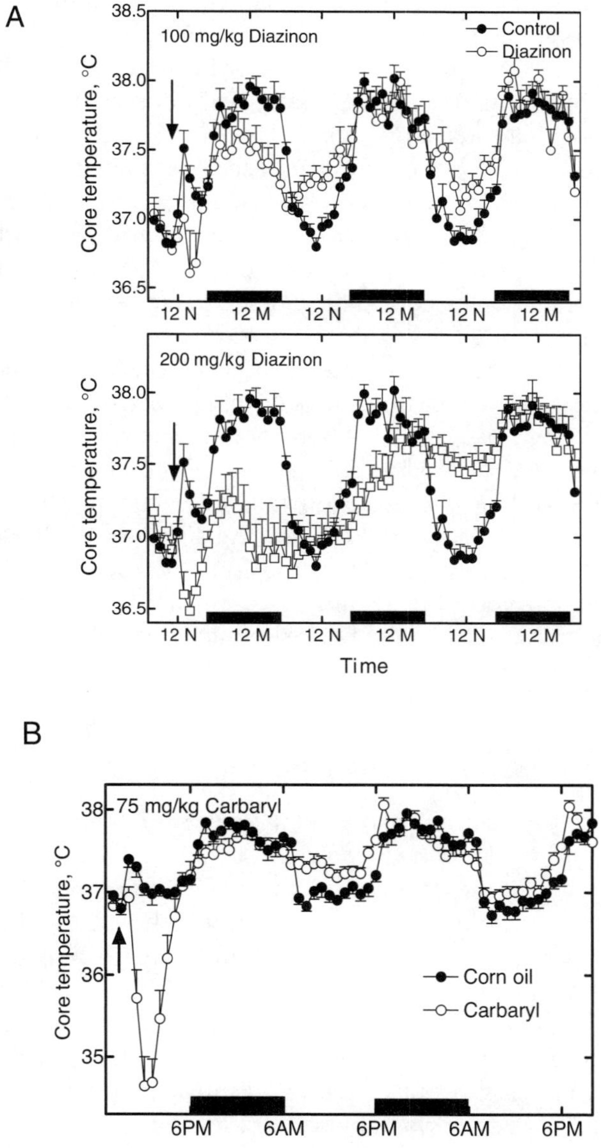


**Figure 6.2 Examples of the acute period of hypothermia and delayed increase in daytime core temperature of the rat dosed with the organophosphate insecticide diazinon and carbamate carbaryl. (Data from Gordon, C.J. and Mack, C.M. (2001). *Toxicology* 169: 93–105; Gordon, C.J. and Mack, C.M. (2003). *J. Toxicol. Environ. Hlth. A* 66: 291–304.)**

The above studies demonstrate a prolonged hyperthermic response following a single exposure to an organophosphate. Interestingly, the hyperthermic response appears to persist with repeated or chronic exposure to organophosphates. Haque et al. (1987) dosed rats repeatedly with the organophosphate malathion for 7 days and found a 1.0°C increase in core temperature on the day after the last dose. Female rats dosed daily in the afternoon with chlorpyrifos showed a vigorous hyperthermic response the day after injection for 4 consecutive days of administration (Rowsey and Gordon, 1997). Rats allowed to feed on a diet containing chlorpyrifos underwent a ~0.2°C elevation in core temperature during the daytime but not during the night over a 15-day treatment period (Gordon and Padnos, 2002). Such a subtle elevation in core temperature can only be detected by monitoring undisturbed animals with radiotelemetry. The dosage of chlorpyrifos to elicit this response was only 7 mg/kg/day but nonetheless resulted in a 87% inhibition in serum cholinesterase activity. This is the first demonstration using radiotelemetry to show a relatively small increase in core temperature following subchronic, dietary treatment with an anti-ChE insecticide. While the hypothermic effects of anti-ChE agents abate with repeated dosing as a result of the development of tolerance, the few studies on the hyperthermic effects of organophosphates suggest little development of tolerance (see Section 3.7.3).

### 6.3.1 Fever Versus Hyperthermia

It is important to distinguish between fever and other types of hyperthermia. This topic is also covered in detail in Chapter 2. There are many ways in which the core temperature of a homeotherm can increase, including (*a*) fever-induced hyperthermia, which is a regulated rise that is defended by regulatory mechanisms; (*b*) exercise-induced hyperthermia, which is a result of the additional heat load produced by contracting skeletal muscle; (*c*) hyperthermia as a result of insufficient ability to dissipate heat, as would occur by exposure in a hot environment; and (*d*) hyperthermia as a result of pathological or pharmacological impairment (Stitt, 1979). Fever was always considered to be unique in that it was the only hyperthermia alleviated with antipyretics such as aspirin. We now have behavioral, physiological, and pharmacological evidence suggesting that the delayed elevation in core temperature following exposure to organophosphate and carbamate insecticides is regulated and similar in some ways but distinct in others to that of an infectious fever.

### *6.3.2 Evidence That Anti-ChE Hyperthermia Is a Fever*

In anti-ChE–induced hyperthermia, the increase in temperature following the acute period of hypothermia might suggest to some that the response is essentially a compensatory rebound or overshoot in body temperature in the course of the recovery from hypothermia. That is, as the rat recovers from being hypothermic for 10 or more hours, heat gain and conserving mechanisms overwhelm heat loss processes and there is a transient overshoot in body temperature. However, rebound hyperthermia does not seem to explain the anti-ChE–induced hyperthermia. Blocking the chlorpyrifos-induced hypothermia by increasing ambient temperature from 22 to 31°C during the period when the rat would be hypothermic led to an increase in the magnitude of the hyperthermic response (Figure 6.3). Moreover, rebound hyperthermia may never be as robust as one might expect in the rat. For example, 6 h of prolonged hypoxic-induced hypothermia in a rat with a mean core temperature of 34°C results in just a meager rebound hyperthermia lasting just a few hours after recovery from hypoxia (Gordon, 1997a; see Figure 5.2).

The responses of thermoeffectors during the period of the hyperthermic response also suggest the development of a regulated hyperthermia. Rats placed in a temperature gradient and allowed to behaviorally thermoregulate prefer normal temperatures during the period of the chlorpyrifos- and DFP-induced fever (Gordon, 1994, 1997). If the elevation in temperature was a forced hyperthermic response as a result of peripheral vasoconstriction and increase in metabolism, then one would expect the rat to prefer cooler temperatures when given the option to behaviorally thermoregulate. In other words, the rats behave in a manner suggesting that they do not feel hot in spite of being hyperthermic. Chronic measurements of tail skin temperature during the period of chlorpyrifos-induced fever show no indication of vasodilation of tail skin blood flow to dissipate the excess heat (Gordon and Padnos, 2002). Finally, the fever persists during the day and disappears at night but then may return the following day. Such a prolonged and reoccurring elevation in temperature would suggest a resetting of the set-point for temperature regulation.

Pharmacological data also reveal unique aspects of the anti-ChE fever. Reducing body temperature with the cholinergic agonist oxotremorine does not result in a delayed elevation in temperature as is seen with chlorpyrifos and carbaryl (Gordon and Mack, 2001). While the induction of the anti-ChE–induced hypothermia is mediated primarily by stimulation of muscarinic pathways in the CNS (see Chapter 3), the development of the fever appears to involve mechanisms not directly related to cholinergic

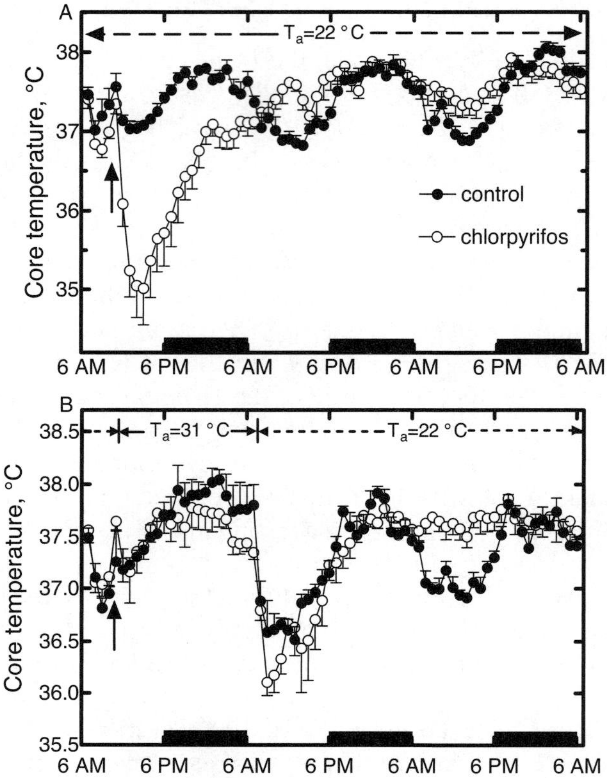


**Figure 6.3 (A) Female rats dosed with 25 mg/kg chlorpyrifos undergo hypothermia and delayed increase in daytime temperature for 48 h after dosing. (B) Blocking the hypothermic effect of chlorpyrifos by raising ambient temperature from 22 to 31°C leads to exacerbation in hyperthermic response at 48 h after dosing. Note difference in temperature scale between (A) and (B). (From Gordon, C.J. (1997). *Toxicology* 124: 165–171.)**

pathways. The fever is not blocked by the muscarinic antagonist scopolamine (Gordon and Grantham, 1999). In fact, while scopolamine and atropine are very effective at blocking the hypothermic effects of anti-ChE–induced hypothermia (see Chapter 3), scopolamine exacerbates the rise in temperature when given during the febrile period. This would suggest that the elevated temperature is being maintained by mechanisms other than the stimulation of cholinergic pathways. The delayed hyperthermia from chlorpyrifos, DFP, and diazinon (Gordon, 1996a; Gordon et al. 1997; Gordon and Mack, 2003) has been found to be effectively blocked with the antipyretic sodium salicylate (Figure 6.4). It is interesting to note that as the effectiveness of sodium salicylate wanes, body temperature returns to a febrile level. This provides further evidence that the fever is

A

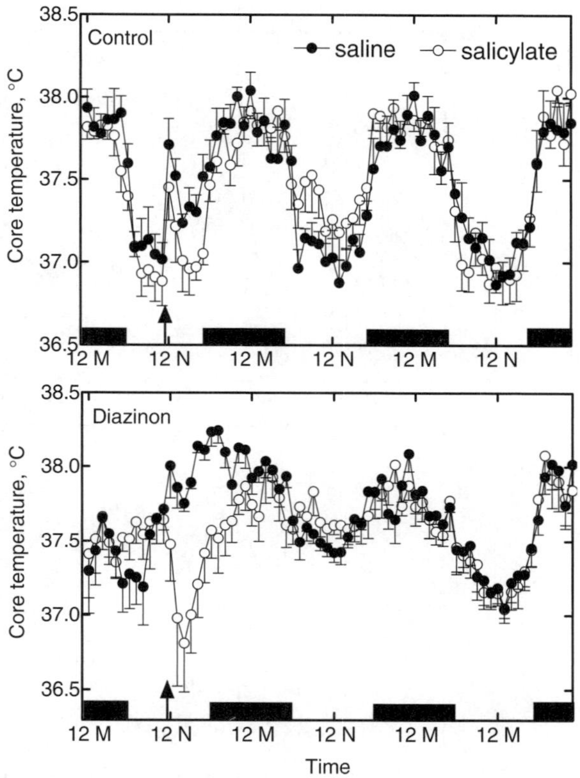


**Figure 6.4 Example of the antipyretic effect of 200 mg/kg sodium salicylate administered intraperitoneally 48 h after rats were dosed orally with the organophosphate insecticide diazinon (A) and 24 h after chlorpyrifos (B). (Modified from Gordon, C.J., Grantham, T.A., and Yang, Y. (1997). *Toxicology* 118: 149–158; Gordon, C.J. and Mack, C.M. (2003). *J. Toxicol. Environ. Hlth.* A 66: 291–304.)**

maintained via central modulation of thermoregulatory pathways, a part of which involves the cyclo-oxygenase system (cf. Figure 6.1). In view of the aforementioned evidence, it is reasonable to define the delayed hyperthermia resulting from exposure to anti-ChE agents as a fever.

In view of the similarity of the anti-ChE–induced fever to infectious fever, it is reasonable to assume that the immune-neural pathways involved in infectious fever are also operative in anti-ChE fever. TNFα and IL-6 are two of several cytokines involved in the mediation of infection-mediated fevers. Gordon and Rowsey (1999) measured plasma levels of these cytokines at several time points following exposure to chlorpyrifos in male and female rats. The only significant findings were an increase in plasma levels of TNFα but no change in IL-6 when measured 48 h after acute exposure to chlorpyrifos (i.e., during the time when a fever develops).

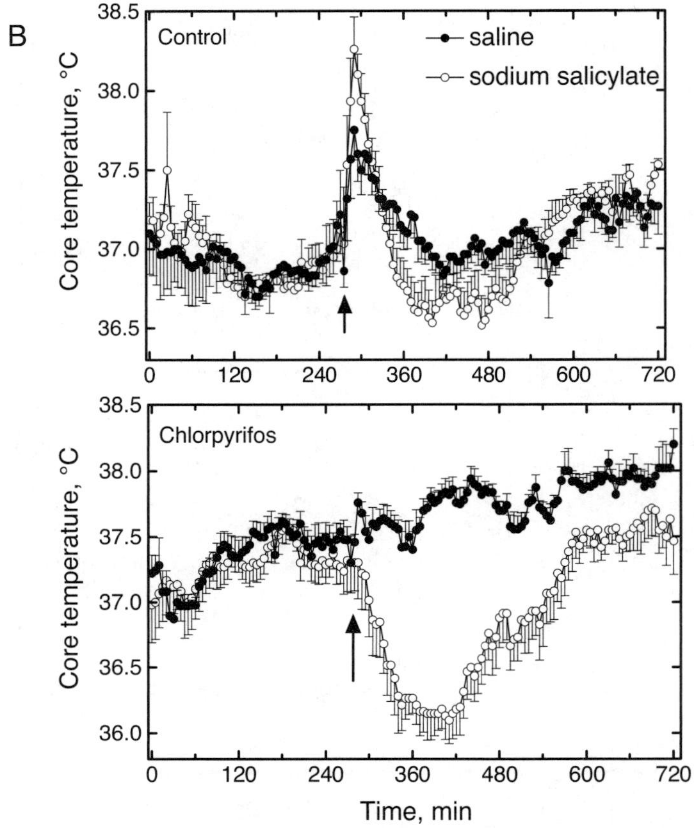


**Figure 6.4 (continued)**

Moreover, there were no remarkable changes in these cytokines at 4, 18, or 24 h after exposure that would be associated with the early phase and full development of the fever. An 89% increase in TNFα levels in female rats 48 h after chlorpyrifos exposure was associated with a febrile state. However, this increase in TNFα was relatively small when compared to the greater than 10-fold rise in blood levels of this cytokine in animals subjected to an LPS fever (Gordon and Rowsey, 1999; also see Kluger et al., 1995). Moreover, IL-6 increased by 1,000-fold in the blood of rats subjected to an LPS fever but changed little following chlorpyrifos administration. In summary, data collected so far do not show the robust elevations in circulating cytokines as are seen with an infectious fever. However, this does not rule out participation of the neuro-immune fever response in the mediation of the anti-ChE fever.

There are numerous pro-inflammatory cytokines that could be operative in anti-ChE fever that have yet to be identified. In addition, the studies have been limited mainly to chlorpyrifos and cytokine responses, which may differ from other organophosphate agents. Localized production of cytokines in the CNS may also be a possible mechanism of action that has yet to be studied. Overall, the anti-ChE agent–induced fever appears to be a unique thermoregulatory response that may involve the participation of multiple physiological systems.

### *6.3.3 Manifestation of Fever: Day Versus Night*

The febrile response to anti-ChE agents in rats is manifested primarily during the daytime, a period when the rats tend to be less active. Core temperature of control and treated rats is usually similar at night when rats are more active (cf. Figure 6.2). This day versus night pattern is commonly seen in infectious fevers in rodents and represents yet another similarity between infectious and anti-ChE fever. It is thought that there is a ceiling core temperature that generally limits the maximum controlled core temperature from day to night as well as during a fever (Feng et al., 1989). For example, during the night, rats maintain a regulated core temperature of 38°C. A pyrogen such as LPS or $PGE_2$ is considered to increase the set-point temperature from 37 to approximately 38°C. This means that if the pyrogen is given during the daytime when core temperature is low, there will be a robust increase in core temperature. If given at night, the febrile reaction is initiated but there is no apparent increase in core temperature because the rat is regulating at a hyperthermic level (Figure 6.5). As the next light cycle approaches and core temperature of afebrile rats decreases, the effect of the pyrogen is again manifested as an elevated core temperature. Feng et al. (1989) noted that fever represents an *increase in the regulated temperature* and not a *regulated increase in temperature*. This explains why the increase in body temperature during a fever will be higher during an animal's period of inactivity

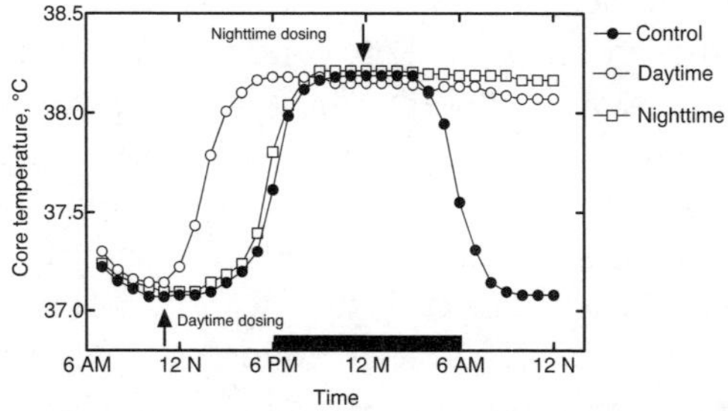


**Figure 6.5 Depiction of the ideal time-course of a fever in nocturnal species depending on whether the pyrogen is administered during the day or night.**

when its core temperature is normally low. A similar pattern of thermoregulation appears to be the case during exposure to organophosphate agents. For a nocturnal species such as the rat, the symptoms of organophosphate poisoning are evident at night (e.g., reduced motor activity) without the expression of a fever. For a diurnal species, an opposite pattern of core temperature would be expected with an elevated core temperature at night and little indication of a fever during the daytime. Thus, interpreting the fever with an origin from infection, toxic agents, or other causes requires that one consider the species' circadian rhythm characteristics.

## 6.4 FEVER AND HYPERTHERMIC RESPONSES IN HUMANS

### *6.4.1 Responses to Anti-ChEs*

Humans are acutely exposed to anti-ChE agents as a result of accidental exposures in the home and places of work and from their use as suicidal or homicidal agents. The data collected during treatment from these poisonings account for much of our understanding of the thermoregulatory effects of anti-ChEs in humans. However, the analysis of these data is hampered by the innumerable uncontrolled variables in these studies, such as variations in time to treatment in the emergency room, age, gender, ambient conditions, pharmacological treatments, lack of information on the patient's normal body temperature, and dose of the anti-ChE agent. The effects of these poisoning incidents are often recorded during emergency room treatment, a period associated with marked stress on the thermoregulatory system. In spite of these drawbacks, these clinical reports provide insight into the thermoregulatory responses of humans exposed to anti-ChEs as well as other toxicants.

Thermoregulation is one of many autonomic processes affected by exposure to organophosphate and carbamate insecticides (Table 6.1). There are also incidences of hypothermia, especially in subjects when first admitted to the emergency room when their symptoms of cholinesterase inhibition are severe. A hyperthermic or febrile response is often seen as a complication in the recovery from exposure to the anti-ChE agent. Namba et al. (1971) conducted one of the first studies to document the occurrence of fever in humans acutely exposed to organophosphate insecticides (Table 6.2). They noted an incidence between the degree of inhibition in serum ChE activity and incidence of fever in patients exposed to the organophosphate insecticides parathion and methyl parathion. In the Namba study, it was noted that mild fevers could persist for over 1 week in some cases.

**Table 6.1 Incidence of Clinical Manifestations in 70 Patients Treated for Acute Exposure to Organophosphate and Carbamate Insecticides**

| *Clinical Manifestation* | *Number of Patients* | *Percentage* |
|---|---|---|
| Muscarinic | | |
| Miosis | 60 | 86 |
| Nausea | 51 | 73 |
| Salivation and bronchial constriction | 51 | 73 |
| Diarrhea/urinary incontinence | 17 | 24 |
| Nicotinic | | |
| Muscular twitching | 31 | 44 |
| Tremor | 9 | 13 |
| CNS | | |
| Headache/dizziness | 44 | 63 |
| Coma | 20 | 29 |
| Cardiac | | |
| Sinus tachycardia | 25 | 36 |
| Sinus bradycardia | 20 | 28 |
| Hypertension | 15 | 22 |
| Hypotension | 12 | 17 |
| Other | | |
| Fever | 34 | 49 |

*Source*: Saadeh, A.M., Al-Ali, M.K., Farsakh, N.A., and Ghani, M.A. (1996). *Clin. Toxicol.* 34: 45–51.

**Table 6.2 Relationship between Degree of Inhibition of Serum Cholinesterase Activity (Expressed as Percent of Normal Activity) and Some Symptoms of Parathion Poisoning in Humans**

| | *Amount of ChE Inhibition* | | |
|---|---|---|---|
| *Symptoms* | *Severe (0–10%)* | *Moderate (11–20%)* | *Mild (21–50%)* |
| Sweating | 14 | 10 | 20 |
| Salivation | 13 | 8 | 10 |
| Weakness | 14 | 11 | 22 |
| Muscular fasciculations | 14 | 6 | 0 |
| Fever | 9 | 6 | 0 |

*Source*: Data taken from Namba, T., Nolte, C.T., Jackrel, J., and Grob, G. (1971). *Am. J. Med.* 50: 475–492.

Others have also reviewed case reports of exposure to anti-ChE insecticides and found a high incidence of fever. Hirshberg and Lerman (1984) performed a retrospective analysis of 236 cases of organophosphate or carbamate poisoning in Israel between 1958 and 1979. They found that a fever was a prevalent symptom in 25% of cases labeled as a "complication or delayed effect" that usually appeared more than 24 h after exposure. Fever was defined as a condition where core temperature was >37.5°C. More recently, Saadeh et al. (1996) described the clinical and sociodemographic facets of 70 adults exposed acutely to organophosphate or carbamate insecticides and found that 49% had a "low grade fever" of 37.5 to 38.5°C with no evidence of infection. The onset and recovery from fever occurred between one and several days after exposure, with nearly half of the fevers persisting for 3 days (Figure 6.6). They noted that patients with fevers had received significantly more atropine than those without

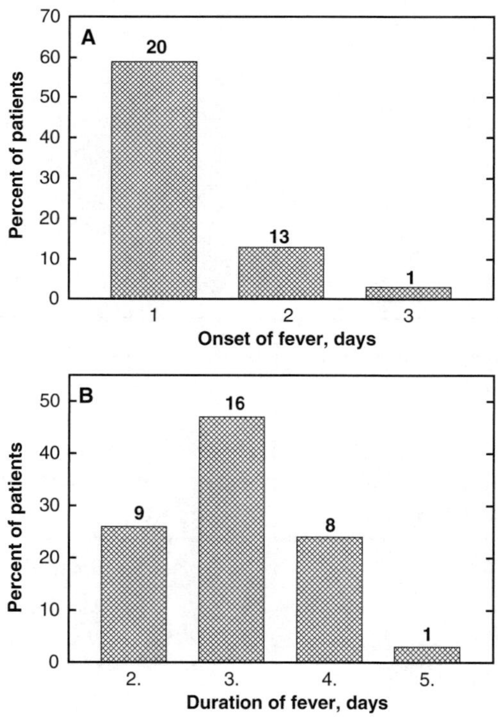


**Figure 6.6 Survey of the development (A) and duration (B) of fever in humans exposed to carbamate and organophosphate insecticides. Of 70 subjects, 49% displayed a fever with no evidence of infection. Numbers above bars indicate actual number of patients. (Data modified from Saadeh, A.M., Al-Ali, M.K., Farsakh, N.A., and Ghani, M.A. (1996). *Clin. Toxicol.* 34: 45–51.)**

fevers. One might consider if the administration of atropine is in some way responsible for the fever since the muscarinic antagonist has well-known hyperthermic effects in humans (Christoph, 1989). However, the authors noted that the fever persisted in 27% of the patients for 4 days in spite of discontinuing atropine therapy.

A presumed terrorist attack with the nerve gas sarin in Matsumoto, Japan led to the poisoning of approximately 600 residents (Morita et al., 1995). A mild, low-grade fever was reported in some subjects for up to 1 month after exposure, and one man showed a low-grade fever at 6 months after exposure. It is interesting to note that the aforementioned papers do not mention if standard antipyretics such as aspirin were prescribed for alleviating the fever.

From the above discussion, it would appear that fever is the most frequent thermoregulatory response in humans acutely exposed to anti-ChEs. However, an acute hypothermic effect of organophosphate and carbamate insecticides has periodically been reported in humans poisoned with anti-ChEs. For example, in one instance, a 16-year-old male suffering from organophosphate poisoning was admitted with a high blood pressure (152/102) but a rectal temperature of only 34.5°C (Cupp et al., 1975). One hour after a 17-year-old male was admitted to the hospital following ingestion of a large amount of malathion, his blood pressure was also high (170/80) and he was mildly hypothermic with a core temperature of 36.2°C (Hassan et al., 1981). A 58-year-old woman ingesting a large amount of diazinon had a low blood pressure and was markedly hypothermic with a core temperature of 34.4°C 1 h after poisoning (Hassan et al., 1981). In another series of case reports of diazinon poisoning, there were incidences of transient hypothermia (34.4°C) and delayed fever (38.9°C) (Klemmer et al., 1978). A 39-year-old man was admitted following malathion poisoning, with high blood pressure (140/90), profuse sweating, and a core temperature of 34°C (Meller et al., 1981). By 15 h after admission, core temperature had recovered to normal. In the same study, an 81-year-old woman was also admitted with high blood pressure (180/100) and a core temperature of 35°C. It is possible that the very young and old may be more susceptible to the hypothermic effects of these pesticides, but this has not been studied in detail.

The hypothermic effects of the anti-ChEs in humans are most likely mediated by the stimulation of muscarinic cholinergic pathways that lead to profuse sweating and peripheral vasodilation. However, it is difficult to say if the hypothermic response is regulated as it appears to be in rodents because, whenever the hypothermia is reported during the course of treatment in the emergency room, measures are usually taken to raise the subject's core temperature to normal. Information on the subject's state of thermal comfort during the onset of poisoning would be useful information in order to judge if the thermal set-point is reduced in the same manner as in rodents subjected to acute anti-ChE exposure. That is, if the subjects felt warm during the period of the hypothermia, then it would suggest, as is evident in the behavioral thermoregulatory response of rodents, that the anti-ChEs induce a regulated hypothermia. In addition, because of the large body mass and small surface area to body mass of adult humans, the ability to lower core temperature in response to a toxicant is going to be markedly attenuated with increasing body mass (see Chapter 5). As the muscarinic effects of anti-ChE poisoning abate, either naturally or due to prophylactic intervention, it would appear that thermoeffectors for increasing heat gain and reduction in heat loss take over, leading to a prolonged albeit low-grade fever.

## 6.4.2 Response to Other Toxicants

### 6.4.2.1 Chlorinated Hydrocarbons

There are incidences of fever or hyperthermia in humans exposed to other types of toxicants, but the mechanism of action is not well understood. The chlorinated hydrocarbon chlordane appears to induce a fever-like response in humans. In one case, a 56-year-old male was subjected to chlordane exposure through either a cutaneous or oral route and showed a spiking core temperature of 38.8 to 39.4°C (Furie and Trubowitz, 1976). A 20-month-old male accidentally ingested chlordane and was hypothermic upon admission (35.6°C) and then became hyperthermic (38.9°C) by 12 h after exposure (Curley and Garrettson, 1969). Imidacloprid, a relatively new class of neonicotinoid insecticides that selectively stimulate nicotinic receptors, was associated with a fever (38.1°C) 24 h after a 64-year-old male ingested a large amount of the insecticide (Wu et al., 2001).

### 6.4.2.2 Oxidative Phosphorylation Uncouplers

The pentachlorphenol (PCP)- and dinitrophenol (DNP)-based agents uncouple oxidative phosphorylation, increase metabolic rate, and cause hyperthermia in rodents and humans (Chapter 3). The PCPs have antifungal properties and have been used for many years as a wood preservative. Occupational exposures in the timber industry were common, with exposures occurring via cutaneous and inhalation routes. Dangerous, prolonged elevations in core temperature, accompanied by profuse sweating, have been reported in workers subjected to acute PCP exposure (Wood et al., 1983). For example, a 22-year-old male exposed to PCP on the work site was admitted to the emergency room with a core temperature of 41.8°C; within 22 h core temperature had increased to 42.2°C and the subject died (Wood et al., 1983). One finds that the hyperthermia in these studies is termed fever or hyperpyrexia. However, assuming that the CNS control of body temperature is unaffected by PCP, then it can be concluded that the rise in core temperature is essentially a forced hyperthermia and not a true fever. It is interesting to find the repeated occurrence of fever and sweating disorders in former workers exposed to PCPs (Walls and Glass, 1998; Gorman et al., 2001). Epidemiological analysis of these workers has not resolved the mechanism of action of PCP exposure that would lead to such long-term effects on temperature regulation. In many instances, the PCP solutions were found to be contaminated with dioxins (Gorman et al., 2001), a factor that may explain these unusual long-term health effects of PCP exposure on thermoregulation.

Oxidative uncouplers such as DNP cause a marked hyperthermic response in mice and rats with relatively rapid recovery. However, adult humans display more prolonged hyperthermic responses. For example, a 32-year-old male accidentally exposed to DNP during the day had a core temperature of 38.8°C that evening, which further increased to 40°C by the next morning in spite of taking antipyretics and applying ice baths (Leftwich et al., 1982). His core temperature exhibited spikes to 39.4°C twice daily, in the morning and evening for 6 days of hospitalization. Although not well studied, the larger body mass and thermal inertia in adult humans may explain the differences in response to these uncouplers between rodents and humans. In the study of the thermoregulatory effects of exercise, it is recognized that small mammals (i.e., <1 kg) are capable of dissipating most of the heat burden by nonevaporative mechanisms because of their large surface area to mass relationship (Taylor, 1977). In large mammals (>10 kg), nonevaporative mechanisms are insufficient, and greater reliance is placed on sweating to dissipate the heat load from exercise. A similar pattern is likely with the heat load produced from exposure to oxidative uncouplers. Hence, in the extrapolation of such effects from small to large mammals, one must be cognizant of the mechanisms by which the heat is dissipated and of potential stress on physiological systems such as water balance in large as compared to small mammals. That is, dehydration is not likely as critical in a small mammal that can dissipate large amounts of heat at ambient temperatures below thermoneutrality without relying so heavily on evaporation. (see Section 5.5)

### *6.4.2.3 Arsenic*

There is apparently little known about the thermoregulatory effects of arsenic in spite of the numerous poisonings each year. Arsenic is an uncoupler of oxidative phosphorylation, but it also has other profound toxic effects not seen with administration of oxidative uncouplers. Chronic exposure to arsenic in the drinking water or by inhalation leads to peripheral circulatory disorders (see Chapter 7). Large doses of arsenic in rat and rabbit lead to hypothermia (Chapter 3). However, CNS administration (intraventricular) of small doses of arsenic in the rabbit ranging from 0.0001 to 0.1 μg leads to a dose-dependent hyperthermia beginning 2 h after injection and persisting for approximately 6 h (Saxena et al., 1991). The hyperthermic response is preceded by a period of respiratory stimulation and an increase in ear skin temperature, an effect that persists for approximately 1.5 h. During this time body temperature is stable or may decrease slightly. Interestingly, there is no evidence of shivering or peripheral vasoconstriction as the hyperthermia ensues, but the hyperthermia is rapidly blocked with intraperitoneal administration of aspirin. Hence, the arsenic-induced hyperthermia does not appear to be regulated as based on the thermoeffector response yet appears to be mediated in part by induction of prostaglandin synthase. Humans are exposed to small amounts of arsenic from some drinking water sources, but there has never been any assessment of the potential impact of low-level exposure on thermoregulation, particularly hyperthermia. In one incident of an acute exposure to arsine gas (arsenous hydride), fever (39°C), headache, vomiting, and other symptoms of acute sickness appeared 1 to 12 h after exposure (Wilkinson et al., 1975).

#### *6.4.2.4 Turpentine*

Localized injection of turpentine to elicit inflammation and fever has been a useful tool in the study of fever (Luheshi et al., 1997; Kozak et al., 1997). The mechanism of action of turpentine on the fever pathway may be relevant to understanding some of the febrile effects of other xenobiotic chemicals. The localized inflammatory response of turpentine provides some advantages over the use of the typical systemic injection of a pyrogen such as LPS because the localized fever is elicited without the exogenous pyrogen circulating throughout the body, which will alter the function of other physiological processes. These studies also show how a toxicant could lead to a fever without entering the CNS such as occurs with metal fume fever (see Section 6.7).

An intramuscular or subcutaneous injection of a small amount of turpentine (~100 μl) elicits a localized inflammatory response and fever that is accompanied by the typical acute phase response. In the rat, the turpentine-induced fever is associated with an approximately 30-fold increase in plasma IL-6 levels (Figure 6.7). Knockout mice lacking the receptor for TNF show an exacerbated febrile response to LPS because TNF serves as a natural cryogen to prevent excessive elevations in core

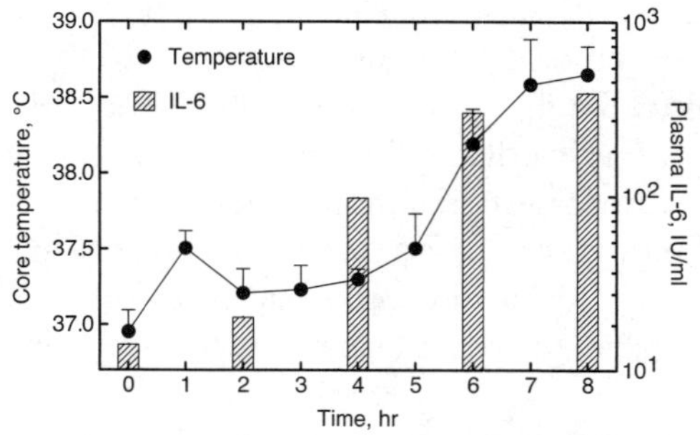


**Figure 6.7 Time-course of core temperature and plasma levels of IL-6 in the rat following intramuscular injection of turpentine (10 μl/rat). Note that rise in IL-6 precedes development of fever. (Data modified from Luheshi, G.N., Stefferl, A., Turnbull, A.V., Dascombe, M.J., Brouwer, S., Hopkins, S.J., and Rothwell, N.J. (1997). *Am. J. Physiol.* 272: R862–R868.)**

temperature (Leon et al., 1997). However, TNF receptor knockouts exhibit a fever to turpentine that is similar to that of their wild-type counterparts that have functional TNF receptors. This further illustrates the apparent unique aspect of the turpentine-induced fever and may shed light on the possible mechanisms of other types of toxicant-induced fever. There is apparently no information on the possible effects of environmental exposure to turpentine and the development of fever.

## 6.5 ALCOHOL: REBOUND HYPERTHERMIA OR FEVER?

The acute hypothermic response to ethanol and other alcohols in rodents is well known (see Chapter 3). The advent of radiotelemetry in rodent studies and the careful monitoring of core temperature in intoxicated humans have shown that ethanol induces a delayed elevation in core temperature. Gallaher and Egner (1987) first noted in telemetered rats dosed intraperitoneally with 4 to 6 g/kg ethanol an acute period of hypothermia that persisted for several hours followed by an elevation in the daytime core temperature that lasted for approximately 72 h after dosing. Interestingly, the rats' nocturnal core temperature was unaffected by ethanol treatment, while the daytime temperature remained significantly elevated above control levels for over 3 days after dosing. It will be recalled that this chronotoxicological effect of ethanol on core temperature is similar to that of organophosphate-induced fever (see Section 6.3.3). The delayed elevation in core temperature following acute ethanol intoxication was coined *rebound hyperthermia* because it appeared to reflect an overcompensating thermoregulatory response to the acute hypothermic effects of ethanol (Gallagher and Egner, 1987). The delayed hyperthermia from ethanol in rats is mediated by either intraperitoneal or oral dosing but is curiously suppressed in animals maintained in sound-attenuated chambers and sheltered from the stress of external stimuli (Sinclair and Taira, 1988). That is, during the recovery from ethanol intoxication, the hyperthermic response may represent a type of stress response exacerbated by the ethanol. However, there is clear evidence that stress augments the hypothermic effects of ethanol (Chapter 7).

*Rebound hyperthermia* may be an inappropriate term to describe this phenomenon, which appears to be a regulated increase in core temperature following acute ethanol exposure. First, blood alcohol levels recover to zero at about the same time as the recovery from the hypothermia (Gallaher and Egner, 1987), suggesting that the hyperthermic response is maintained by mechanisms other than direct effects of ethanol on thermoregulatory processes. Second, the hyperthermia persists for several days after treatment, and only during the daytime, suggesting that the hyperthermia is not simply a rebound response. Third, it is apparently not known if antipyretics are effective at blocking the hyperthermic effects of ethanol; however, since ethanol is known to induce the synthesis of $PGE_2$ (and other prostaglandins) and behavioral effects of ethanol are blocked with COX inhibitors (George and Collins, 1985), one might expect the delayed hyperthermia to be blocked with antipyretics. If this is the case, then one could view the rebound hyperthermia as a fever, possibly similar to that observed with the anti-ChE agents discussed earlier.

Recent studies in human subjects have shown a similar type of delayed, ethanol-induced hyperthermia (Danel et al., 2001). Human subjects on bed rest and dosed with ethanol over a 36-h period to maintain a blood alcohol level of 0.19 to 0.71 g/l showed a small, transient hypothermia followed by a sustained 0.2°C increase in their nighttime core temperature (Figure 6.8). This subtle hyperthermic response was manifested because the core temperature of the treated subjects did not decrease to the normal nocturnal levels as is seen during control periods. In other words, ethanol does not mediate an increase in core temperature as one would expect with typical forms of hyperthermia but rather interferes with the nocturnal decrease in core temperature. This thermoregulatory response is qualitatively the same as that of the aforementioned rat studies with the hyperthermic effects of ethanol and the anti-ChE agents; the hyperthermia is manifested during the period when core temperature for the species is normally at its lowest point.

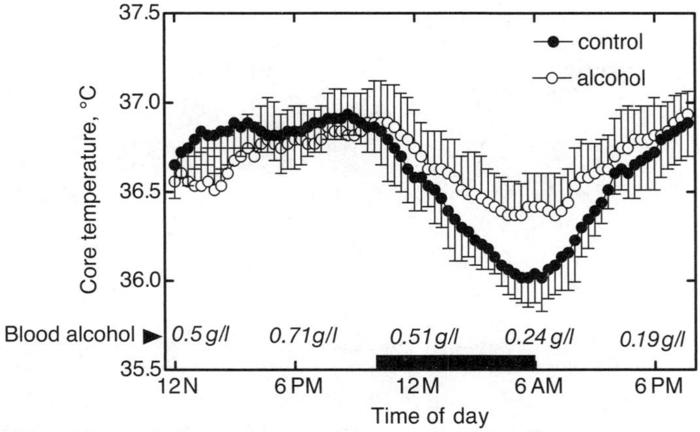


**Figure 6.8 Circadian profile of core temperature of nine healthy men given a control fluid or 256 g of alcohol administered regularly over a 36-h period. Note that the significant hyperthermic effect of ethanol occurs only during the night. Also note transient hypothermia at start of alcohol exposure. Blood alcohol levels measured at selected hours are given in italics. (Data modified from Danel, T., Libersa, C., and Touitou, Y. (2001).** ***Am. J. Physiol.*** **281: R52–R55.)**

## 6.6 CARBON MONOXIDE: TOXICANT AND ENDOGENOUS MEDIATOR OF FEVER

Carbon monoxide is a well-characterized pollutant and byproduct of incomplete combustion that is responsible for a large number of deaths annually throughout the world. Whole-body exposure to carbon monoxide leads to hypometabolism and hypothermia as a result of impaired ability of hemoglobin to bind oxygen (Chapter 3). However, endogenously produced carbon monoxide has recently been found to serve as a neuromodulatory agent and may play a role in the development of fever (Steiner et al., 1999). Carbon monoxide is generated in the brain as a result of the enzyme heme oxygenase that catalyzes the metabolism of heme to biliverdin, free iron, and carbon monoxide (Figure 6.9). Administration of an inhibitor of heme oxygenase activity to normal animals has no discernable effect on normal body temperature. However, inhibiting heme oxygenase in rats made febrile by administration of LPS leads to a marked attenuation in the fever. Administering carbon monoxide–saturated saline into the CNS reverses the effect on the fever and has been found to raise core temperature of afebrile rats (Jang et al., 2002). Interestingly, the carbon monoxide–heme oxygenase pathway raises core temperature during fever in a manner that is independent of the prostaglandin pathway (Steiner and Branco, 2000). All together, these studies indicate that carbon

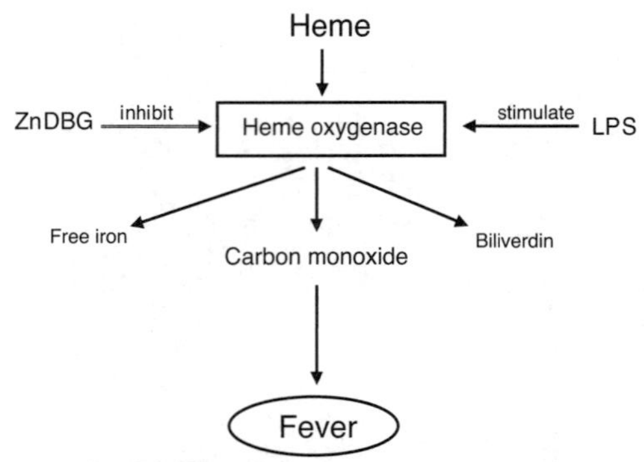


**Figure 6.9 Proposed mechanism of the carbon monoxide–heme oxygenase pathway during LPS-induced fever. Intraperitoneal LPS activates heme oxygenase in the CNS, an enzyme that metabolizes heme to biliverdin, free iron, and carbon monoxide. Carbon monoxide is a neuromodulatory substance that activates thermoregulatory processes leading to fever. ZnDPBG (zinc deuteroporphyrin 2,4-bis glycol) is an inhibitor of heme oxygenase and attenuates LPS-induced fever. (Modified from Steiner, A.A., Colombaria, E., and Branco, L.G. (1999). *Am. J. Physiol.* 277: R499–R507.)**

monoxide is a neuromodulator that is apparently not involved in the maintenance of basal thermoregulatory processes but does serve as a mediator during fever. One should consider how carbon monoxide exposure and the heme oxygenase pathway may operate during carbon monoxide poisoning. That is, is the hyperthermia commonly observed in carbon monoxide poisoning a result of not only impaired oxygen-carrying capacity but also modification of the heme oxygenase-fever pathway? This pathway may also be relevant in the manifestation of other types of toxicant-induced fevers discussed in this chapter.

## 6.7 METAL FUME FEVER

Metal fume fever has been recognized for over 100 years in occupations involving exposure to fumes of metal oxides (Gordon and Fine, 1993). It is interesting to note that metal fume fever may be the first observation of an environmental exposure to a xenobiotic that mediates an integrated, thermoregulatory response. Inhalation of freshly formed metal oxides in welding, smelters, and galvanizing operations leads to a pattern of symptoms characteristic of an immunological reaction including cough, headache, fatigue, tachycardia, and fever. The symptoms are manifested 4 to 12 h after exposure, and zinc oxide appears to be the most toxic agent to induce metal fume fever. Oxides of other metals also appear to have roles in metal fume fever, including aluminum, antimony, cadmium, copper, magnesium, manganese, and tin. Most of the data on this syndrome come from human studies (Gordon and Fine, 1993). There are apparently no data in rodent models to show the occurrence of metal fume fever.

If metal fume fever activates the classic mechanisms of fever as shown in Figure 6.1, then the elevations in body temperature should occur concomitantly with a rise in circulating cytokines involved in fever. Fine et al. (1997) characterized the relationship between metal fume fever and the elevation in circulating levels of IL-6 in humans subjected to controlled exposures to zinc oxide (Figure 6.10). Subjects inhaling zinc oxide for 2 h underwent delayed elevations in core temperature accompanied by the typical symptoms of metal fume fever (e.g., cough, myalgia, and fatigue). A small increase in plasma IL-6 levels was seen at 3 h after exposure, and a significant elevation was evident by 6 h, whereas core temperature increased slowly for the first 8 h and then peaked at 11 h after exposure

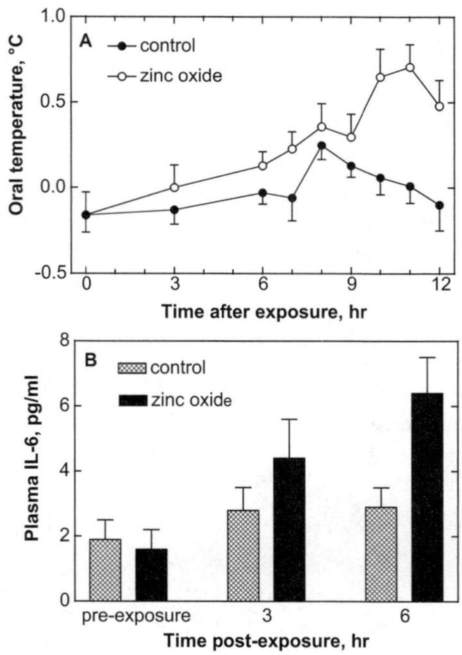


**Figure 6.10 (A) Time-course of the change in core temperature of 13 human subjects (8 male) following 2 h of inhalation exposure to zinc oxide fumes (5 mg/m$^3$) or control air. (B) Plasma levels of IL-6 of the same subjects when measured at 3 and 6 h after exposure. Note peak elevation in core temperature at 11 h after initial exposure to zinc oxide fumes. (Data modified from Fine, J.M., Gordon, T., Chen, L.C., Kinney, P., Falcone, G., and Beckett, W.S. (1997). *J. Occup. Environ. Med.* 39: 722–726.)**

(Figure 6.10). Plasma levels of TNFα were unaffected by zinc oxide inhalation. Repeated, daily exposure to zinc oxide in humans leads to tolerance with the fever reduced by more than 50% by the third day of exposure (Fine et al., 2000). Since IL-6 is a key mediator in the CNS activation of fever (Figure 6.1), it seems clear that metal fume fever from zinc oxide involves activation of the fever pathway. However, the increase in plasma levels of IL-6 is significant but relatively small and would normally be insufficient to evoke a fever. This would suggest that metal fume fever involves the participation of more than one pyrogenic agent in the development of the fever (Fine et al., 2000).

## 6.8 INFLAMMATION, FEVER, AND THE P-450 PATHWAY

Recent studies on the effects of P-450 enzyme activity[1] on the manifestation of inflammatory responses and fever raise interesting questions about how anti-ChE insecticides and other toxicants could modulate or induce a fever. The mechanism of action of typical antipyretic drugs such as aspirin, ibuprofen, indomethacin, and others is the inhibition of COX activity and the subsequent reduction in $PGE_2$ levels (Figure 6.1). The COX pathway is just one of several avenues for the metabolic conversion of arachidonic acid. Once arachidonic acid is liberated from membrane phospholipids, it can be oxygenated by three principal enzyme systems: COX, lipoxygenases, and cytochrome P-450 (Figure 6.11A). The expression of these enzymatic pathways has been demonstrated in the brain, liver, kidney, and lung (see Kozak et al., 1998, for review). Interestingly, the products of these enzymatic pathways play a critical role in the magnitude of fever as well as in the modulation of a variety of inflammatory processes (Figure 6.11B).

The production of $PGE_2$ by COX-1 and COX-2 is clearly one of the final steps in the mediation of fever (see Figure 6.1). However, the production of the epoxyeicosatrieonic acids (EETs) by P-450 pathways appears to lower the set-point and counter the febrile effects of $PGE_2$ (Figure 6.11B). In view of the reciprocal relationship between the products of COX and epoxygenase, it is evident that drugs that alter P-450 activity can modulate the magnitude of a fever. It is well known that inducers of P-450 such as bezafibrate and dehydroepiandrosterone reduce the fever following exposure to LPS (Kozak et al., 2000). However, inhibitors of P-450 activity such as proadifen have been found to augment the magnitude of an LPS-induced fever (Kozak et al., 2000a).

[1] Modernized nomenclature refers to the P-450 enzymes as a part of the cytochrome P-450 (CYP) enzymes (see Bolt et al., 2003).

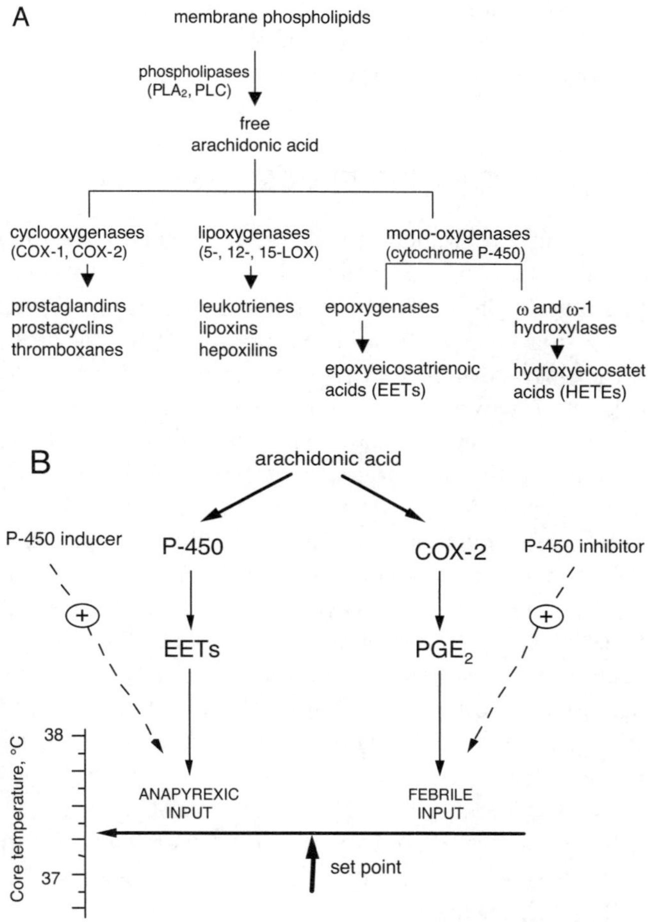


**Figure 6.11 (A) Summary of the three principal pathways for the metabolism of arachidonic acid. (B) Postulated mechanism showing how induction or inhibition of P-450 activity can alter the magnitude of a fever. (Drawings modified from Kozak, W., Kluger, M.J., Tesfaigzi, J., Kozak, A., Mayfield, K.P., Wachulec, M., Dokladny, K. (2000b).** ***Ann. NY Acad. Sci.*** **917: 121-134; Tesfaigzi, Y., Kluger, M., and Kozak, W. (2001).** ***Respir. Physiol.*** **128: 79–87.)**

Altogether, it would seem that there is a possible mechanism of action between the P-450 function, the metabolism of toxicants, and fever. The P-450 enzymes, particularly CYP2E1, are critical in the oxidation and excretion of a variety of toxicants (Bolt et al., 2003). Substrates for this enzyme include acetone, ethanol, chloroform, acrylamide, and many others. CYP2E1 is induced and inhibited by a variety of chemicals and it shows distinct genetic polymorphisms in human populations (Bolt et al., 2003). Since the P-450 enzyme expression is affected by a multitude of toxicants and influences the febrile and inflammatory response, it behooves one to take a closer look at the impact of toxicants on the mediation of inflammation and fever.

## 6.9 IS TOXIC-INDUCED FEVER ADAPTIVE?

The febrile response to infection has been shown to be adaptive and improves the recovery and survival to many but not all infectious agents (Hart, 1988; Kluger, 1986). The combination of a high temperature and reduction in plasma levels of iron and zinc during fever has been shown to suppress the growth of certain pathogens, thus affording protection to the infected host. However, fever is metabolically costly, and animals must expend valued energy stores to maintain an elevated body temperature at the same time that the sick animal is lethargic and unable or unwilling to forage or eat. Considering that the acute phase response to an infection has evolved at many levels and is metabolically costly, it is likely to be very advantageous in the recovery from and survival of infection.

Does a fever from exposure to pesticides, metal fumes, ethanol, and other toxicants afford any kind of protection or is it detrimental? There is little if any data to suggest that treating the fever or allowing it to progress has any consequences on overall health. There are several questions that come to mind in regard to treating the fever from certain classes of toxicants. For example, does the fever from acute organophosphate poisoning provide any benefit in the recovery of AChE activity? Does the elevated body temperature from metal fume fever stimulate the immune system and recovery of pulmonary inflammation? Answers to these questions should help clinicians to decide if treating the fever of a poisoning victim improves or impedes recovery.

# Chapter 7

# Environmental Stress

## 7.1 INTRODUCTION

Toxicologists traditionally study the biological response to chemical toxicants under ideal environmental conditions that result in little physiological or psychological stress. On the other hand, environmental physiologists deal with the dynamics of the physical environment and how physiological responses are affected by one or more physical factors. Stress is an unavoidable facet of life, and it follows that all organisms must deal concurrently with environmental stress and toxicant exposure.

Environmental physiology is broadly defined as the study of the physiological mechanisms that allow animals to cope with and adapt to changes in temperature, humidity, atmospheric pressure, and other natural factors of their physical environment (Folk, 1974). The physiological response to a toxicant is determined by three factors (Casarett and Doull, 1975): the nature of the poison or toxic agent, the exposure situation, and the subject. Moreover, the factors that influence the subject are subdivided into *internal* (i.e., factors inherent in the subject such as species, nutritional status, age, etc.) and *external* to the subject (i.e., environmental factors). Toxicologists rarely study how the external environmental factors modulate the physiological response to toxic agents. Likewise, only a handful of environmental physiologists have an interest in toxicology. In this chapter, I have attempted to meld the fields of environmental physiology and toxicology to assess the impact of environmental stress on the thermoregulatory response to chemical toxicants.

## 7.2 ROLE OF ENVIRONMENTAL PHYSIOLOGY IN TOXICOLOGY: A BRIEF HISTORY

It is evident throughout this book that there has been an interest in assessing how environmental factors influence toxicity. Temperature is probably the most frequently studied of all environmental variables. There is a remarkable amount of research beginning in the early 20th century that showed how ambient and body temperature could modulate the toxicity of a variety of toxicological and pharmacological agents (see Chapter 4). Dr. Anna Baetjer was one of the first to make notable contributions in the fields of environmental physiology and toxicology. Using anti-ChE insecticides with a specific mode of action that affects the neural control of body temperature (see Chapter 3), she was the first to develop a comprehensive picture of how environmental heat and cold stress interact with the physiological responses to a toxicant (Baetjer and Smith, 1956). Her work showed a direct relation between the toxicity of an organophosphate insecticide (parathion) and changes in body and ambient temperature (see Chapter 4 for further discussion of her work).

Dr. John Doull succinctly summarized the role of environmental physiology and toxic response in a 1972 review, "The Effect of Physical Environmental Factors on Drug Response." This critical review focused on the effects of radiation, ambient pressure, and temperature on the toxicity of drugs and xenobiotics. In the *Textbook of Environmental Physiology*, Dr. G. Edgar Folk (1974) devoted a chapter to the impact of environmental stress and pollution on the quality of life. Dr. Steven Horvath and his colleagues at the Institute of Environmental Stress were the first to address how variations in the physical environment, particularly temperature and humidity, affect the physiological response to airborne toxicants in humans during rest and exercise (Drinkwater et al., 1974; Gliner et al., 1975). Researchers from the Institute of Environmental Medicine in Natick, MA, have made numerous contributions to the role of environmental stress, exercise, and response to anti-ChE that date back to the 1970s. In the 1980s our laboratory and others showed how the thermoregulatory system responds to toxicants and drugs in an integrated fashion to lower body temperature and enhance survival to the toxic agent (Gordon et al., 1988; Gordon, 1993).

## 7.3 THE PHYSICAL ENVIRONMENT

Why should toxicologists and pharmacologists be concerned with environmental stress? There are many natural as well as artificial (i.e., man-

**Table 7.1 Natural Physical Factors of the Environment**

- Heat
- Cold
- Humidity
- Air movement
- Barometric/water pressure
- Visible light
  - Wavelength
  - Intensity
  - Photoperiod
- Geomagnetism
- Ultraviolet radiation
- Airborne dust and particulate matter
- Noise

*Source*: Modified from Folk, G.E., Jr. (1974). *Textbook of Environmental Physiology*. Philadelphia: Lea and Febiger.

made) physical factors in the environment that are often beyond our control that can affect the physiological response to a toxicant or drug (Table 7.1). While all the factors listed in Table 7.1 are important, temperature and humidity are most critical in the study of the effects of stress on toxicological response. The geographic distribution, health, and survival of all animal and plant life are essentially linked to environmental temperature and humidity.

Humans, agricultural species, and wildlife inhabit a variety of environments that can be relatively mild or very stressful (Figure 7.1). Living at lower latitudes and in tropical and subtropical areas means there will be exposures to very warm and humid environments during the summer months and relatively mild winters. Life in the desert is associated with extremely hot and dry conditions with intense solar radiation during the day and cold temperatures at night. Living at higher latitudes and at altitude means exposure to extremely cold temperatures during the winter months. Humidity is also a factor to consider, but the seasonal variations do not appear as great as those of ambient temperature (Figure 7.1B). Not shown are daily changes in humidity that can be extreme depending on geographic location.

The data in Figure 7.1 suffice for one to consider how temperature and humidity can affect a toxicological response. Moreover, toxicologists should consider how acclimatization to the daily and seasonal changes in temperature and humidity may affect a particular physiological response. There are innumerable environmental scenarios that could affect the response to a toxicant. Some of the data presented later in

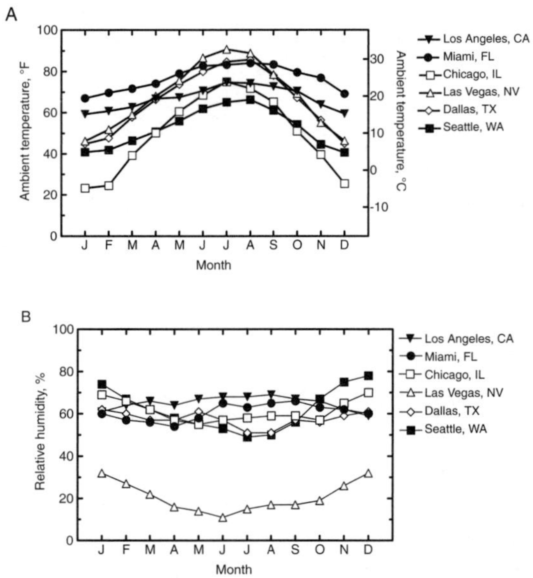


**Figure 7.1 Mean 24-h air temperatures (A) and afternoon relative humidity (B) averaged by month in selected cities in the United States. Air temperature data averaged from 1985 to 1989 from the University of Dayton. Relative humidity data represent monthly averages from tables provided by the U.S. Weather Service (see Gordon, 2003).**

this chapter should prompt one to wonder how adaptation to a particular climate and geographic location might affect susceptibility to a toxicant. For example, would a resident of Miami who is acclimatized to a mean yearly temperature of 24.8°C and subjected to an annual temperature swing of 9.5°C respond to a toxicant differently than a resident of Chicago who is acclimatized to 9.7°C and experiences an annual temperature variation of 29°C?

The process of acclimatizing to a new environment may also affect a toxic response. Would a native of a temperate zone be more susceptible if relocated to a subtropical or desert climate? This question is pertinent to the thousands of migrant agricultural workers who deal with pesticide exposures while acclimatized to a variety of extreme climates. Although the data in Figure 7.1 are for U.S. cities, the same relative climatic patterns are seen worldwide. It should also be noted that these data are mean 24-h temperatures and do not reflect the day-to-night variations in ambient temperature that also affect the distribution and physiology of animal and plant life. While much of the above discussion focuses on human health, the same rationale is relevant for studying toxicological responses of agricultural species and wildlife (see Chapters 8 and 9).

**Table 7.2 Ideal Environmental Conditions Typically Used for Testing Physiological Response of Rodents and Other Laboratory Species to Toxic Agents**

- $T_a$ of 20 to 26°C with clean, insulative bedding (temperature dependent on species)
- 40 to 60% relative humidity
- Still or calm air movement
- 12:12 light:dark cycle
- Fluorescent lighting
- Resting, confined conditions with no option for work or exercise
- No psychological stress
- *Ad libitum*, nutritionally balanced food and water
- Filtered air exchanged ~10 times per hour
- Sea level

### 7.3.1 Selecting an Appropriate Laboratory Test Environment

From the above discussion, it is obvious that toxicologists and pharmacologists should consider how variations in the natural environment may alter a physiological response. However, the majority of laboratory rodent studies have been performed in animals acclimatized to environmental conditions that are usually considered ideal (i.e., no stress) and are standard across most laboratories (Table 7.2). These conditions are generally characterized by an ambient temperature associated with a minimal (i.e., basal) metabolism, relative humidity of 50%, no work or psychological stress, 12 to 12 light to dark photoperiod, still air, *ad libitum* food and water, filtered air, and atmospheric pressure near or equal to sea level (see National Research Council, 1996). Note that the ambient temperature of rodent facilities is generally below thermoneutrality, but the animals are usually provided with bedding material to facilitate thermal comfort. In most instances, rodents in animal housing facilities are maintained at a temperature that is considered a slight to moderate cold stress (but is comfortable for laboratory personnel). Test animals are almost always sedentary with no option for exercise. Overall, most quality laboratory investigations require a well-controlled environment for animal research, but this is not representative of the dynamic natural environment encountered by humans and other species on a day-to-day and seasonal basis. Alterations in one or more of the environmental variables in Table 7.2 are likely to alter the physiological response to a toxic chemical.

## 7.4 TEMPERATURE AND WORK: THEIR IMPACT ON A TOXIC RESPONSE

The thermoregulatory system operates continuously to maintain a relatively constant body temperature in the face of daily and seasonal fluctuations of the internal (e.g., work) and external heat loads and heat sinks (e.g., ambient temperature, solar radiation, and wind). Under ideal environmental conditions (e.g., thermoneutral environment), a stable core temperature is maintained with minimal strain on physiological systems. However, in the face of marked changes in ambient temperature, relative humidity, and work load, a constant core temperature is maintained but at the expense of activation of thermoeffectors and physiological stress. In a detailed review of the effects of heat stress alone and heat stress plus exercise, it was shown that exercise in a hot environment exacerbates plasma levels of various stress hormones, including catecholamine, corticosteroids, and growth hormone (Brenner et al., 1998). These hormonal responses are indicative that the physiological response to toxicants will be altered and likely exacerbated in a working and especially a working plus heat-stressed environment. As will be shown below, a stressed thermoregulatory system with activation of thermoeffectors can alter the physiological response to toxicant agents in many ways.

In a global perspective, the function of the thermoregulatory system can be viewed as a focal point that interacts with biotic and abiotic facets of the ecosystem (Figure 7.2). Temperature homeostasis is in fact linked in many ways with the ecosystem, including environmental temperature, humidity, work and stress, and the physiological response to toxic agents. The subcomponents of this figure are touched on briefly here and discussed in more detail later in this chapter (see Sections 7.4 and 7.10). Temperature and humidity interact to affect the production and degradation of toxicants in air, land, and water. The accumulation of greenhouse gases exacerbates the formation of some toxicants, raises environmental temperature, and thus affects thermoregulation. These processes eventually affect the occurrence and possible exposure to toxicants. Heat and humidity also affect thermoregulation, leading to activation of thermoeffectors (respiration, sweating, and skin blood flow) that have a direct effect on the potential dosage of a toxicant. Moreover, work, exercise, and other forms of stress exacerbate the effects of heat and humidity on thermoregulation.

### 7.4.1 Thermal Stress and Entry of Toxicants into the Body

It is interesting to consider how the processes that affect the biological dosage of an environmental toxicant are also linked to thermal homeo-

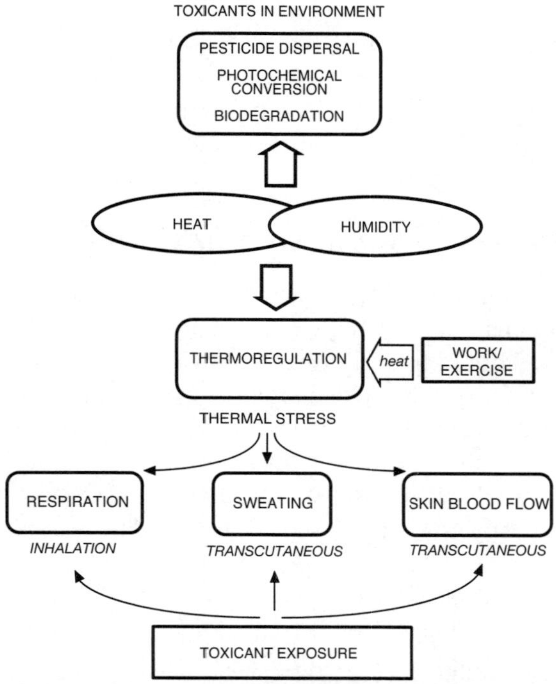


**Figure 7.2 Central role of thermoregulation in the physiological response to environmental toxicants. (Modified from Gordon, C.J. (2003). *Environ. Res.* 92: 1–7.)**

stasis. Toxic agents can enter the body by three principal routes: respiratory surfaces, gastrointestinal tract, and transcutaneously (Casarett and Doull, 1975). These routes of entry into the body can be affected by exposure to heat or cold stress. The surfaces of the respiratory tract and skin are integral for the operation of thermoeffectors for evaporative and dry heat loss (see Chapter 2). Hence, when a homeotherm is in an environment where it must actively dissipate heat, it is likely to be more susceptible to lower doses and concentrations of a toxicant. However, the increased demand for heat production in a cold environment results in an elevation in respiratory rate, thus increasing the intake of airborne toxicants and also raising susceptibility.

The thermoregulatory system responds to heat stress and exercise by activating three key systems to dissipate excess heat: cardiovascular, respiratory, and sudomotor (sweating). The combination of peripheral vasodilation to increase skin blood flow and raise skin temperature along with sweating results in an effective mechanism to dissipate a heat load (Folk, 1974; Blatteis, 1998). True panting animals exhibit marked increases in breathing frequency during heat stress. Nonpanting homeotherms, including humans and rodents, also exhibit increases in breathing frequency and minute ventilation that contribute to a modest increase in evaporative water loss when heat stressed (Ingram and Mount, 1975). The added heat load of exercise will further increase ventilation and augment the total intake of airborne pollutants (Mautz, 2003).

### 7.4.2 Sweating and Absorption of Toxicants

Sweating is the principal thermoeffector response in heat stressed humans and some other mammals (Chapter 2). In humans, eccrine sweat glands are activated by cholinergic pathways; activation of these pathways generally occurs concurrently with an increase in skin blood flow. The increased flow of warm blood from the core to the surface combined with evaporative cooling from sweating is an effective means of dissipating excess body heat. However, the combination of moisture, warm temperatures on bare skin, and increased skin blood flow provides an ideal environment to accelerate the transcutaneous absorption of many types of pesticides (Chang et al., 1994; Wester et al., 1996). It is not clear how sweat production in species other than humans affects the absorption of pesticides. The proportion of eccrine and apocrine sweat glands utilized for thermoregulation differs considerably between primates and other species (Ingram and Mount, 1975; Folk, 1974). In species other than humans, the sweat is produced in densely furred skin, and thus it is not clear how pesticides would be absorbed as compared to bare skin. Agricultural species that sweat may well be susceptible, but this has apparently not been assessed.

Overall, both *in vitro* and *in vivo* studies suggest that activation of thermoeffectors during heat stress and exercise to dissipate heat will accelerate pesticide absorption in humans. An *in vitro* model of cutaneous absorption of parathion has been used to show how temperature, blood flow, and relative humidity affect the absorption of parathion. A small section of porcine skin positioned over a flow-through diffusion cell provides an ideal means to control air temperature, relative humidity, perfusate temperature (i.e., an indication of body temperature), and flow of the perfusate (i.e., an indication of the potential effects of blood flow) while studying the percutaneous absorption of a pesticide (Chang and Riviere, 1991). The absorption of radiolabeled parathion across porcine skin increases dramatically with an elevation in air and perfusate temperature (Figure 7.3A). For example, a 5°C increase in air and perfusate temperature leads to more than a twofold increase in parathion absorption. It is possible that skin warming can raise lipid fluidity and permeability of the dermal tissues, leading to increased penetration of the pesticide. The cutaneous absorption of parathion is directly affected by relative

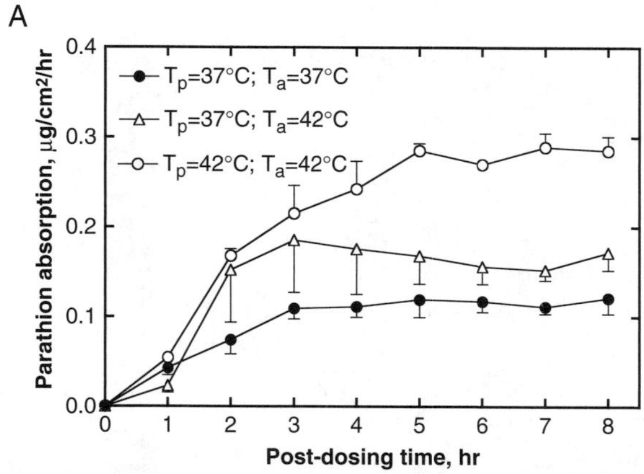


**Figure 7.3 (A) Effect of air temperature ($T_a$) and perfusate temperature ($T_p$) on the passive absorption of parathion (dose = 40 µg/cm²) *in vitro* in porcine skin. (From Chang, S.-K. and Riviere, J.E. (1991). *Fund. Appl. Toxicol.* 17: 494–504.) (B) Overall effects of perfusate temperature, relative humidity, and perfusate flow on parathion absorption. (Data modified from Chang, S.-K., Brownie, C., and Riviere, J.E. (1994). *J. Vet. Pharamcol. Ther.* 17: 434–439.)**

humidity and perfusate flow (Figure 7.3B). The effects of humidity are profound, suggesting that increased moisture on the skin raises the permeability to parathion. Parathion is a lipophillic molecule, and it is thus not clear why percutaneous absorption would increase with additional moisture on the skin. It is also interesting to note the marked effect of humidity in the pig, which is a nonsweating animal. One can only surmise that species that sweat to thermoregulate in the heat would be especially susceptible to pesticide absorption in a warm and humid environment.

Studies of humans have shown how perspiration can accelerate the cutaneous absorption of organophosphate agents. Human volunteers were exposed to ambient temperatures of 14, 21, 28, and 40.5°C while their hand and arm were exposed to a 2% parathion dust for 2 h (Funckes et al., 1963). The absorption of the insecticide was estimated by the quantity of paranitrophenol, a metabolite of parathion, that was excreted in the urine. The dermal absorption of parathion was mildly affected by skin warming at low temperatures and markedly affected at warm temperatures. Parathion absorption increased by 25% when the temperature of exposure was raised from 14 to 21°C; however, from 21 to 28°C, parathion absorption increased by just 17%. Raising the ambient temperature from 28 to 40.5°C led to a 180% increase in absorption (Funckes et al., 1963). Although the rate of sweating was not quantified, it is clear that subjects perspired profusely at the warmest ambient temperature.

It is also interesting to note the increase in parathion absorption at the lower ambient temperatures in spite of a lack of sweating. The warmer skin

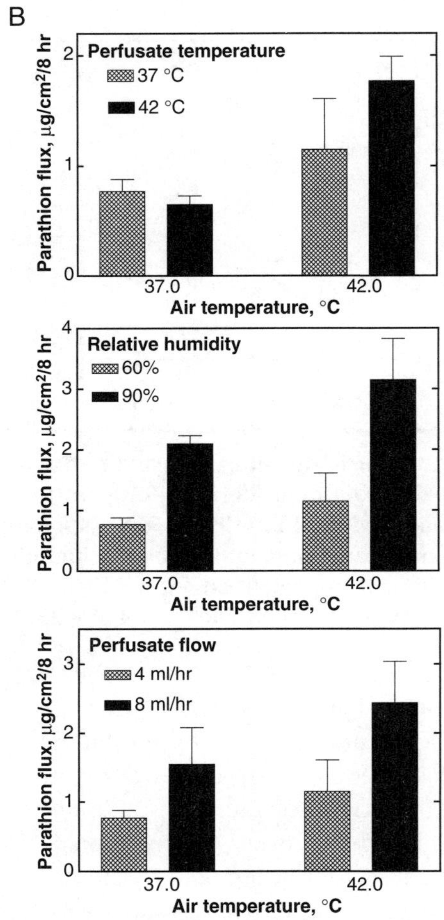


**Figure 7.3 (continued)**

temperature is likely to be a critical factor affecting parathion absorption even without sweating. The dose of parathion was relatively low because red blood cell and plasma cholinesterase activity were unaffected by the treatment. Since sweat glands are activated by cholinergic stimulation, parathion and other anticholinesterase pesticides should directly stimulate sweating through inhibition of cholinesterase activity. Hence, the dosage from exposure to anti-ChE pesticides should be exacerbated in a warm and humid environment because of cholinergic stimulation of sweating combined with greater transcutaneous absorption across moist skin. The authors also noted that subjects exposed to parathion at the highest ambient temperature exhibited sweating from the exposed area of skin for several days after pesticide decontamination (Funckes et al., 1963).

In another human study, volunteers had small amounts of the nerve gas VX [S-(2-diisopropylaminoethyl) o-ethyl methylphosphonothioate] applied topically to their cheeks and forearms at ambient temperatures of −18, 2, 18, or 46°C. The VX was left on the skin for 3 h and its penetration into the body was estimated by measuring the inhibition of red blood cell ChE activity (Craig et al., 1977). The transcutaneous absorption of VX was directly dependent on ambient temperature. The decimal fraction of penetration of VX on the cheek was 0.04 at −18°C and 0.32 at 46°C, whereas for the forearm the penetration was 0.004 at 18°C and 0.029 at 46°C. Overall, penetration across the cheek was much more effective than the forearm, with greater than 50% inhibition in ChE activity occurring at a dose of less than 10 μg/kg. It was postulated that after exposure to VX or a comparable agent, cooling the skin would delay absorption, thus allowing for safer decontamination of an exposed subject.

### *7.4.3 Sweating and Toxicant Excretion*

From the above discussion, it is clear that heat stress leads to an augmentation in the cutaneous absorption of some toxicants in humans. However, sweating also provides a means of excreting some toxicants including heavy metals as well as trace metals essential for life. It has been well known that significant amounts of iron can be excreted in sweat. In fact, iron deficiency and anemia in some individuals in tropical and other hot environments have been attributed to excessive sweating (Prasad et al., 1963). In addition to iron, appreciable quantities of trace metals and toxic heavy metals are secreted with sweat, including aluminum, cobalt, copper, lead, manganese, mercury, molybdenum, nickel, tin, and zinc (for review, see Hohnadel et al., 1973).

Subchronic heat exposure has been shown to elicit significant reductions in circulating levels of trace metals. For example, a 7-day exposure to a sauna (30 min at 80°C twice per day) in humans led to a decrease in serum copper (15 to 13.5 μmol) and zinc (13.8 to 9.8 μmol; Uhari et al., 1983). In a comparison of metal excretion in men and women subjected to heat stress in a sauna, concentrations of nickel, copper, zinc, and lead were approximately doubled in the sweat of women; however, the total volume of sweating in men was three times that measured in women (Table 7.3). It is of interest to compare the blood levels of these metals to the concentrations found in the sweat. In the case of nickel, the concentration of the metal in the sweat is at least an order of magnitude greater than that in the blood. Overall, sweating plays a significant role in the homeostatic control of trace metals and can be a viable therapeutic means of excreting significant quantities of toxic metals. Volatile organic

**Table 7.3 Trace Metals (mean ± s.d.) in the Sweat and Blood of Healthy Adults Collected During a Dry Heat Sauna (15 min at 93°C)[a]**

| | *Metal Concentrations in Sweat, Φg/l* | | | |
|---|---|---|---|---|
| | *Nickel* | *Copper* | *Zinc* | *Lead* |
| Men | 52 ±36 | 550 ±350 | 500 ±480 | 51 ±42 |
| Women | 131 ±65 | 1480 ±610 | 1250 ±770 | 118 ±72 |
| | *Metal Concentrations in Blood, Φg/l* | | | |
| | *Nickel* | *Copper* | *Zinc* | *Lead* |
| Men | 3.1 ±1.0 | 1120 ±100 | 950 ±210 | 160 ±49 |
| Women | 2.7 ±1.2 | 1170 ±150 | 860 ±60 | 96 ±25 |

[a] Sweat volume for men and women was 23 ± 12 and 7 ± 3 ml, respectively.

*Source*: Data from Hohnadel, D.C., Sunderman, F.W., Jr., Nechay, M.W., and McNeely, M.D. (1973). *Clin. Chem.* 19: 1288–1292.

compounds such as acetone, ether, ethanol, and toluene are excreted in sweat (Naitoh et al., 2002).

## 7.5 INTERACTION BETWEEN HEAT STRESS, WORK, AND TOXICANT EXPOSURE

### *7.5.1 Carbon Monoxide*

Some motor sports present an ideal field study for one interested in toxicology and environmental physiology. The interaction between heat stress, work, and response to airborne toxicants is a concern to race car drivers where ventilation and environmental control are severely compromised. In certain types of races, the thermal environment inside a car can be extremely taxing to the thermoregulatory system, with air temperatures of over 50°C measured in the summertime. The required protective clothing worn during a race places further limits on heat loss. Added to the heat stress is the driver's exposure to carbon monoxide, which can easily reach levels of 200 ppm (Walker et al., 2001). Clearly, this could be a major factor in the health and safety of the drivers during a race. Additionally, drivers must perform motor skills and maintain an alertness throughout a race; the consequences of a mental error can lead to severe injury or death.

Physically fit drivers were evaluated in a race car simulator maintained under typical summertime racing conditions (50°C and 200 ppm carbon monoxide) while core temperature and other thermoregulatory parameters were monitored (Figure 7.4). Subjects were exercised for 15 min prior to

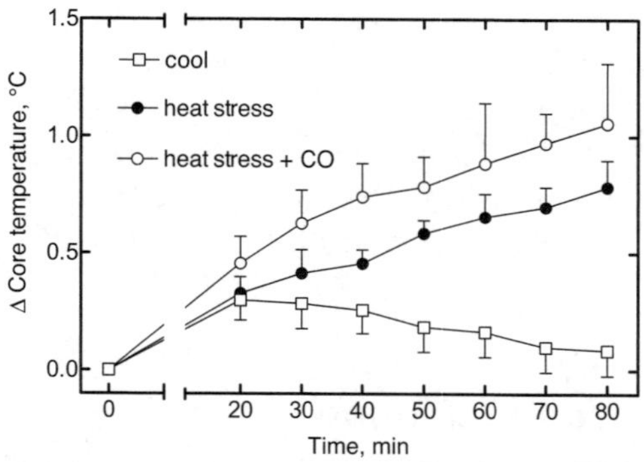


**Figure 7.4 Time-course of core temperature in fit human subjects while inside an interactive race car simulator and subjected to cool conditions (20°C), hot conditions alone (50°C), and hot conditions while exposed to carbon monoxide (CO) resulting in blood carboxyhemoglobin levels of 10 to 12%. (Data modified from Walker, S.M.., Ackland, T.R., and Dawson, B. (2001). *Comp. Biochem. Physiol. Part A* 128: 709–718.)**

operating a simulator under racing conditions for 60 min. Under relatively comfortable ambient conditions of 20°C, the rise in core temperature while performing in the simulator was minimal. Exposing the drivers to heat stress led to a marked increase in core temperature and sweating. However, the combination of heat stress and carbon monoxide exposure (10 to 12% carboxyhemoglobin) led to a significant increase in core temperature, stored body heat, sweating, and a number of driving errors when compared to heat stress alone. Since this study was performed on well-trained athletes, it would be of interest to determine the susceptibility of untrained individuals to these types of environmental conditions. That is, there may be other exposure scenarios and occupations where susceptible individuals (e.g., aged) could be subjected to heat stress and carbon monoxide exposure.

Carbon monoxide exposure leads to hypothermia in rodents when they are maintained at ambient temperatures below thermoneutrality (Chapter 3). It is not clear in these human studies why carbon monoxide exacerbates the hyperthermic response to heat stress. Nielsen (1971) found that mild to moderate carbon monoxide poisoning in exercising humans led to an increase in the plateau level that core temperature reached during work. A plateau may in fact be equivalent to an elevated set-point temperature. Lower skin temperatures were noted during carbon monoxide poisoning in human subjects. This thermoeffector response would lead to higher core temperature with all other factors being equal (Nielsen, 1971). In view of the recent discovery of the role of carbon monoxide as a possible neuromodulator that mediates a fever (Chapter 6), it would be important to reevaluate how heat stress and exercise affect the thermoregulatory effects of environmental exposures to low levels of carbon monoxide. There is surprising little known about how low levels of carbon monoxide affect the regulation of body temperature in experimental animals and humans. Ideally, the incorporation of radiotelemetry into the rodent studies would allow one to detect subtle thermoregulatory changes that might help to explain the overall thermoregulatory effects of carbon monoxide.

## 7.5.2 Cholinesterase Inhibitors

### 7.5.2.1 Animal Studies

Short-term exercise results in marked elevations in core temperature, heart rate, and blood pressure and is likely to alter the sensitivity to toxicants when administered through a variety of routes. Exercising animals are especially susceptible to exposure to airborne toxicants as a result of the increase in minute ventilation and subsequent rise in the uptake of a toxicant (Mautz, 2003). A general paradigm for studying these processes in experimental animals is to put a rodent on a treadmill that has an electrical shock device to reinforce running activity, sometimes until the point of exhaustion (Mautz, 2003). Rats can be trained and will improve their running performance on the treadmill, but it is not always clear how such data can be extrapolated to human studies. With forced running, the experimental animal may respond physiologically to the activity as a stress and not simply as volitional running behavior.

Exercise performance and thermoregulation are inexorably linked. One important factor that limits running performance in rodents is ambient and core temperature. Spontaneous running activity or performance on a treadmill is reduced with a rise in ambient temperature. The thermoregulatory system is unable to maintain a stable core temperature during exercise as ambient temperature rises. Rats on a treadmill will cease running activity at a critical brain temperature, a response that prevents a debilitating or lethal hyperthermia (Gordon, 1993). Hence, a drug or toxicant could interfere with exercise performance by directly affecting thermoeffectors for heat gain and loss, altering the set-point core temperature, or lowering the temperature threshold that suppresses running activity.

There is some evidence that anti-ChE agents can impair exercise performance in rodents. Physostigmine administered to rats 15 min before placement on a treadmill (11 m/min; 6° incline) led to a more rapid increase in core temperature and impairment in running performance as indicated by a shortened time to exhaustion (Matthew, 1993). However, the effects were only seen at ambient temperatures of 15 and 26°C but not at a very cold temperature of 10°C or a warm temperature of 30°C. Pyridostigmine also led to a reduction in time to exhaustion and elevation in heating rate in rats forced to run at a very warm temperature of 35°C (Francesconi et al., 1984). This deficit in exercise and thermoregulatory performance was associated with a 64% reduction in plasma ChE activity. Pyridostigmine given orally to rats for 7 days resulted in a 23% reduction in plasma ChE activity (Francesconi et al., 1986). While this treatment had no effects on running endurance in the heat, it did result in a significant increase in evaporative water loss. Repeated administration of the organophosphate malathion to the rat (7.5 mg/kg for 4 days), resulting in a 35% reduction in blood ChE activity, also had no remarkable effects on running performance or core temperature in the heat (Francesoconi et al., 1983). In spite of some of the negative findings, the data suggest that exposure to relatively low doses of anti-ChE agents is apparently ineffective under relatively mild ambient conditions but can impair thermoregulation and exercise performance in warm and hot environments.

### 7.5.2.2 Human Studies

Pyridostigmine has been routinely administered to military personnel as a prophylactic against the possible exposure to nerve gases that are potent anticholinesterase inhibitors. Nerve gases such as sarin and VX enter the body via inhalation or cutaneously and cause rapid inhibition in central and peripheral ChE activity. Pyridostigmine slows the toxicity of the nerve gas agent by reversibly binding to the cholinesterase active site, thus preventing the organophosphate agent from forming an irreversible bond (see Chapter 3). While pyridostigmine is generally considered to be safe under ideal environmental conditions, there has been a recent surge in studies to determine if there are adverse effects on thermoregulation under stressful conditions. This research was prompted by the incidence of sickness in many veterans in the Gulf War (see Section 7.9). In addition to its use in the military, pyridostigmine is also prescribed in the treatment of myasthenia gravis.

When pyridostigmine is administered daily with oral tablets to healthy human subjects, red blood cell ChE activity is decreased by approximately 40%. This degree of inhibition of circulating ChE activity may lead to subtle but significant changes in thermoregulation. For example, after 50 h of pyridostigmine treatment, resting heart rate was reduced by 11 beats/min and core (esophageal) temperature decreased by 0.23°C in subjects maintained at an ambient temperature of 35°C (Kolka and Stephenson, 1992). The reduced core temperature compared to controls was maintained even with a vigorous workout. Another study found that pyridostigmine treatment was effective at lowering heart rate during exercise at ambient temperatures of 29 and 36°C but not at 22°C. Skin blood flow during exercise was also reduced at warmer temperatures by pyridostigmine treatment (Kolka and Stephenson, 1990). In a hot and dry environment (42°C; 20% relative humidity), pyridostigmine treatment in young, healthy exercising subjects led to increases in evaporative water loss and lowered chest skin temperature, along with a mild reduction in heart rate (Wenger et al., 1993). A similar regimen of pyridostigmine treatment in healthy males had no effect on heart rate and core temperature while exercising in a hot environment but did alter the proportions of dry and evaporative heat loss (Epstein et al., 1990).

Pyridostigmine administered to humans in cold climates appears to have little effect on thermoregulation under basal conditions and during exercise (Roberts et al., 1994). Peripheral cholinergic stimulation is likely responsible for the autonomic effects (decreased skin blood flow and heart rate and increased sweating) of pyridostigmine, but the drug has no effect on heat production, nor does it impair the ability to regulate core temperature during exercise in the heat (although a low core temperature may in fact be considered a deficit in regulation). It is important to note that these studies have all been performed in very healthy and physically fit subjects. One should wonder how the aged or young would respond under similar conditions. This could be especially poignant considering today's threats from terrorist attacks with nerve gas and the possibility that the general public could be routinely prescribed with pyridostigmine under a wide range of ambient conditions.

## 7.6 AGRICULTURAL WORKERS AND PESTICIDE EXPOSURE

Many occupations in agriculture involve potential exposure to pesticides under dire thermal conditions. Pesticide applicators must frequently don protective clothing to minimize exposure to anticholinesterase insecticides and many other hazardous agents. The protective clothing is uncomfortable because it impedes heat loss and can lead to marked hyperthermia when worn in the summer months. Indeed, studies have shown that many workers will forgo the required protective clothing when spraying pesticides during the warm season while risking potentially hazardous exposure (Gunther and Gunther, 1977).

Studies that assess the potential hazards of insecticides are generally performed in thermally comfortable environments. The potential hazards are understood in a nonstressed thermoregulatory system in spite of the fact that insecticides, herbicides, and other agents are often applied on crops during hot and humid conditions. Agricultural workers and the general public can be exposed to various agents when their thermoregulatory system is actively dissipating heat. Plasma ChE activity of pesticide applicators in Florida was shown to decrease dramatically from winter to the summer months, corresponding with intensive application of organophosphate-based insecticides (Figure 7.5). Plasma ChE activity of applicators decreased to below 50% of normal from January to August (Yeary et al., 1993). Of course, ambient thermal conditions change drastically in Florida and other southern states from January to August, characterized by marked elevations in temperature and humidity (cf. Figure 7.1A). Thus, there is significant likelihood of pesticide workers being exposed to organophosphate insecticides and sustaining significant reductions in ChE activity when they are heat stressed.

Aerial spraying is another very risky occupation that involves exposure to pesticides and heat stress. This is a very demanding job that requires pilots to endure long, daily work loads that can be very stressful, as they must fly very low to the ground in sprayed fields with exposures to noise, vibration, and gravitational forces (Richter et al., 1981). Spraying often occurs in very warm environments, and the donning of protective clothing can lead to thermoregulatory stress. Pilot error and crashes are likely to be exacerbated by hyperthermia and exposure to the pesticides (see Section 7.5). Overall, testing pesticide safety in laboratory rodents housed at ambient temperature below their lower critical temperature does little to mimic the commonly stressful conditions of many agricultural workers exposed to pesticides.

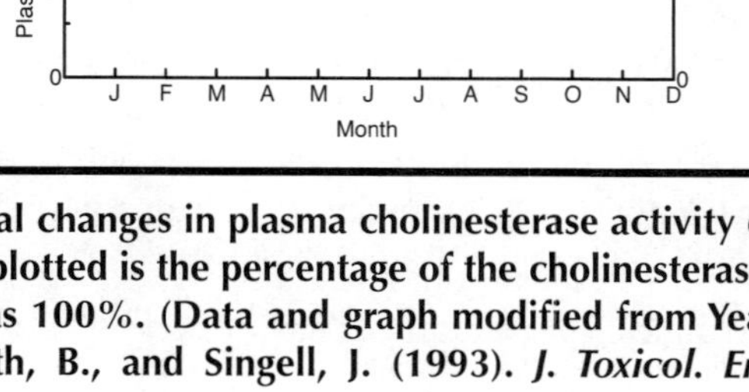


**Figure 7.5 Annual changes in plasma cholinesterase activity of lawn care workers in 1987. Also plotted is the percentage of the cholinesterase activity using the value of January as 100%. (Data and graph modified from Yeary, R.A., Eaton, J., Gilmore, E., North, B., and Singell, J. (1993). *J. Toxicol. Environ. Health* 39: 11–25.)**

## 7.7 TRAINED VERSUS SEDENTARY MODELS OF TOXICANT SUSCEPTIBILITY

There are many endogenous factors that can compromise the use of laboratory rodents as a model to study human toxicology. The levels of fitness and activity are two factors that are rarely evaluated. The thermoregulatory response of a physically fit subject is bound to differ from that of a sedentary subject. Nearly all toxicological and pharmacological research is performed in sedentary animals. Moreover, administration of drugs and toxicants and the monitoring of a physiological response are generally made during the daytime, when rodents are inactive and spend a majority of their time sleeping. Humans display a wide range of fitness and activity levels, and it is important to understand if the level of fitness and circadian timing would affect the physiological response to a toxicant or drug.

The thermoregulatory system is one of many physiological systems that are affected by long-term exercise training. Hamsters, mice, and rats (especially females) will exhibit vigorous nocturnal running behavior when given access to a wheel. Female Sprague-Dawley rats run an average of 7.6 km per night when given free access to a running wheel. The additional heat production from running results in a marked increase in core temperature compared to sedentary animals (Figure 7.6A). After several weeks of running on a wheel, rats display a small albeit significant rise in the daytime core temperature in spite of the fact that the animals are inactive and generally sleeping. This mild hyperthermia during the daytime is thought to be a result of a nonspecific immune response involving increased levels of circulating cytokines known to raise core temperature, including IL-1 and IL-6.

Rowsey et al. (2001) determined if exercise training would modulate the thermoregulatory response to an organophosphate insecticide. After training for 8 weeks, female rats were dosed with corn oil or 25 mg/kg chlorpyrifos while core temperature was monitored by radiotelemetry. Both sedentary and trained animals showed a similar hypothermic response to chlorpyrifos; however, the delayed fever that normally occurs 24 to 48 h after exposure to an organophosphate was essentially absent in the trained animals (Figure 7.6B). In another study, the hypothermic effects of smaller doses of chlorpyrifos (10 mg/kg) administered daily for 4 days were attenuated by exercise training, and the delayed fever was abolished (Rowsey et al., 2003). The inhibition in plasma ChE activity was unremarkable between the sedentary and trained animals following exposure to chlorpyrifos. Hence, the trained animals undergo the same degree of cholinergic stimulation and experience the same hypothermic response to acute chlorpyrifos but possess a mechanism to resist the delayed fever.

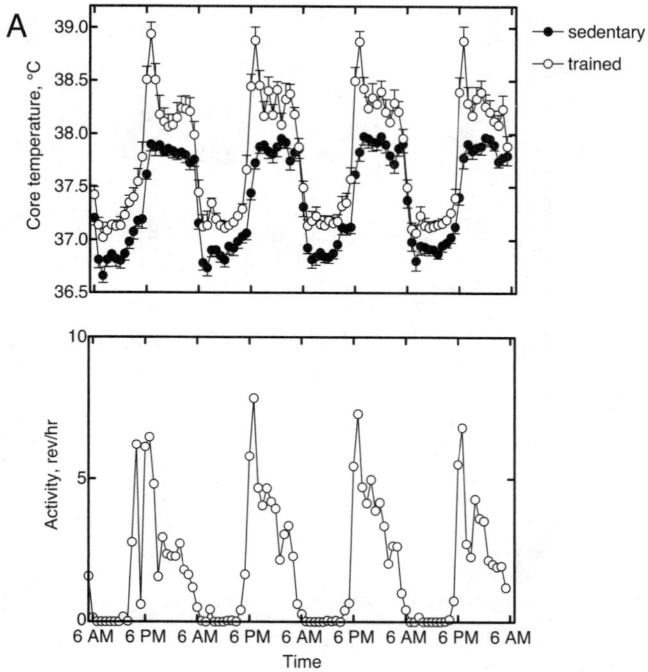


**Figure 7.6 (A) Comparison of core temperature in sedentary and trained female rats that were allowed access to a running wheel for 8 weeks. Note elevated core temperature in trained animals during the night when they are active as well as during the day when activity is minimal.**

Chronic, treadmill exercise training has been found to alter the sensitivity of rats to anti-ChE agents and other drugs (McMaster and Carney, 1986; McMaster and Finger, 1989). Rats were trained to run daily on a treadmill for 8 to 10 weeks such that they could maintain a speed of 1 mph for 1 h. The sedentary and trained animals were assessed for their sensitivity to physostigmine and other cholinergic drugs using an operant task as a sensitive end point. The tests were performed in rested animals as well as in animals that had recently been forced to exercise on the treadmill. Although core temperature was not monitored in these studies, it can be assumed that the exercise training will be associated with marked changes in thermoregulation. Exercise training lowered the potency to the muscarinic antagonists atropine and scopolamine by 10- and 40-fold, respectively (McMaster and Carney, 1986). Training led to a marked desensitization to physostigmine when animals were evaluated under conditions of rest and immediately after exercise (Figure 7.7). When the rats were evaluated 5 min after recovery from a bout of running (0.5 mph for 15 min), untrained rats displayed increased sensitivity to physostigmine,

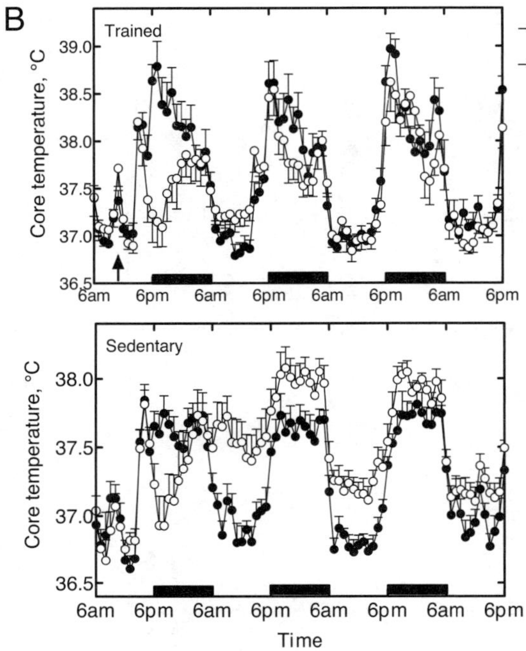


**Figure 7.6 (B) Thermoregulatory response to control vehicle or 25 mg/kg chlorpyrifos (oral dosing at arrow) in sedentary and trained rats. Note prolonged fever during the daytime in sedentary animals. (Data from Rowsey, P.J., Metzger, B.L., and Gordon, C.J. (2001). *Biol. Res. Nursing* 2: 267–276.)**

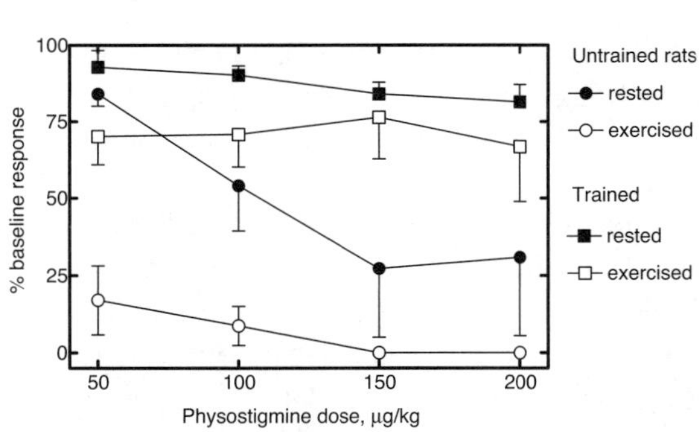


**Figure 7.7 Effects of exercise training for 10 weeks on the operant response to physostigmine when administered to resting animals and animals that had just been forced to run on a treadmill. Untrained animals are more susceptible to the behavioral effects of physostigmine. (Data from McMaster, S.B. and Finger, A.V. (1989). *Pharm. Biochem. Behav.* 33: 811–813.)**

characterized by a marked decrease in response with increasing dosage of the drug. Exercise-induced increases in heart rate, blood flow, and body temperature could be responsible for the increase in sensitivity to physostigmine in untrained rats. It is possible that an attenuated hyperthermic response to forced exercise in trained animals could be partly responsible for altered drug sensitivity.

## 7.8 STRESS AND MODULATION OF THERMOREGULATORY RESPONSE

From the above discussion, it is clear that the hyperthermic effects of exercise and consequences on thermoregulation can alter an animal's sensitivity to toxicants. Exercise studies in rodents may or may not be considered a stress. That is, exercise on a treadmill is most likely a stress if the activity is forced, whereas volitional running activity on a wheel is most likely not a stress. Assessing the effects of exercise and work on a toxicological response is therefore hampered by whether or not the activity is interpreted by the test animal as a stress. However, there are external factors that are undoubtedly a stress to rodents that impart changes in body temperature by altering the set-point and thermoeffector activity for control of heat production and heat loss. Like exercise, these thermoregulatory responses will alter the physiological response to toxicants and drugs.

### *7.8.1 Psychological Stress*

An open field environment is psychologically stressful to rodents because it is associated with vulnerability to predation. Rodents avoid open areas and seek the safety of a confined or sheltered environment. Open field exposures are often used to study physiological and behavioral responses to psychological stress in rodents. When placed in an open field environment such as a well-illuminated white box, rodents undergo a marked increase in core temperature along with other physiological responses associated with activation of the sympathetic–adrenal axis. Open field hyperthermia is mediated at least partially by cytokines and is blocked with administration of antipyretics, suggesting that the elevation in core temperature is akin to a fever involving an elevation in the set-point (Singer et al., 1986; Kluger et al., 1987). The hyperthermic response to open field stress is mediated by both an increase in metabolic rate and a transient reduction in tail blood flow.

Open field hyperthermia thus represents a challenge to the thermoregulatory system that is normally not observed under basal, resting conditions. One might expect that a toxicant's effect on a thermoregulatory parameter is easier to detect when the animal is stressed. The possible effects of a toxicant on vasomotor control or metabolic rate may only be manifested when these effector systems are activated during stress. For example, core temperature of male rats administered the organophosphate chlorpyrifos was unaffected by a dose of 10 mg/kg (Gordon and Yang, 2001; also see Chapter 3). However, when rats were placed in an open field environment 3 h after exposure to the same dose of chlorpyrifos, the stress-induced hyperthermia was attenuated relative to controls (Figure 7.8). It is of interest to note the stress-mediated hyperthermia from handling and administration of the control vehicle and chlorpyrifos. Rats dosed with chlorpyrifos showed an attenuated hyperthermic response to handling and injection that was apparent immediately after dosing.

The hyperthermic response to open field stress is apparently controlled by activation of peripheral cholinergic pathways. Administration of methyl scopolamine, a muscarinic antagonist that does not cross the blood brain barrier, led to an attenuation of open field hyperthermia (Rowsey et al., 2002). However, pyridostigmine at doses that do not affect baseline core temperature augmented the open field hyperthermia. Thus, peripheral cholinergic stimulation appears to enhance the hyperthermic response, whereas blockade of muscarinic receptors attenuates the hyperthermic response. This would suggest that relatively low doses of organophosphates or carbamates can affect the open field hyperthermia through modulation of the activation of peripheral cholinergic pathways.

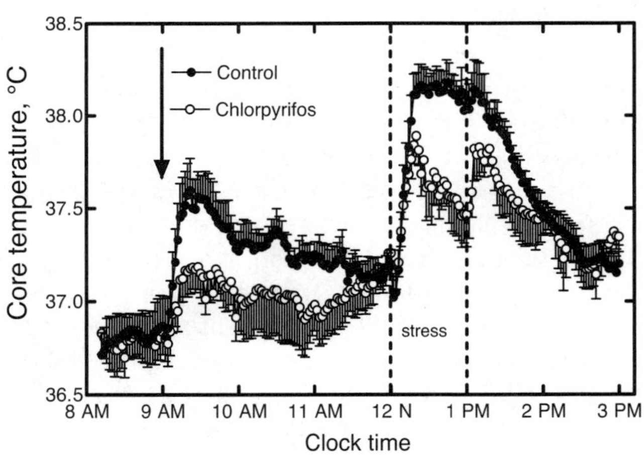


**Figure 7.8 Example of how the hyperthermic response to open field stress is attenuated following oral dosing with 10 mg/kg chlorpyrifos. Note that core temperatures of control and treated groups are nearly equal prior to and following recovery from exposure to stress. (Modified from Gordon, C.J. and Yang, Y.-L. (2001). *J. Thermal Biol.* 26: 313–318.)**

Open field stress testing has also been used as a sensitive tool to detect adverse effects of pre-natal exposures to a toxicant. Fewell et al. (2001) subjected rats to pre-natal nicotine exposure by implanting osmotic minipumps that delivered a constant dose of nicotine tartrate (6 mg/kg/day) beginning from day 6 to 7 of gestation to parturition. When the offspring reached an age of 7 to 8 weeks, they were subjected to open field stress for 1 h while their core temperature was monitored by radiotelemetry (Figure 7.9). Pre-natal exposure to a dose of nicotine in the rat that was equivalent to the blood levels of nicotine in a "heavy smoker" led to a marked reduction in the hyperthermic response to open field stress. Baseline body temperature, body weight, and the overall health of the animals appeared normal. The same laboratory further showed that the ability to mount a fever in response to central injection of prostaglandin $E_1$ was unaffected by pre-natal nicotine exposure (Fewell and Eliason, 2002). Hence, a pre-natal toxic insult that would appear to be benign during rest was manifested as a thermoregulatory deficiency in the adult by challenging the animal's thermoregulation system with open field exposure. Overall, the activation of a stress response in a rodent model using thermoregulation as an end point may be an ideal way of assessing potential toxicity of environmental contaminants that otherwise would not be detected under resting conditions.

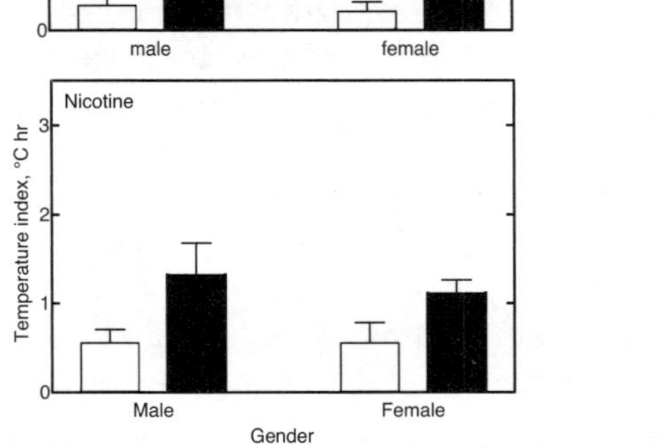


**Figure 7.9 Effect of pre-natal exposure to nicotine tartrate on the hyperthermic response of adult rats to open field stress. Data expressed as the mean + S.E. of the temperature index, an integration of the increase in temperature with time over the 3-h exposure to open field stress. (Data modified from Fewell, J.E., Eliason, H.L., and Crisanti, K.C. (2001). *Physiol. Behav.* 74: 595–601.)**

## 7.8.2 Restraint and Handling Stress

Many types of toxicology studies necessitate the use of a restraining device for rodents and other species. Restraint devices are often used to restrict the movement of rats and mice used in nose-only inhalation studies (see Chapter 3). Restraint is also needed for physiological studies in which the animal is instrumented to monitor cardiovascular end points and other physiological parameters. Moreover, blood pressure measurements using the tail cuff technique in rats can only be done in restrained animals. The development of radiotelemetry over the past 25 years has revolutionized the monitoring of autonomic parameters in unrestrained rodents and other species. Although the equipment is expensive, it is well worth the investment in that the data are free of artifacts from handling and restraint. From this author's perspective, restraint has such a profound effect on thermoregulation and, hence, essentially all physiological responses of rodents, that the topic merits a thorough discussion in this book.

### 7.8.2.1 Thermoregulation and Restraint

The stress of handling and restraint has a profound influence on thermoregulation that imparts significant effects on the efficacy of a drug or toxicant. Simply picking up a mouse or rat and measuring its temperature with a colonic probe leads to an elevation in core temperature that will persist for several hours (see Gordon, 1990, 1993, for review). Note that a measurement made initially within a few seconds after picking up an undisturbed animal provides an accurate measurement of the true baseline temperature because the stress response does not affect body temperature for several minutes. However, repeated measurements made at regular intervals (e.g., <1 h) lead to a stress-induced hyperthermia. It is noteworthy that much of the thermoregulatory data presented in this book were collected by investigators who had to rely on repeated handling to measure core temperature. The core temperature of control rats is often found to be above 38°C and occasionally close to 39°C, whereas the minimal, daytime core temperature of the rat is equal to or slightly below 37°C (when measured with radiotelemetry; see Chapter 2). The presence of personnel near a rodent can lead to significant elevations in core temperature (see Gordon, 1990, for review). How can one accurately interpret the effects of a toxicant on body temperature or any parameter that is dependent on temperature when the control and treated animals are forced to sustain a profound hyperthermia? Furthermore, how does restraint in a rodent compromise the extrapolation of a toxicological response to a rested, unstressed human subject?

The thermoregulatory response to restraint is dependent on the species, type of restrainer, conditioning and training to the restraint, and ambient temperature. *Emotional hypothermia* and *emotional hyperthermia* in rodents and rabbits are often reported, but the mechanism of action is not well understood. Rats placed in a tight-fitting wire-mesh restrainer at an ambient temperature of 25°C undergo an increase in heat production, while heat loss remains unchanged, resulting in a net heat storage and a marked rise in core temperature (Nagasaka et al., 1979). In general, restraint leads to hypothermia when rodents are maintained below a threshold ambient temperature and to hyperthermia when maintained above the threshold temperature (Thornhill et al., 1979). Mice often become hypothermic with restraint at a standard room temperature of 22°C (Miller and O'Callaghan, 1994), whereas golden hamsters undergo a 0.9°C increase in core temperature with restraint in nose-only inhalation tubes (King-Herbert et al., 1997). In the cold, restrained rodents are unable to shiver and rapidly lose body heat. In excessively warm environments, restraint impedes the ability to groom saliva on the fur to dissipate heat (see Gordon, 1993, for review). The thermoregulatory responses to restraint can be blunted with repeated training in the devices (Thornhill et al., 1979), but this does not necessarily mean that the animal's response to a drug or toxicant will be unaffected by adaptation to restraint relative to a free-moving, undisturbed state. In a recent study on the adaptation to restraint stress, mice and rats placed in nose-only restrainers daily for 4 h per day required 2 weeks before heart rate and core temperature returned to basal levels (Narciso et al., 2003). The novel environment of the restraining device is also a confound in the thermoregulatory response. Overall, ambient temperature is clearly one of the most important factors that will govern whether the emotional response to restraint leads to hypothermia or hyperthermia.

#### *7.8.2.2 Restraint and Response to Drugs and Toxicants*

The effects of restraint on the thermoregulatory response to drugs and other pharmacological agents are profound and the results are relevant to studies with xenobiotics. The toxicity of a chemical or drug will be affected by a toxicant, even if core temperature does not change during restraint. For example, rats placed in plastic or wire-screen restrainers have similar baseline core temperatures (37.6 to 38.3°C) when compared to rats housed in a confined but free-moving state. These body temperatures are high, even for the free-moving rats, compared to core temperature monitored with telemetry. The presence of laboratory personnel near the rodents likely contributes to an emotional hyperthermia. In spite of the similar baseline core temperatures, the hypothermic effects of a high dose of morphine were influenced by restraint (McDougal et al.,

1983). The hypothermic effect of morphine was accentuated by restraint with the reduction in temperature greatest in rats restrained in a wire-screen as compared to a plastic restrainer. Using telemetry to monitor tail skin and core temperature, it was shown that restraint leads to a lower tail skin temperature but does not affect the vasomotor response to isoproterenol or angiotensin II (Wright and Katovich, 1996).

Most investigators would expect that the stress from restraint would exacerbate chemical toxicity because of the expectations that restraint equals hyperthermia in rodents. While this is generally true for the rat at standard test temperatures, the thermoregulatory response of the mouse to restraint is more unpredictable. The core temperature of mice is more labile, and they often become hypothermic when restrained at a standard room temperature of 22°C (Miller and O'Callaghan, 1994). Since hypothermia affords protection to a variety of chemical toxicants (see Chapter 4), restraint-induced hypothermia could be protective. The CNS cytotoxicity of the substituted amphetamine MDMA was found to be attenuated by restraint in the C57 mouse. Administration of MDMA to unrestrained mice led to hyperthermia and a significant accumulation of glial fibrillary acidic protein (GFAP) in the striatum (a biomarker of CNS damage; see Chapter 4). In comparison, the restrained mouse became hypothermic with or without MDMA administration, and the accumulation of GFAP was significantly reduced (Figure 7.10). Overall, the novel environment and limited movement of a restraining device are stressful to rodents and thermoregulation is compromised. Depending on the species and ambient temperature, restraint can raise or lower body temperature and modulate toxicity and drug efficacy.

Other forms of psychological stress can modulate the thermoregulatory response to a toxicant. There has been considerable work on the effects of stress on the thermoregulatory response to ethanol. Stressors other than restraint that normally raise body temperature will exacerbate the hypothermic effects of ethanol (Cunningham and Bischof, 1987). Rats that are handled in a manner that a researcher would use to measure rectal temperature exhibit a greater hypothermic response to ethanol as compared to nonhandled rats. Repeated measurements of rectal temperature in mice resulted in a marked increase in lethality of mice following administration of soman, atropine, and oxotremorine (Table 7.4). There was 100% survival in mice given the drugs and handled but without repeated insertion, whereas mice that were probed to measure temperature and given the organophosphate and drugs sustained 90% mortality after 4 days. The repeated insertion of the rectal probe may have damaged the intestinal lining, increasing the animals' susceptibility to microbial infection (Clement, 1993a).

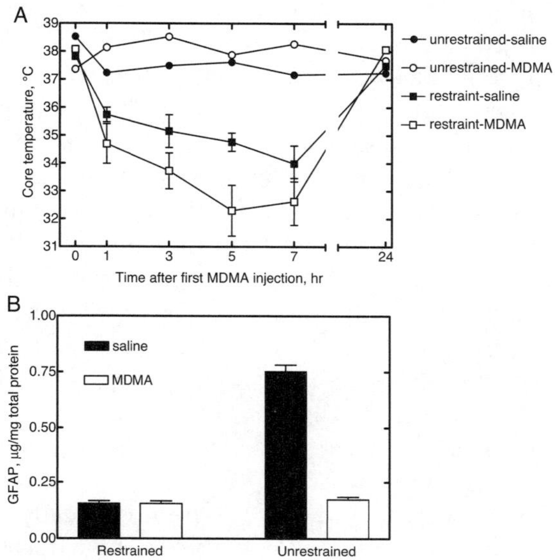


**Figure 7.10** **(A) Effect of restraint on core temperature of mice injected with saline or the substituted amphetamine MDMA (20 mg/kg × 4 doses). (B) Striatal levels of glial fibrillary acidic protein (GFAP), a biomarker of CNS damage, measured 72 h after treatment. (Data from Miller, D.B. and O'Callaghan, J.P. (1994). *J. Pharmacol. Exp. Ther.* 270: 752–760.)**

**Table 7.4 Effect of Rectal Probing to Measure Core Temperature on Percent Mortality of Mice Dosed with Soman (100 µg/kg; SC) on Day 0 Followed by Daily Injections of Atropine (2.2 mg/kg; IP) 5 min Before Oxotremorine (625 µg/kg; IP)[a]**

| | *Treatment* | | | | *Percent Mortality* | | | |
|---|---|---|---|---|---|---|---|---|
| *Group* | *Soman* | *Atropine* | *Oxotremorine* | *Rectal Probe* | *Day 1* | *Day 2* | *Day 3* | *Day 4* |
| 1 | + | + | + | + | 0 | 20 | 70 | 80 |
| 2 | + | + | + | + | 0 | 20 | 50 | 70 |
| 3 | – | + | + | + | 0 | 20 | 70 | 90 |
| 4 | + | + | + | – | 0 | 0 | 0 | 0 |

[a] Mice either had rectal temperature measured or were held in the same manner but not probed.

*Source*: Data from Clement, J.G. (1993). *Lab. Anim. Sci.* 43: 381–382.

Psychological stressors such as housing on a wire-grid floor and a strobe light exacerbate, whereas electric foot shock attenuates ethanol's hypothermic effect. Foot shock, handling, and strobe light stressors all cause hyperthermia in the absence of ethanol, but they have differential effects on the thermoregulatory response to ethanol. It is interesting to note the opposite effects of handling stress on heart rate and body temperature in the rat. Handling stress was found to augment ethanol-induced tachycardia but exacerbated the hypothermic effects of ethanol (Peris and Cunningham, 1986). Activation of the opiate system in response to stress, while critical in the physiological response to stress, may in fact exacerbate the hypothermic effect of a toxicant such as ethanol (Cunningham and Bischof, 1987).

In spite of the reliance on the use of restraint in toxicological and pharmacological studies, we know relatively little about how restraint affects the thermoregulatory response to a drug or toxicant. A biological end point in a toxicological study, even one not directly associated with the thermoregulatory system, may nonetheless be affected by restraint. How can the mechanism of action be understood if the stress response dampens or exacerbates the physiological response? The understanding of peripherally versus centrally active drugs and toxicants and the impact of stress is also not well developed. For example, in a recent study from this laboratory it was shown that stress from cage switch, handling, open field exposure, and vehicle injection exacerbated the hypothermic effect of methyl scopolamine in the rat (Rowsey and Gordon, 2000; Gordon and Yang, 2001). Methyl scopolamine is a muscarinic antagonist that does not cross the blood brain barrier and was thought to have no remarkable effects on body temperature. The continued development and advances of radiotelemetry should lead to marked improvements for the study of *in vivo* toxicological responses by eliminating handling and other stressors. Minimizing stress will lead to more sensitive assessment of toxicological insults in laboratory animals. Where the use of restraint is unavoidable, better attempts should be made to measure core temperature and assess how the toxicant and stress interact in the control of heat gain and heat loss thermoeffectors.

### 7.8.3 Metallothionein Induction and Stress

The induction of metallothionein is influenced by a variety of stressors (Oh et al., 1978). This protein is induced in response to exposure to heavy metals and is involved in their metabolism and detoxification. Acute heat and cold stress was found to induce the synthesis of metallothionein in liver of the rat. Exercise stress (forced swimming for 3 h) was also a very effective inducer, whereas a severe skin burn had no effect (Oh et al.,

1978). Metallothionein has been found to have a role in brown adipose tissue thermogenesis (see Chapters 2 and 3). Its induction by a variety of stressors that also affect thermoregulation may be linked to the control of nonshivering thermogenesis.

## 7.9 GULF WAR SYNDROME

Many veterans from the 1991 Gulf War have been diagnosed with unexplained illnesses that are commonly referred to as the Gulf War syndrome. The syndrome has received considerable attention over the past decade, and a variety of factors have been considered in its etiology, including stress from combat, exposure to pesticides and warfare agents, heat stress, and pyridostigmine administration (Spencer et al., 2001). The possible interaction between heat stress and exposure to organophosphate nerve gas agents as a possible cause of the Gulf War syndrome has received considerable attention in the mass media and scientific literature.

Simulating some of the environmental conditions of desert warfare in an animal model may provide a better understanding of the etiology of the Gulf War syndrome. In one approach, rats were exposed to thermoneutral or heat stress conditions while subjected to nose-only exposure to the organophosphate sarin (Henderson et al., 2002; Conn et al., 2002). Housing the rats an ambient temperature of 32°C led to a sustained 1°C elevation in core temperature. These ambient conditions represent a significant heat stress, but the rats could still thermoregulate. Exposure to low or high doses of sarin (0.2 or 0.4 $mg/m^3$) for 1 h per day for 5 or 10 days had no effect on body temperature, and there were no marked clinical effects under thermoneutral or heat stressed conditions. However, the inhibitory effects of sarin on AChE activity in the hippocampus were more pronounced in rats exposed to heat stress. There were also long-lasting combined effects of heat stress and sarin exposure as indicated by down-regulation of type $M_1$ and up-regulation of type $M_3$ muscarinic receptors in certain parts of the CNS (Henderson et al., 2002). These neurochemical effects persisted for at least 30 days after a 5-day exposure to sarin and heat stress. In some CNS sites such as the cerebral cortex and striatum, heat stress exacerbated the down-regulation of $M_1$ receptors in rats exposed to low and high doses of sarin. Observations such as these could provide a data base to help explain some of the mental dysfunctions in soldiers during the Gulf War who were subjected to thermal stress and other insults.

Prescribing pyridostigmine tablets to soldiers in the Gulf War has also been explored as a possible cause of the syndrome. As explained earlier (Section 7.5.2.2), pyridostigmine can alter thermoregulatory responses in the heat in spite of the fact that it does not cross the blood–brain barrier. Acute heat stress has been shown to increase the permeability of the blood–brain barrier in rodents; penetration of pyridostigmine into the CNS could account for the manifestation of some symptoms of the Gulf War syndrome. For example, in young rats exposed for 4 h to an ambient temperature of 38°C, a marked increase in permeability of the blood–brain barrier was associated with an increase in core temperature of 3.6°C (Sharma and Dey, 1987). A 1.0°C increase in core temperature after 2 h of heat exposure had no effect on permeability. However, in a study of guinea pigs exposed to acute heat stress with body temperature reaching levels associated with heat stroke (44.3°C), there was no evidence of an increase in permeability of the blood–brain barrier to pyridostigmine (Lallement et al., 1998).

Considering that soldiers in the field are exposed to excessively high temperatures and may be required to wear chemical-resistant suits that impede heat loss, it is prudent to consider the possible interactions between heat stress, blood–brain permeability, and pyridostigmine treatment. Even though there is no indication that permeability is affected by these conditions, the peripheral effects of pyridostigmine on body temperature during rest and stress (see Section 7.8.1) should lead researchers to consider how chronic treatment with a cholinesterase inhibitor may affect human health under stressful combat conditions in the desert. Thus far, there is no definite connection between heat stress and pyridostigmine exposure. In a thorough epidemiological survey of veterans of the first Gulf War, Spencer et al. (2001) concluded that the unexplained illnesses could not be attributed to exposure to anticholinesterase agents.

## 7.10 METEOROLOGICAL CONDITIONS AND ENVIRONMENTAL TOXICOLOGY

Meteorological conditions are integral to human thermoregulation and the physiological responses to toxicants. Taking into consideration this author's bias, the thermoregulatory system can be viewed as a focal point of environmental toxicology (Figure 7.2). This diagram depicts a variety of possible interactions between the thermoregulatory system, meteorological conditions, and exposure to toxicants. A key component of the diagram is to show how temperature and humidity can affect the occurrence and distribution of toxicants in the air, land, and water. Meteorological conditions also affect the physiological response to environmental toxicants through their effects on thermoregulation. The potential exposure to a toxicant and entrance into the body via the skin or respiratory tract are affected by activity of thermoeffectors controlling heat loss, including skin blood flow, sweating, and respiration (see Section 7.4 for details).

Temperature and humidity have direct effects on the distribution, accumulation, and breakdown of pollutants, pesticides, and other environmental toxicants. There are complex effects of temperature and humidity on the generation and degradation of airborne pollutants (Viswanathan and Murti, 1989). Warmer air temperatures increase both the generation and degradation of many air pollutants. Human activity and generation of pollutants are also inexorably linked to temperature and humidity. For example, ozone alerts are common in the summertime, a season when there are hot temperatures and concomitant heavy use of automobiles. In the winter months, people are more likely to be exposed to particulates from the industrial and domestic burning of fossil fuels for electricity and heat (e.g., fireplaces and wood stoves).

The dispersion of pesticides into the air after spraying an agricultural field is also affected by temperature, rain, wind, and other factors (Gunther and Gunther, 1977). Generally, high temperatures accelerate the dispersion and increase the concentration of airborne pesticides within a sprayed area. It is interesting to note that many pesticide workers, to alleviate thermal discomfort, will not utilize the recommended protective clothing when spraying pesticides during the warm summer months. In a roundabout way, high ambient temperatures can increase the dosage of a pesticide through a biological and environmental mechanism. That is, high environmental temperatures can lead to higher pesticide concentrations in a sprayed field and, biologically, the behavioral thermal preference of the personnel causes them to shun the protective clothing, resulting in an increased dosage of the insecticide.

## 7.11 ARSENIC, COLD STRESS, AND RAYNAUD'S DISEASE

Occupational and environmental exposure to arsenic through inhalation and in the drinking water has been associated with the development of Raynaud's disease, characterized by localized reduction in peripheral blood flow. Related peripheral vascular disorders such as black foot disease occur in humans exposed to relatively high concentrations of arsenic in the drinking water. Raynaud's disease can be considered an example of heterothermic thermoregulatory dysfunction in that it is associated with localized cooling and numbing of the appendages (fingers, toes, nose) as a result of reduction in peripheral blood flow. It appears that Raynaud's is more common in cold climates, presumably because of the vasconstrictive effect of cold exposure on skin blood flow. Finger systolic blood pressure is significantly reduced in smelter workers exposed to arsenic but only at relatively cold temperatures of 10 and 15°C (Lagerkvist et al.,

1986). The reduction in finger systolic blood pressure as a result of arsenic exposure appears to be long lasting since the Raynaud's symptoms persisted in smelter workers in spite of the absence of arsenic exposure for 4 to 8 weeks (Lagerkvist et al., 1988). It is also interesting to note that Raynaud's disease is more prevalent in women, but the reason for a gender difference is not known. Overall, the disease represents an example of a profound interaction between environmental cold stress and the manifestation of a thermoregulatory response (i.e., heterothermy) in humans exposed to acute or chronic arsenic.

## 7.12 AMBIENT TEMPERATURE, POLLUTION, AND HUMAN MORTALITY

There is a growing interest in assessing the possible interaction between exposure to air pollutants, human mortality, and thermal stress (Kunst et al., 1993; Michelozzi et al., 1998; Katsouyanni et al., 1993; Mercer, 2003). Young adult and aged humans have a similar baseline body temperature, but their thermoeffector mechanisms differ considerably (Kenney and Munce, 2003). In general, aged humans have deficiencies in their ability to reduce skin blood flow and increase heat production when subjected to cold stress and an impairment to increase sweat output and skin blood flow when subjected to heat stress. In view of the deficits in thermoeffector function, it is not surprising to find an increased incidence of deficiencies in thermal homeostasis in the aged during periods of thermal stress (for review, see Mercer, 2003). The interaction between season, thermoregulatory function, and health of the aged and other susceptible groups should lead one to consider how sensitivity to pollutants and other toxicants may vary as a function of seasonal change.

The association between extreme changes in environmental temperature and mortality would be expected in view of the deficiencies in thermoeffector function in the aged. Indeed, high rates of death and sickness during summer heat waves and winter cold snaps receive considerable attention in the media. However, one should also wonder how the annual changes in temperature such as depicted in the graphs of Figure 7.1A affect human health and the susceptibility to environmental toxicants. Many would take for granted that exposure to the seasons of a temperate climate would have no remarkable effects on human health, but this is not necessarily the case, as will be explained below. Seasonal effects on susceptibility to disease should prompt toxicologists and pharmacologists to consider how sensitivity to a drug or toxicant may be affected by seasonal change.

Many studies have found a dependence of human mortality on seasonal changes in temperature. Kunst et al. (1993) performed a thorough analysis of mortality data in The Netherlands and published one of the most well-delineated relationships between daily ambient temperature and human mortality (Figure 7.11). To say that this relationship should be tantalizing to any student of environmental physiology is an understatement. The authors analyzed a data base that consisted of an average of 324.7 deaths per day from 1979 to 1987 and found a V-shaped relationship between air temperature and mortality. One cannot help but notice the remarkable similarity between the pattern in Figure 7.11 and a typical thermoneutral profile of metabolic rate for a homeotherm (see Chapter 2). Mortality index was minimal at an ambient temperature of 16°C, rose gradually to a temperature of 19°C, and then increased sharply at higher temperatures. At temperatures below the optimum, mortality underwent a gradual and steady increase as temperature decreased below 13°C.

Many studies have shown a V-, U-, or J-shaped relationship between temperature and human mortality (Curriero et al., 2002; Braga et al., 2001). One would expect that seasonal effects on mortality would be dependent on latitude with respect to annual changes in temperature. In a review of mortality data in northern and southern cities in the United States, a relationship between temperature and mortality similar to that of Figure 7.11 was observed (Curriero et al., 2002, 2003). Cold-induced mortality

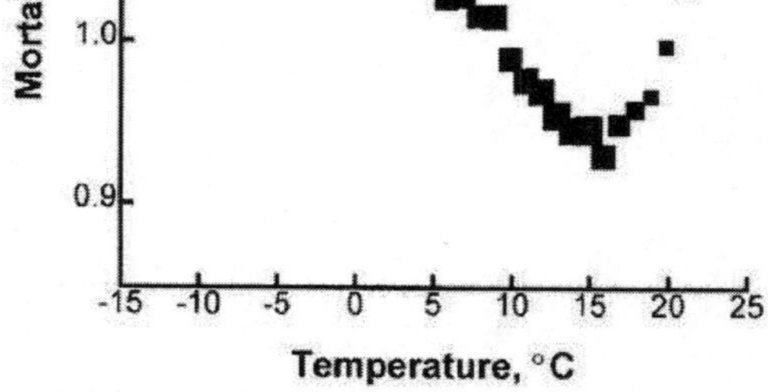


**Figure 7.11 Example of the impact of the average air temperature and daily mortality in humans. These data represent the mortality (324.7 deaths per day) in The Netherlands from 1979 to 1987. The size of each block is proportional to the sample size. Data and graph modified from Kunst, A.E., Looman, C.W., and Mackenbach, J.P. (1993). *Am. J. Epidem.* 137: 331–341.)**

had a greater impact on people living in warmer, southern cities, whereas heat-induced deaths were more numerous in individuals living in northern cities. A comparison between Chicago and Miami illustrates how acclimatization to upper or lower latitudes can affect the temperature–mortality relative risk functions (Figure 7.12). Residents of Chicago experienced mean summer and winter temperatures of 71.9 and 25.6°F as compared to Miami with temperatures of 82.3 and 68.7°F, respectively. The estimated lower threshold temperature of the temperature–mortality function was 42°F for Chicago as compared to 71°F for Miami. The upper threshold temperature was 75.5°F for Chicago and 81.9°F for Miami. The significant relationship between temperature and mortality is remarkable when one considers the uncontrolled biotic and abiotic variables in these studies. Most subjects would be expected to be sheltered much of the time from heat and cold stress, while variables such as humidity, levels of physical

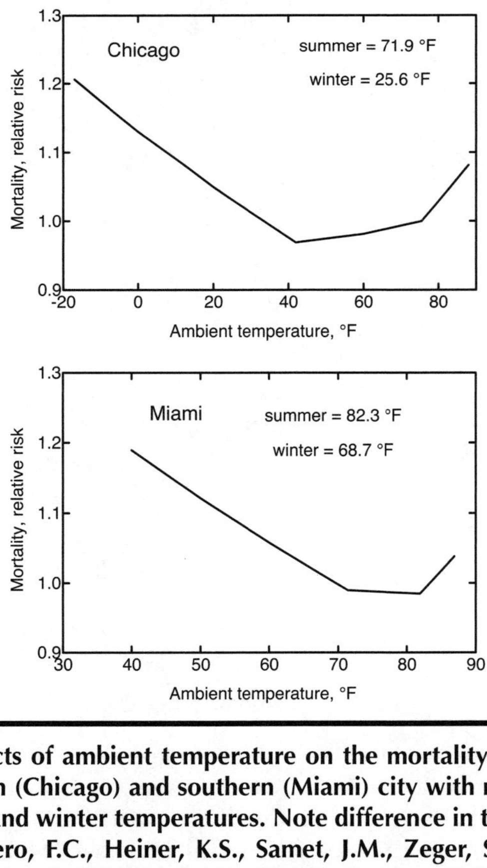


**Figure 7.12 Effects of ambient temperature on the mortality relative risk function for a northern (Chicago) and southern (Miami) city with marked differences in mean summer and winter temperatures. Note difference in temperature scales. (Data from Curriero, F.C., Heiner, K.S., Samet, J.M., Zeger, S.L., Strug, L., and Patz, J.A. (2002). *Am. J. Epidemiol.* 155: 80–87; Curriero, F.C., Samet, J.M., and Zeger, S.L. (2003). *Am. J. Epidemiol.* 158: 93–94.)**

activity (fitness), and diet would also affect mortality. There is also a wide range in genotypes, socioeconomic status, susceptibility to disease, and possible exposure to environmental contaminants that could affect these functions.

Statistical analysis of epidemiological data necessitates sophisticated models, the details of which go far beyond the scope of this book. That is, in view of the tremendous variability in biotic and abiotic factors that have a role in human health as described above, there may be subtle effects of climate that can only be detected with very sensitive analytical methods. For example, in a study of the role of heat stress and pollution on human mortality in Toronto, a small but significant effect of air pollution on heat stress–related mortality was detected (Rainham and Smoyer-Tomic, 2003). Frank and Tankersley (2002) have recently developed a hypothesis to explain the link between fluctuations in ambient particulate matter (PM) pollution and human deaths that is based on a decline in thermoregulation and other homeostatic processes. Using telemetric monitoring of core temperature and electrocardiogram in the aging mouse, they suggested that sudden death from exposure to PM pollutants is related to increased susceptibility of individuals, as their ability to regulate body temperature and heart rate declines with aging. Variations in ambient temperature and other climatic factors will also have a critical role in the ability to maintain normal body temperature and heart rate with aging. If mortality is a harbinger of other serious health effects of pollutants and high temperatures, then it can only be assumed that adding exercise and work to these conditions would increase the likelihood of serious health effects that are linked to thermoregulation.

### 7.12.1 Greenhouse Effect and Thermoregulation

The essence of this chapter is that life as well as exposure to toxic chemicals do not occur under ideal environmental conditions. The mechanism of action of toxic chemicals is best understood in animals maintained under ideal environmental conditions (e.g., Table 7.2). Less is known about how environmental stress alters the mechanism and efficacy of toxicants. Natural and man-made perturbations in the climate along with the continued escalation in production and use of pesticides and other contaminants should stimulate interest to understand how environmental stress affects the physiological response to toxic agents in humans, domestic animals, and wildlife.

We have a limited understanding of how short- and long-term alterations in one or more physical factors of the environment will affect the health and well being of humans, as well as all animal and plant life.

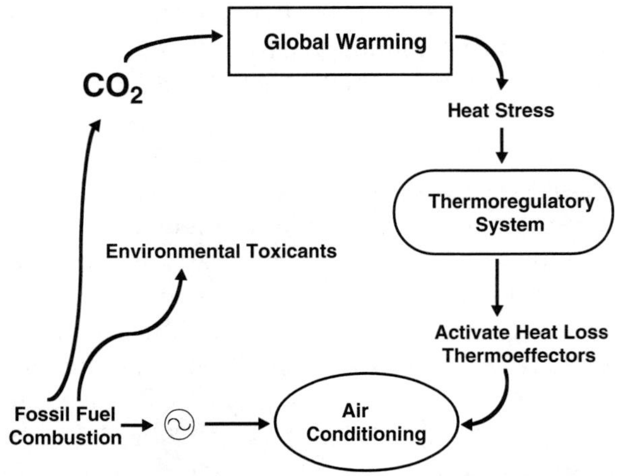


**Figure 7.13 A proposed model of human thermoregulation and global warming. Massive amounts of fossil fuels are burned to provide power to operate air conditioning (i.e., a behavioral thermoeffector). Burning fuels generates toxicants and $CO_2$, leading to more global warming and a greater need for air conditioning.**

Global warming seems inevitable as a result of the atmospheric accumulation of greenhouse gases that, in most cases, has been a result of human activities (Turco, 2002). The greenhouse effect is controversial, complex, and difficult to predict. Nonetheless, the buildup of carbon dioxide, nitrogen oxide, methane, and other gases that are generated in the combustion of fossil fuels is a leading cause of the greenhouse effect. To this end, it is interesting to consider the positive feedback cycle that has developed between the thermoregulatory system and the generation of greenhouse gases (Figure 7.13). It is theorized that our need for thermal comfort is in fact a major source of pollution, including the generation of greenhouse gases. How could such a cycle come about? Human thermoregulation is exquisitely sensitive to ambient temperature and relative humidity. Peoples of western countries and other developed nations that have the financial and technological resources are able to use air conditioning to behaviorally thermoregulate and maintain an ideal level of thermal comfort in their homes, automobiles, and work places. In this simplified scenario, the electrical power needed to operate air conditioning systems leads to the generation of greenhouse gases (and other environmental toxicants), leading to more global warming, increasing the demand for more air conditioning, and so forth. It would be of interest to consider how severe the greenhouse effect would be if not for our demand for an ideal thermal environment. Moreover, as humans continue to burn fossil fuels to maintain their microenvironments for ideal thermal comfort, one should question how such behavior will contribute to the greenhouse effect and exposure to toxicants.

The impending consequences of the greenhouse effect and global warming have spurred interest in understanding how future changes in the climate will impact human health (Martens, 1999). Global warming occurs in the process of producing pollutants that, in some cases, are more toxic at warmer temperatures. The human race is likely on a road to an atmosphere that is warmer and has more pollutants, an environment that is partially driven by our need for thermal comfort. Future research should address the health effects in humans as well as the ecological effects of environmental toxicants and the possible exacerbating role of global warming.

# Chapter 8

# Comparative Physiological Responses

## 8.1 INTRODUCTION

Laboratory rodents have been the animals of choice in most biomedical and toxicological research endeavors. Indeed, most of the data discussed in this book come from studies using commercially bred mice and rats. Most biomedical researchers would contend that toxicological research should focus on experimental mammals since they are the closest genetic match to humans. In fact, limiting our toxicological research to mammals while disregarding comparative responses is likely to hamper our understanding of human responses. In an eloquent thesis on the problems of ultraspecialization in biomedical research written in 1975, the eminent physiologist Allan C. Burton elaborated on the perils of researchers that specialize on one group of organisms. That is, we "learn more and more about less and less" (Burton, 1975). He went on to say that the quality and originality of biomedical research will decline as we focus solely on mammalian responses and much of what we need to know about mammalian (and human) responses will come from research on lower life forms.

Comparative responses to toxicants are not only critical for one to study ecotoxicology but provide valuable insight into the mechanism of toxicity in mammals, including humans. In many instances, the molecular and cellular mechanisms of action of toxicants and pesticides have been best understood by using invertebrate models, and these data eventually led to a better understanding of mammalian responses. Ballatori and Villalobos (2002) published a thorough analysis and review of comparative animal models in toxicology research and point to three key developments that have led a resurgence in the use of comparative models: (*a*) the trend by research and regulatory agencies to fund studies that reduce the number of mammals used in research; (*b*) better recognition of the power of nonmammalian models to study toxicology with attributes such as high rates of fecundity, unique anatomy, exaggerated biochemical pathways (e.g., acetylcholinesterase activity in electric eel), and cells and tissues that are easy to study (e.g., giant lobster axons, *Xenopus* oocytes, etc.); and (*c*) the discoveries from genomic sequencing projects that have demonstrated a remarkable similarity of the human genome to the so-called lower organisms.

With a resurgence in comparative physiology in toxicology, it is paramount to assess the role of temperature in the toxic responses of species other than the commonly used laboratory rodents, including invertebrates, poikilothermic vertebrates, and bird and mammalian wildlife. This chapter endeavors to assess the thermoregulatory responses of fish, amphibians, invertebrates, and other species to toxicant exposure and to provide an assessment, from an ecological perspective, of how temperature affects toxicological responses.

## 8.2 ECOTOXICOLOGY

The thermoregulatory responses to toxicants and the effects of temperature on chemical toxicity in wildlife, lower vertebrates, and invertebrates are an essential facet of the study of ecotoxicology. The field of ecotoxicology consists of three components: ecology, environmental chemistry, and toxicology — all of which are affected by temperature, humidity, wind, radiation, and other facets of the climate (Bourdeau et al. 1989; Gordon, 2003). Temperature is by far the most important climatic factor that limits an organism's ability to inhabit and survive in any habitat. With the expansion of human populations, the vulnerability of ecosystems to toxicant contamination is expected to increase precipitously as animal and plant life are forced into more limited habitats (Everts, 1997). The interactions between thermal stress and toxicology on the susceptibility of ecosystems will continue to be a crucial issue as human activities cause wildlife and other organisms to be pushed into less than optimal thermal zones. Everts (1997) notes that arid and semiarid zones that encounter marked temperature extremes in space and time could be especially critical in this assessment of climate and toxicant susceptibility. Of course, wetlands and other climatic zones that are densely populated by humans are also crucial. Moreover, understanding the thermal physiological response to toxicants from a comparative point of view is essential in ecotoxicological risk assessment.

## 8.3 EFFECTS OF TEMPERATURE ON TOXICITY IN AQUATIC ORGANISMS

Excluding birds and mammals and a few fish species such as tuna, all aquatic organisms are poikilothermic, meaning that their body temperature is essentially equal to that of the ambient temperature (see Chapter 2). This trait is advantageous in the study of the toxicology in these organisms because it is not necessary to measure body temperature. Measurement of the water temperature is sufficient to provide an accurate estimate of the organism's body temperature in most cases. However, terrestrial ectothermic vertebrates, including many species of amphibians and reptiles, are capable of homeothermic-like characteristics when placed in environments where they can use behavior to maintain a core temperature that is partially independent of changes in ambient temperature. For an excellent and thorough review on the thermoregulation of nonmammalian vertebrates and invertebrates, the reader is referred to a series of books on the comparative physiology of thermoregulation edited by Whittow (1970).

The potential toxicity of pesticides and other environmental contaminants in aquatic organisms can be influenced by a complex array of environmental and biological factors (Figure 8.1A). In addition to interspecies differences in vulnerability, factors such as water temperature, oxygen availability, pH, activity level, and stage of development can exert a significant impact on toxicity. The diagram in Figure 8.1A serves as a general framework to illustrate the possible interaction of biotic and abiotic variables in the toxic response of fish and other aquatic organisms. A considerable portion of this research was performed several decades ago and has been thoroughly reviewed in papers by Cairns et al. (1975, 1975a).

Water temperature must always be considered as the key variable in ecotoxicological studies. Because of the $Q_{10}$ effect (see Chapters 1 and 4), a rise in temperature of 10°C results in an approximate doubling in metabolism for most aquatic organisms. Oxygen availability is usually a factor in toxicological studies on mammals and birds. However, oxygen solubility in water decreases with rising temperature, resulting in reduced levels in dissolved oxygen with increasing temperature. Thus, in the case of chemicals that impede respiratory gas exchange or block cellular respiration, one would expect their toxicity to be exacerbated at higher temperatures because of combined factors of (*a*) increases in metabolic rate and oxygen demand and (*b*) lower levels of dissolved oxygen. For example, exposing the fathead minnow to high temperatures and low dissolved oxygen leads to an inhibition in acetylcholinesterase activity, resulting in greater susceptibility to exposure to an organophosphate-based insecticide (Baer et al., 2002). One other key factor in aquatic biology is the interaction between acclimation temperature and lethal temperature,

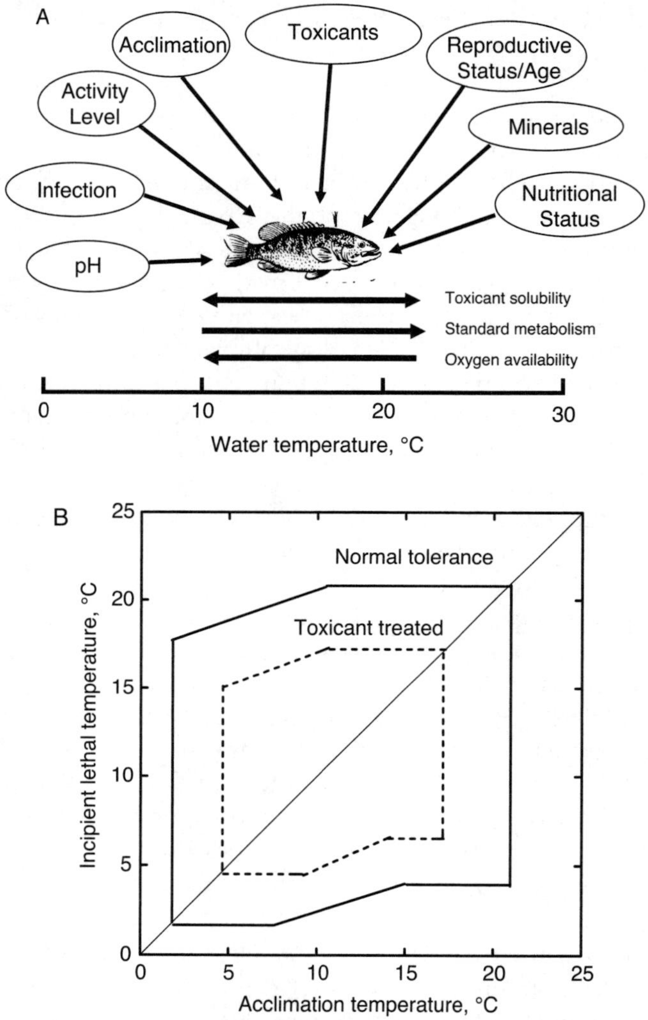


**Figure 8.1 (A) Overall concept of how water temperature, oxygen availability, and toxicant solubility along with other biotic and abiotic factors could interact to affect the toxicity in fish and other aquatic organisms. (B) A temperature tolerance polygon that depicts the relationship between temperature of acclimation and lethal temperature as affected by toxicant exposure and the interaction of the other factors depicted in Figure 8.1A. The smaller area of the polygon means reduced environmental fitness.**

termed a tolerance polygon (Figure 8.1B). Exposure to toxicants is likely to alter the ability of aquatic organisms to survive extreme thermal environments. The thermal tolerance, reflected by the area within the polygon, provides a broad assessment of the ability of the organism to adapt to environmental stress (Cossins and Bowler, 1987). Overall, the complex effects of temperature and toxicity in fish and other aquatic organisms are obvious when assessing the array of biological and environmental factors, including the organism's limits of thermal tolerance, temperature acclimation, activity levels, reproductive status and gender, nutritional state, and pH (Figure 8.1A).

Acute toxicity testing in the aquatic organisms, including the lower vertebrates and invertebrates, typically involves determining the lethal concentration of a toxicant in the water of organisms acclimated to a constant set of environmental conditions and exposed to the toxicant for a given period of time. In view of the effects of temperature on respiration and oxygen availability, one would generalize that toxicity of chemicals in aquatic organisms increases with temperature. In general, the lethal dose or concentration of the toxicant in the water resulting in 50% mortality (i.e., $LC_{50}$) will decrease with a rise in temperature (Table 8.1). However, this is not always the case, and one finds that toxicity and temperature are influenced by the toxicant, species, acclimation conditions, and other factors. For example, it is well known that pesticides such as DDT and

**Table 8.1 Summary of the Effects of Temperature on Chemical Toxicity (i.e., Lethality) in Aquatic Organisms**

| *Species* | *Toxicant* | *Thermal Effect* |
|---|---|---|
| Trout | Ammonium ion | More toxic with warming |
| Fish | Cyanide | More toxic with warming |
| Fish | Copper, zinc, nickel | Variable |
| Trout | Mercury compounds | More toxic with warming |
| Trout, blue gill | Methoxychlor | More toxic with cooling |
| Trout, blue gill | DDT | More toxic with cooling |
| Fish, fathead minnow | Profenofos | More toxic with warming |
| Fish | Endrin | More toxic with warming |
| Fiddler crab | Mercuric chloride | More toxic with cooling |
| Trout, blue gill | Parathion, malathion | More toxic with warming |
| Trout | Phenols | More toxic with cooling |
| Tadpole | Carbaryl | More toxic with warming |
| Tadpole | Cadmium | More toxic with cooling |
| Crayfish | Fluvalinate (pyrethroid) | More toxic with warming |
| Crayfish | Cadmium | More toxic with warming |
| Paramecium | Sodium azide | More toxic with warming |

*Source*: Most of the data taken from review by Cairns et al. (1975) except for studies on paramecium (Malvin et al., 1994), fluvinate in crayfish (Sogorb et al., 1988), carbaryl in tadpole (Boone and Bridges, 1999), cadmium in tadpole and crayfish (see Figure 8.2), and profenofos in fathead minnows (Baer et al., 2002).

pyrethroids are generally more toxic at lower temperatures (see Chapter 4). In addition, there are instances where toxicants can be more toxic with increasing temperature for some species but less toxic for others. The 96-h $LC_{50}$ for cadmium in the crayfish decreases from 58.5 to 18.4 mg/l with an increase in temperature from 20 to 28°C, but the $LC_{50}$ for cadmium in the tadpole rises with an increase in water temperature from 20 to 25°C (Figure 8.2). The goldfish is a commonly used species in studies on the effects of toxicants on water quality. It is interesting to note that the sensitivity to toxic chemicals in the goldfish as a function of temperature is well defined as compared to other principal freshwater species (Smith and Heath, 1979).

## 8.3.1 Critical Thermal Maximum and Minimum

Lethality is a common end point in the study of the effects of toxicity and temperature in aquatic organisms. The lethal temperature of an organism that is exposed to or has a previous history of exposure to a toxicant can also be a valuable parameter. A change in an organism's lethal temperature induced by exposure to a toxicant provides a gross measure of its ability to inhabit and survive in seasonally stressful environmental conditions. However, assessing the point of death can be uncertain in many instances because it is difficult to define the exact point of time when the organism is dead. A thermal parameter is needed that defines in ecological terms a body temperature that is essentially lethal.

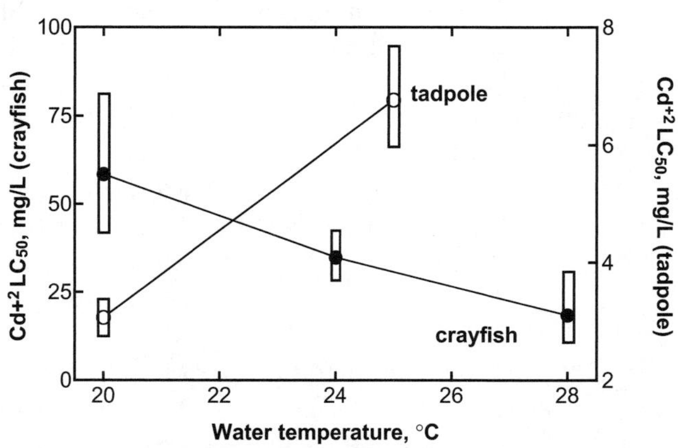


**Figure 8.2 Comparison of the differential sensitivity to cadmium in crayfish and tadpoles (prometamorphic stage) as a function of water temperature. Lethal concentration ($LC_{50}$) measured after 96 h of exposure and plotted as mean ±95% confidence interval. (Data from Del Ramo, J., Diaz-Mayans, J., Torreblanca, A., and Núñez, A. (1987). *Bull. Envrion. Contam. Toxicol.* 38: 736–741; Ferrari, L., Salibián, A., and Muiño, C. (1993). *Bull. Environ. Contam. Toxicol.* 50: 212–218.)**

To this end, Cowles and Bogert (1944), studying the thermal tolerance of reptiles, developed the term *critical thermal maximum* (CTMax), defined as "the thermal point at which locomotor activity becomes disorganized and the animal loses its ability to escape from conditions that will promptly lead to its death." The point of disorganized locomotor activity is usually defined as the body or ambient temperature where there is a loss of a righting reflex in terrestrial vertebrates or loss of equilibrium in aquatic vertebrates. Determining the lower limits of thermal tolerance is hampered by the uncertainty of hypothermic death. Heart contractions and breathing may progress for hours at low temperatures while the organism is essentially immobile and vulnerable to predation. Hence, the counterpart of the CTMax, the critical thermal minimum (CTMin), is used to define the point of ecological death in a cold environment, typically defined as the temperature of loss of a righting reflex or equilibrium. An increase in the CTMin and decrease in the CTMax translates to a smaller area of the temperature tolerance polygon (see Figure 8.1B) and reduced ability to adapt to stressful environments.

Studies have shown that some pesticides and other toxicants lower the CTMax and increase the CTMin (Table 8.2). Such a response could have an ecological impact by limiting the distribution of fish into certain habitats. For example, red shiners acclimated to 23 or 30°C and then exposed to atrazine for 14 days show a significant reduction in their CTMax (Figure 8.3). In this study, the fish were placed in a beaker of water that was heated at a rate of 1°C/min. The temperature at which the fish lost equilibrium and became motionless was defined as the CTMax (Messaad et al., 2000). Control fish acclimated to 30°C had a CTMax that is 1.8°C higher than that of fish acclimated to 23°C. CTMax decreased with increasing dose of atrazine, with fish acclimated to 30°C showing markedly greater sensitivity. A similar pattern was seen with the insecticide terbufos as well as with mixtures of atrazine and terbufos. Fathead minnow larvae exposed to high concentrations of carbofuran (50% of the $LC_{50}$) exhibited a reduction in the CTMax but no change in the CTMin, an effect that persisted for 10 days after the 4-day exposure period (Heath et al., 1997). Muskellunge fry exposed to sublethal concentrations of arsenic also showed a decrease in their CTMax (Paladino and Spotila, 1978). Fathead minnows exposed to the pyrethroid cyfluthrin for 96 h sustained a decrease in their CTMax and an increase in their CTMin (Heath et al., 1994). DDT exposure was found to raise the lower lethal temperature but had no effect on the upper lethal temperature of Atlantic salmon (Anderson, 1971).

**Table 8.2 Summary of Effects of Pesticides and Other Environmental Contaminants on the Critical Thermal Maximum (CTMax) and Minimum (CTMin) in Fish**

| *Species* | *Toxicant* | *CTMax* | *CTMin* | *Reference* |
|---|---|---|---|---|
| Stoneroller minnow (*Campostoma anomalum*) | Phenol | Decrease | — | Chagnon and Hlohowskyj (1989) |
| Trout (*Salmo gairdneri*) | Nickel | Decrease | — | Becker and Wolford (1980) |
| Muskellunge fry (*Esox masquinongy*) | Arsenic | Decrease | — | Paladino and Spotila (1978) |
| Channel catfish (*Ictalurus punctatus*) | Nitrite | Decrease | — | Watenpaugh and Beitinger (1985) |
| Fathead minnow (*Pimephales promelas*) | Cyfluthrin (pyrethroid) | Decrease | Increase | Heath et al. (1994) |
| Fathead minnow | Carbofuran | Decrease | No change | Heath et al. (1997) |
| Red shiner (*Cyprinella lutrensis*) | Atrazine | Decrease | — | Messaad et al. (2000) (see Figure 8.3) |

## 8.4 FISH BEHAVIORAL THERMOREGULATION

Thermoregulatory behavior is clearly the most critical mechanism fish possess to avoid harmful thermal environments. Other than tuna, which are capable of regulating body temperature by endothermy, all fish regulate their body temperature solely through behavior. Thermoregulatory behavior in fish has a major effect on their natural distribution, feeding, reproduction, development, and growth. Temperature preference behavior can determine fish distribution, and it has been argued that an alteration in thermoregulatory behavior brought on from a toxicant could disorient and alter the distribution of fish in their natural habitats (Domanik and Zar, 1978). Hence, the presence of pollutants in water and bioaccumulation of toxicants through the food chain could have profound effects on the biology of fish species if these toxicants affect thermoregulatory behavior (Domanik and Zar, 1978).

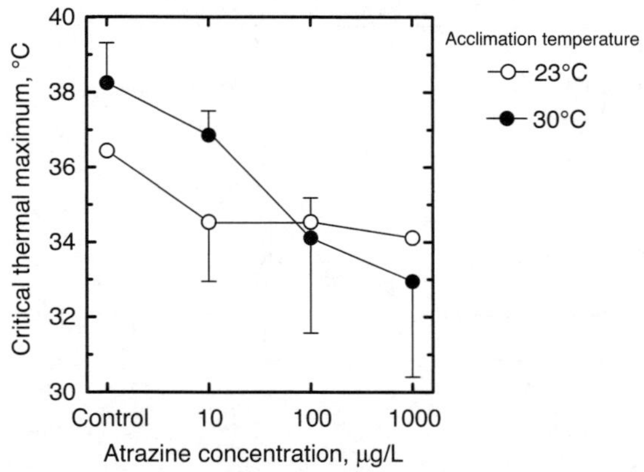


**Figure 8.3 Effect of 14-day exposures to atrazine at two acclimation temperatures on the critical thermal maximum of red shiners (*Cyprinella lutrensis*). (Data from Messaad, I.A., Peters, E.J., and Young, L. (2000). *Bull. Environ. Contam. Toxicol.* 64: 748–754.)**

The behavioral response of fish placed in a horizontal or vertical temperature gradient has been used to assess the effects of various pesticides and toxic chemicals on thermoregulation. The typical temperature gradient is a relatively simple device that is made up of a long, shallow trough with fresh water that has heaters and chillers placed at appropriate distances to maintain a gradient of temperatures (Figure 8.4). Water depth in the gradient has to be maintained shallow to avoid thermal stratification. Many of the older studies utilized an observer to record the position of the fish in the gradient at various time intervals. A mirror was positioned such that the fish could be watched without the observer being detected by the fish. Modernization of gradient systems with automated monitoring of the position of the fish in the gradient has improved the accuracy and efficiency for performing these types of studies (e.g., see Crawshaw et al., 1989). Measuring the upper and lower escape temperatures can also be utilized to assess effects of toxicants on thermoregulatory behavior. In such a system, the water is warmed or cooled, and the point at which the fish exits into a more neutral water temperature is determined (Green and Lomax, 1976). The temperature gradient is ideal for studies in fish and other species because it represents a paradigm of natural thermotropic behavior and the animal can move to its optimal thermal environment and remain there without having to perform a motor or operant task. A toxicant or drug that impairs motor activity could affect the performance of an animal in an operant system, thus altering the true thermoregulatory behavior (see Gordon, 1993).

A plethora of studies in the 1960s and 1970s showed that thermoregulatory behavior of fish was affected by a variety of toxic chemicals,

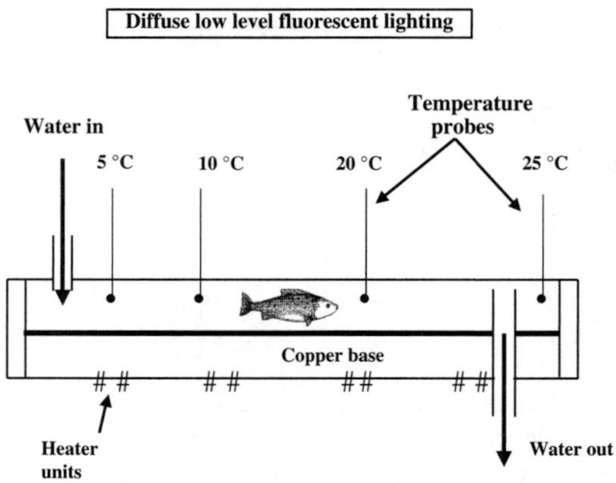


**Figure 8.4 Diagram of a temperature gradient used to measure selected water temperature of fish. Fish are placed in an aquarium for 24 h while exposed to a toxicant and then housed in a gradient for periods of several hours while their position in the gradient is monitored regularly. (Drawing modified from Gardner, D.R. (1973). *Pest. Biochem. Physiol.* 2: 437–446.)**

including DDT and other organochlorines, anti-ChE insecticides, metals, and alcohols (Anderson, 1971; Peterson, 1976; Table 8.3). Organophosphate insecticides were found to elicit a reduction in selected temperature of young fish when they were exposed to the toxicant for 24 h and then placed in a temperature gradient for observation (Figure 8.5). The organophosphate insecticides chlorpyrifos (0.1 ppm) and guthion (0.002 ppm) were very effective at lowering the selected temperature. However, exposure to the carbamate carbaryl up to a dose of 1.0 ppm had no effect on selected temperature (Peterson, 1976). Not all xenobiotics elicit an effect on selected temperature, even when administered at very toxic doses close to the $LC_{50}$. For example, some metals such as copper sulfate were effective at lowering selected temperature, whereas others such as zinc sulfate and cadmium sulfate had little effect on selected temperature of young Atlantic salmon (Table 8.3).

Acclimation temperature and age are critical factors in the study of the effects of pesticides on behavioral thermoregulation in fish. Temperature acclimation leads to marked biochemical and physiological alterations that affect behavior, metabolism, and the potential uptake and toxicity of aquatic toxicants in aquatic organisms (Prosser, 1986; Cossins and Bowler, 1987). The effects of malathion on the thermoregulatory behavior of the common shiner were shown to be dependent on acclimation temperature and age (Figure 8.6). Fish were exposed to the pesticide for 24 h and

**Table 8.3 Survey of the Effects of Various Toxicants on the Selected Water Temperature of Fish When Placed in a Temperature Gradient and Effects of Temperature on Chemical Lethality Expressed as the $Q_{tox}$. Concentrations of Toxicants Represent Maximal Effects on Selected Temperature**

| *Chemical* | *Concentration (ppm)* | *Δ Sel. temp (°C) (species)* | *$Q_{tox}$ (species)* |
|---|---|---|---|
| *pp* DDT | 0.03–0.05 | +4–5 (*Salmo salar*) | 0.7 (*Salmo gairdneri*) |
| Methoxychlor | 0.04–0.05 | +4–5 (*Salmo salar*) | 0.7 (*Salmo gairdneri*) |
| Aldrin | 0.05–0.15 | –3 (*Salmo salar*) | 2.6 (*Salmo gairdneri*) |
| Dieldrin | 0.1–0.025 | –1 (N.S.) (*Salmo salar*) | 3.6 (*Salmo gairdneri*) |
| Heptachlor | 0.005–0.25 | ~–1.0 (*Salmo salar*) | 1.3 (*Salmo gairdneri*) |
| Lindane | 0.01–0.03 | +1 (*Salmo salar*) | 1.2 (*Lepomis macrochirus*) |
| Sodium pentachlorphenol | 0.05–0.10 | –5 (*Salmo salar*) | 0.8 (*Salmo salar*) |
| Azinphosphorethyl | 0.002–0.003 | –4 (*Salmo salar*) | 1.8 (*Salmo gairdneri*) |
| Malathion | 0.025–0.050 | –2 (*Salmo salar*) | 0.8 (*Salmo gairdneri*) |
| Fenitrothion | 0.30–1.00 | –2 (N.S.)(*Salmo salar*) | 1.9 (*L. macrochirus*) |
| Chlorpyrifos | 0.10–0.25 | –4 (*Salmo salar*) | 7.9 (*Salmo gairdneri*) |
| Naled | 0.15–0.30 | –4 (*Salmo salar*) | 4.6 (*Salmo gairdneri*) |
| $CdSO_4$ | 0.002 | –1 (*Salmo salar*) | — |
| $CuSO_4$ | 0.015–0.03 | –2 (*Salmo salar*) | — |
| $ZnSO_4$ | 0.4 | 0 (*Salmo salar*) | 1.9 (*Salmo gairdneri*) |
| Potassium cyanide | 0.02–0.04 | –5 (*Salmo salar*) | 3.8 (*Gillichthys*) |
| Phenol | 10 | –3 (*Salvelinus fontinalis*) | 0.4 (*Salmo gairdneri*) |

*Source*: Data from Peterson, R.H. (1976). *J. Fish. Res. Board Can.* 33: 1722–1730.

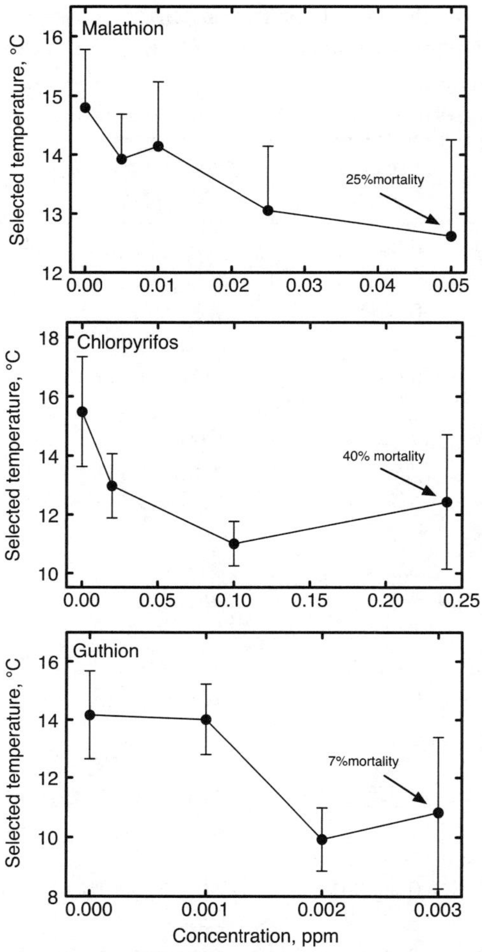


**Figure 8.5 Dose–response relationship for three organophosphate insecticides on selected water temperatures of juvenile Atlantic salmon. Salmon were exposed for 24 h to insecticides and then placed in the temperature gradient. (Data from Peterson, R.H. (1976). *J. Fish. Res. Board Can.* 33: 1722–1730.)**

then placed in the temperature gradient while behavior was monitored for 20 min. Two-year-old fish were more sensitive than the young of the year and one-year-olds, displaying a decrease in their selected temperature at malathion doses of 0.25 and 1.0 μg/l. However, this effect was only seen in fish acclimated to 17°C and not in fish acclimated to 8°C (Domanik and Zar, 1978). The warmer acclimation temperature may be more effective in the manifestation of malathion toxicity, possibly by increasing the uptake and cellular toxicity. Temperature selection behavior of very young fish may be incompletely developed, thus hampering comparisons between young and mature fish.

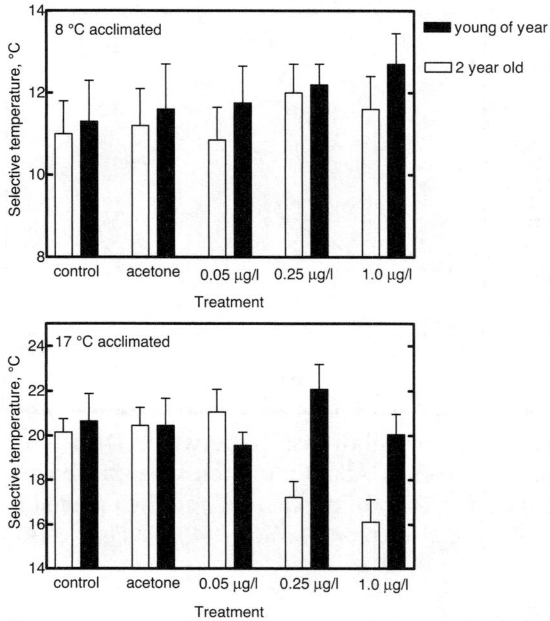


**Figure 8.6 Effect of acclimation to 8 or 17°C on the selected water temperature of common shiners when exposed to malathion for 24 h. (Data from Domanik, A.M. and Zar, J.H. (1978). *Arch. Environm. Contam. Toxicol.* 7: 193–206.)**

The use of DDT has been banned in most countries, but its heavy use and slow degradation and environmental persistence prompted extensive studies on DDT's toxic effects in fish and other aquatic fauna. DDT is an unusual neurotoxicant in that its toxicity increases with decreasing temperature (see Chapter 4). In general, the selected temperature of fish following exposure for 24 h to DDT displays a V-shaped function. In one of the first studies showing that fish behavioral temperature regulation was altered by a toxicant, Ogilvie and Anderson (1965) showed that young Atlantic salmon exposed to DDT for 24 h underwent a preference for slightly cooler temperatures at relatively low concentrations of DDT (5 to 10 ppb) and then selected temperature increased at higher concentrations up to 50 ppb (Figure 8.7). Fish acclimated to warmer temperatures were more sensitive to DDT, showing a preference for warmer temperatures at lower doses of DDT as acclimation temperature increased (also see Anderson, 1971; Javaid, 1972). Most of the fish studies have been performed in horizontal temperature gradients similar to the diagram in Figure 8.4. Vertical temperature gradients, which reflect the variation in temperature in nature in static water columns, have also been used to show that fish exposed to DDT and other chlorinated hydrocarbons prefer warmer temperatures near the water surface (Peterson, 1973).

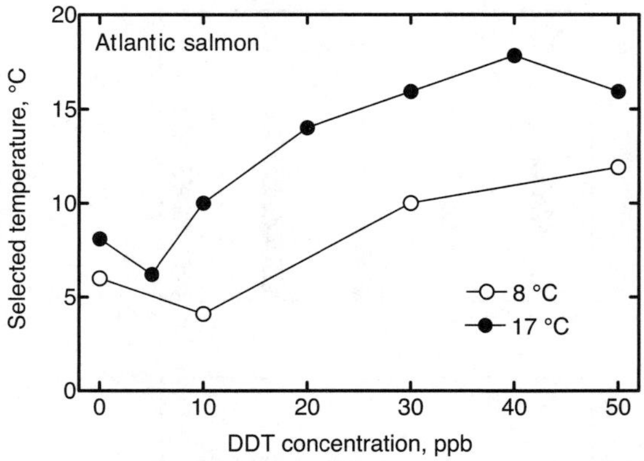


**Figure 8.7 Dose–response relationship between DDT concentration and selected temperature of young Atlantic salmon when acclimated to 8 or 17°C. Salmon were exposed for 24 h to insecticides and then placed in gradient. (Data from Ogilvie, D.M. and Anderson, J.M. (1965). *J. Fish. Res. Board Can.* 22: 503–512.)**

The effects of DDT on behavioral temperature regulation have been shown to be long-lasting. For example, very young brook trout exposed to varying concentrations of DDT for 24 h show a preference for cooler temperatures at a maximal dose of 20 ppb and then prefer warmer temperatures with increasing doses of DDT above 20 ppb (Figure 8.8A). A similar V-shaped dose–response curve was observed for other organochlorine compounds that are structurally similar to DDT (Gardner, 1973). The effects of DDT and other organochlorines on temperature preference appear to be very long-lasting following a single exposure. For example, after brook trout were exposed to 20 ppb DDT for 24 h, they preferred cooler temperatures for at least 9 days in spite of being maintained in clean water (Figure 8.8B). A similar pattern was observed with fish exposed to 20 ppb methoxychlor, although recovery was more rapid, occurring within 5 days after exposure (Gardner, 1973).

## *8.4.1 Endogenous Ethanol and Hypothermia*

The goldfish (*Carassius auratus*) is frequently used in behavioral thermoregulatory studies. When placed in a temperature gradient, goldfish select a temperature of approximately 25 to 27°C (Figure 8.9). When the water in the gradient is replaced with a 1% ethanol solution, the goldfish undergo a prompt 2 to 3°C reduction in selected temperature (O'Connor et al., 1988). Switching between clean water and ethanol in the temperature gradient demonstrates a rapid onset and recovery of selected temperature

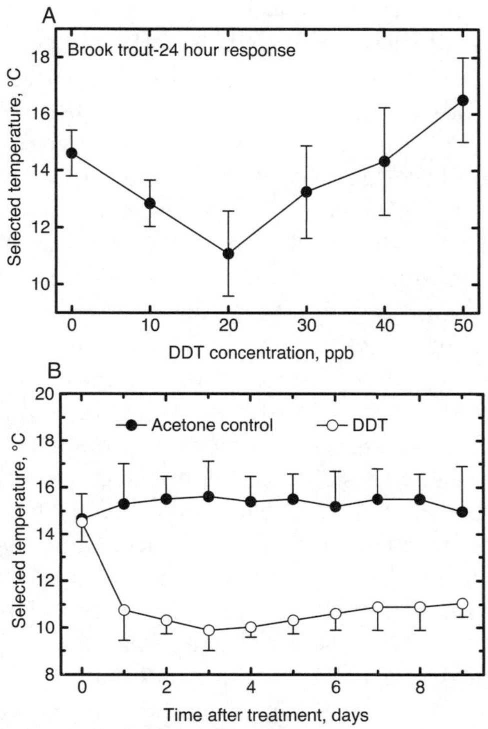


**Figure 8.8 Acute (A) and prolonged (B) effects of exposure to DDT on the selected temperature of brook trout. (Data from Gardner, D.R. (1973). *Pest. Biochem. Physiol.* 2: 437–446.)**

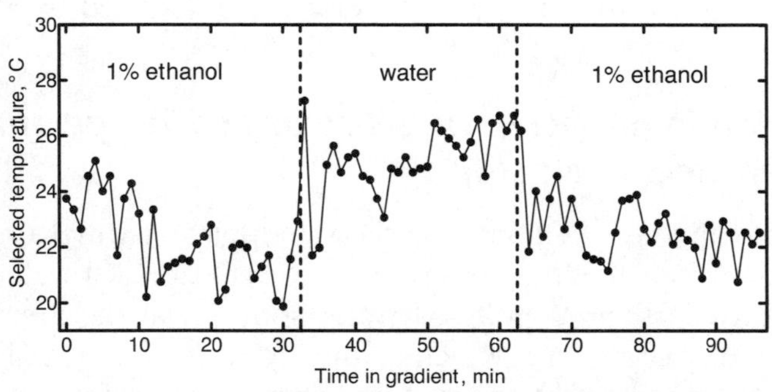


**Figure 8.9 Time-course of selected temperature of goldfish ($N$ = 6) placed in a temperature gradient and exposed to alternating concentrations of 1% ethanol, water, and 1% ethanol. (Data from O'Connor, C.S., Crawshaw, L.I., Bedichek, R.C., and Crabbe, J.C. (1988). *Pharm. Biochem. Behav.* 29: 243–248.)**

in response to a toxic insult. It is interesting to note that ethanol is an effective hypothermic agent in mammals and has been shown to elicit a regulated hypothermic response in mice and rats (see Chapter 3).

Goldfish and other cyprinids possess a unique ability to withstand prolonged periods of anoxia by converting lactic acid to ethanol. Lactic acid, the normal end product of metabolism during anoxia, does not easily cross the respiratory epithelium. Ethanol is extremely permeable and is easily excreted. Hence, the evolution of biochemical pathways for converting lactic acid to ethanol appears to be an ideal means of maintaining metabolism during anoxia without offsetting the acid–base balance (Shoubridge and Hochachka, 1980). Crawshaw et al. (1989) found that microinjecting ethanol into the anterior portion of the nucleus preopticus periventricularis (NPP) of goldfish housed in a temperature gradient resulted in a marked preference for colder temperatures. The goldfish was found to be extremely sensitive to ethanol and responded behaviorally to tissue levels that are orders of magnitude below that needed to lower body temperature in mammals. It was speculated that ethanol is an endogenous neuromodulating agent that controls thermoregulatory reflexes during anoxia. That is, the production of ethanol during anoxia stimulated the NPP to lower selected temperature, thereby improving the ability of the goldfish to survive anoxia (Crawshaw et al. 1989). However, it was later shown that there was no correlation between the levels of ethanol in the CNS and selected temperature (Rausch et al. 2000). That is, the selected temperature of the goldfish decreased prior to a detectible increase in ethanol during anoxia. Nonetheless, in view of the link between acute toxicity and hypoxia in aquatic organisms, it would be of interest to determine what, if any, role endogenously produced substances such as ethanol serve in the manifestation of toxicant-induced hypothermia.

### 8.4.2 Relationship between Behavior and Temperature-Dependent Lethality

Fish are in a unique position from the standpoint that their thermoregulatory behavior can alter the dosage and toxic efficacy of a water pollutant. The location of fish in a polluted water source will be affected by the nature of the toxicant (i.e., whether the fish can detect and avoid the toxicant) and the distribution of temperatures within the water column. That is, the behavior of the fish to move toward or away from the source of a toxicant could be affected by its thermoregulatory behavior. It is also of interest to note that fish develop a fever in response to an infection and use their thermoregulatory behavior to elevate body temperature when administered a variety of febrile agents (Reynolds and Casterlin,

1982). Considering that many toxicants are inflammatory and may lower resistance to infection, the potential febrile response combined with the effects of toxicants on thermoregulatory behavior could be profound.

Because the toxicity of many of the chemicals discussed here and in other chapters (see Chapter 4) is directly dependent on temperature, the thermoregulatory behavior elicited by the toxicant will affect survival. The greater the relative dependence of toxic-induced lethality on water temperature, then the more likely one would expect the toxicant to evoke a change in thermoregulatory behavior. For example, if the lethal concentration for 50% mortality ($LC_{50}$) decreases sharply with a rise in water temperature, then the fish should select cooler temperatures when exposed to the toxicant. This interaction between lethal toxicity and behavioral thermoregulation was analyzed by Peterson (1976). He developed a relationship between temperature-dependent lethality and thermoregulatory behavior using salmonids primarily and a few other species (Table 8.3). The temperature-dependent lethality was expressed using an equation akin to that of the $Q_{10}$ (see Chapter 4), but to avoid confusion it is termed the $Q_{tox}$:

$$Q_{tox} = (LC_{50} \text{ at } T_2/LC_{50} \text{ at } T_1) \exp (10/T_1 - T_2) \quad (8.1)$$

where $T_1 > T_2$; $LC_{50}$ = 24-h concentration to achieve 50% lethality; T = temperature in °C. A $Q_{tox}$ of 2 means that the $LC_{50}$ is halved for a 10°C elevation in temperature. A $Q_{tox}$ of less than 1.0 means the $LC_{50}$ decreases with a reduction in temperature. The $Q_{tox}$ parameter will be affected by a variety of biological and environmental factors; acclimation temperature and species are expected to have a marked effect (Peterson, 1976). For the majority of chemicals and toxicants, $Q_{tox}$ is greater than 1.0 because, as metabolic rate increases with rising temperature, there is greater uptake as well as increased toxicity of the chemical on cells, tissues, and organs.

Just as ambient temperature affects the toxicity of nearly all chemicals in mammals (Chapter 4), water temperature has even more profound effects on toxicity because the body temperature of the fish is essentially equal to water temperature. The aquatic organism's survival when exposed to a toxicant will be dependent on (*a*) the availability of thermal gradients, (*b*) its behavioral thermoregulatory response in the water column, and (*c*) the toxicant's $Q_{tox}$ value. Peterson (1976) first developed a plot to show the intertwining between thermoregulatory behavior and $Q_{tox}$ (Figure 8.10A). The behavior–lethality responses should essentially fall into either quadrant I or III if the fish (or other aquatic organism) is displaying the adaptive thermoregulatory behavior to counter the toxicological effects. That is, for a toxicant with a $Q_{tox}$ >1.0, a decrease in selected temperature will lead to a reduction in toxicity, whereas for a toxicant with a $Q_{tox}$

<1.0, an increase in selected temperature will reduce toxicity. The neurotoxicity of DDT as well as pyrethroids generally increases with decreasing temperature in mammals, other vertebrates, and invertebrates (see Chapter 4). Indeed, DDT and methoxychlor are some of the few examples of toxicants that have a $Q_{tox}$ less than 1.0, and it has been shown that fish prefer warmer temperatures when subjected to relatively high concentrations of these toxicants (Figure 8.10B). A few toxicants such as lindane have a $Q_{tox}$ close to 1.0, and the $LC_{50}$ is slightly affected by temperature. Many pesticides, metals, and other toxicants have a $Q_{tox}$ >1.0, and fish display a preference for cooler temperatures, presumably a response to lower the toxicity.

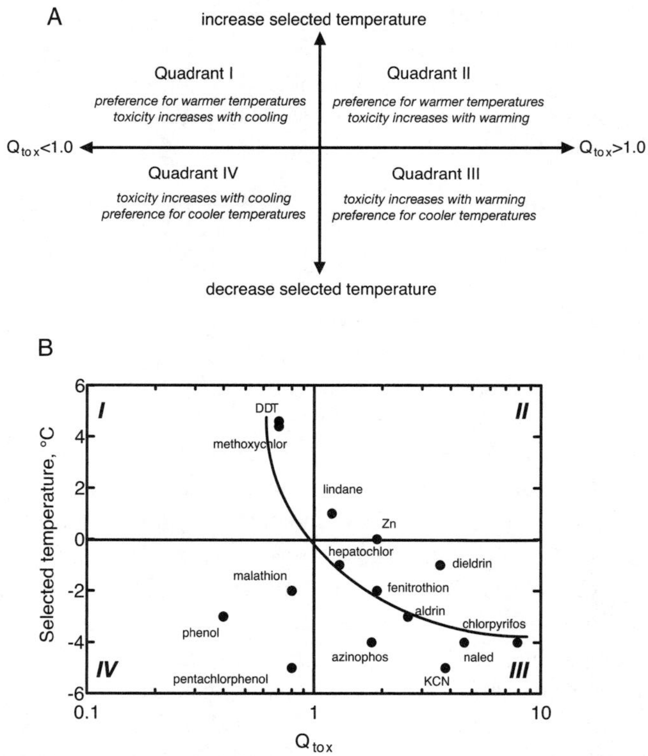


**Figure 8.10 (A) General scheme to show the interaction between temperature-induced lethality of a toxicant ($Q_{tox}$) and behavioral temperature regulation (i.e., selected temperature) of aquatic organisms. Regulatory responses to minimize toxicity fall into quadrants I and III. Thermoregulatory dysfunction falls into quadrants II and IV. (B) Plot of data from Table 8.3 showing the relationship between $Q_{tox}$ and selected temperature for a variety of species of fish. (Graph redrawn from Peterson, R.H. (1976). *J. Fish. Res. Board Can.* 33: 1722–1730.)**

That fish use behavior to attenuate the toxicity of chemicals in quadrants I and III illustrates the adaptability of the CNS control of body temperature. The fish CNS thermoregulatory centers are capable of responding to an incredibly wide range of toxic chemicals and activate the appropriate thermoeffector response to increase their likelihood of survival. However, any responses in quadrants II or IV would be indicative of a toxicant-induced thermoregulatory dysfunction. The ecotoxicological effects of a response in quadrants II or IV could be profound, as aquatic organisms incorrectly respond behaviorally to a pollutant and exacerbate the toxicity. A few of the chemicals in Figure 8.10B fall into quadrant IV, including malathion, pentachlorophenol, and phenol. Could this imply that these toxicants have induced thermoregulatory dysfunction? Such an interpretation should be viewed with caution until more studies are performed on these toxicants. This anomaly could be attributed to inconsistencies in acclimation temperature, exposure duration, and other factors. It is interesting to note that pentachlorophenol is an uncoupler that increases metabolism, while potassium cyanide blocks metabolism, yet they both elicit a reduction in selected temperature (Table 8.3). Another study showed that Atlantic salmon underwent a marked reduction in selected temperature following exposure to potassium cyanide, whereas the oxidative uncoupler dinitrophenol had no effect on behavioral temperature selection (Javaid, 1972).

The interaction between water temperature, toxicity, and behavior during long-term low-level exposures using end points other than lethality would be very important to assess. The relationship between $Q_{tox}$ and selected temperature has been established for relatively brief, acute exposures to toxicants. Lethality has been the primary endpoint of toxicity, but other benchmarks of toxicity that are manifested at lower toxicant concentrations could be used in the calculation of the $Q_{tox}$ variable. In most of the behavioral thermoregulatory studies in fish, the observation period was relatively brief, lasting just a few hours. It would be of importance to assess the effects of toxicants on behavioral thermoregulation for subchronic and chronic toxicant exposures and also evaluate the time-course of recovery. Advances in monitoring behavioral thermoregulation and developments in telemetry devices for tracking and recording pertinent physiological data in laboratory and field studies, along with the development of more sensitive indicators of toxicity such as genomic markers, could be used to get a better understanding of the physiological and ecotoxicological effects of toxicants on aquatic organisms.

## 8.5 AMPHIBIANS

The viability of populations of frogs and other amphibians is considered a critical indicator of water quality (Boyer and Grue, 1995). The perceived worldwide decline in populations of frogs may be attributed to a decline in habitats. In addition, water pollution, extension of agricultural practices, and the resulting eutrophication may have contributed to the decline in amphibian populations. These anthropometric practices have led to warmer water temperatures and higher levels of ammonia and other toxicants in aquatic habitats. Although there are uncertainties in assessments at a global level, thermal stress and pollution could be contributing to the decline in amphibian populations.

Adult frogs and tadpole larvae utilize behavioral temperature preferences to regulate their body temperature (Whittow, 1970). Compared to the data base on fish, there is relatively little known about how toxicants affect thermoregulation in amphibians. Hypoxia leads to a regulated decrease in body temperature in essentially all species ranging from single-celled paramecia to mammals (see Wood, 1991). Since many of the effects of toxicants on aquatic organisms appear to be linked to limitations on respiration, one might expect that many toxicants would elicit regulated decreases in temperature as have been seen in fish and mammals (see Chapter 3). Because the toxicity of many pesticides and other toxicants in amphibians increases with body temperature (Table 8.1), one would hypothesize that amphibians would respond behaviorally to toxicants in a manner that would lessen the toxicity (i.e., selecting lower body temperatures).

Wood and Malvin (1991) first noted that toads responded to hypoxia by selecting lower body temperatures. This behavioral response was deemed beneficial because the selection of a lower temperature improved arterial oxygen saturation and minimized respiratory alkalosis (also see Figure 5.10). Pesticides or toxicants that impair tissue availability of oxygen should exert thermoregulatory effects similar to those of hypoxia. In the toad (*Bufo marinus*), subcutaneous injection of sodium azide or cyanide, which are potent inhibitors of oxidative phosphorylation, led to a rapid reduction in selected temperature (Branco and Malvin, 1996). For example, administration of 0.6 mmol/kg sodium cyanide led to more than an 8°C decrease in body temperature within 1 h, and the decrease in body temperature persisted for at least 24 h. Interestingly, intracerebroventricular injection of cyanide or azide at doses that were 1/1000 that of the subcutaneous injection led to similar reductions in body temperature. This would suggest that systemic exposure to these toxic agents elicits a thermoregulatory response by exerting effects directly on CNS control mechanisms (Branco and Malvin, 1996).

Comparative physiologists and toxicologists are faced with the challenge of designing appropriate experiments to determine if thermal stress and environmental toxicants are contributing to the decline in amphibian populations. Laboratory studies traditionally focus on how a static or stable thermal stress affects toxicity. While it is clear that warmer temperatures alone can antagonize the toxicity of many chemicals, organisms in the field are subjected to a dynamic thermal environment. Fluctuating temperature can represent a significant stressor in toxicological exposures. This is very pertinent in amphibian studies because eggs are deposited and larvae develop in shallow waters that experience daily temperature fluctuations. To address this issue, Broomhall (2002) used a novel approach of exposing tadpoles from the Australian frog *Litoria citropa* to endosulfan, an organochlorine cyclodiene insecticide. The tadpoles were exposed to the insecticide while maintained in a stable thermal environment with a mean temperature of 20°C with a range of ±2°C or a variable temperature of 21°C with a range of ±7.5°C where temperature oscillated with a daily rhythm reflecting natural variations. Although the average temperatures were similar, the 96-h exposure to 0.8 μg/l endosulfan led to significantly greater mortality in tadpoles exposed to the variable water temperature (Figure 8.11A). Moreover, when tested for their ability to evade predation from odonates 24 days after endosulfan exposure, tadpoles exposed to the variable temperatures and the pesticide were significantly more susceptible to predation (i.e., shorter time to capture; Figure 8.11B). Thus, the variable temperature stress, rarely used in laboratory studies, is an ecotoxicologically relevant factor that has been shown to potentially reduce overall fitness by raising vulnerability to predation.

## 8.6 INSECTS

Toxicity of essentially all insecticides exhibits a dependence on temperature. The interaction between environmental temperature and toxicity of pesticides in insects was noted as early as the 1930s. In fact, the concept of a negative temperature coefficient for the toxicity of an insecticide was apparently first coined in the studies of DDT toxicity in insects in the 1940s (for review see Scott, 1995, and Chapter 4). Temperature can affect pesticide toxicity by altering the dispersion, penetration, metabolism, distribution, and target site interactions in insects (Scott, 1995). In general, the dispersion and penetration of the toxicant are directly proportional to ambient temperature, but the metabolism and target site interaction may show positive or negative temperature coefficients. This can make the overall interactions between temperature and toxicity in insects rather complex and difficult to interpret and predict. Less understood are the

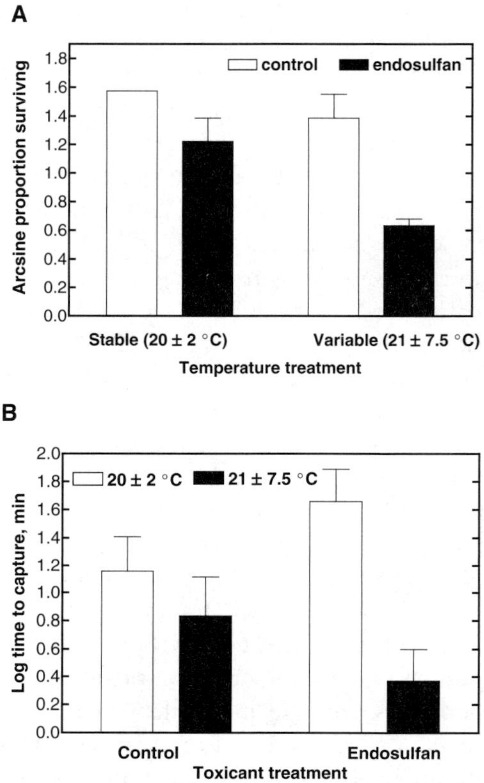


**Figure 8.11 Effects of previous exposure to static or varying temperature while exposed to 0.8 μg/l endosulfan on the 96-h lethality (A) and time to capture by an odonate predator in *Litoria citropa* tadpoles when tested 24 days after endosulfan exposure (B). (Data from Broomhall, S. (2002). *Aquat. Toxicol.* 61: 243–250.)**

interactions between thermoregulatory response and insecticide toxicity. This is important because the thermoregulatory responses can influence the overall toxicity of the insecticides. In addition, assessing the thermoregulatory response should lead to a better understanding of the insecticide's mechanism of action. In cases where some insect species are used as models in health risk assessment of toxic agents, it behooves one to understand the mechanisms of the toxicants on thermoregulation.

The honeybee *(Apis mellifera)* is one of many species of insects that use endothermic mechanisms to regulate internal temperature. Vandame and Belzunces (1998) utilized the endothermic characteristics of the honeybee to study the thermoregulatory effects of deltamethrin and azole fungicides. Internal temperature was estimated by measuring surface thorax temperature with infrared thermography while bees housed in groups of 10 at an ambient temperature of 22°C were sprayed with the insecticides (Figure 8.12). Endothermy was evident by the fact that control bees

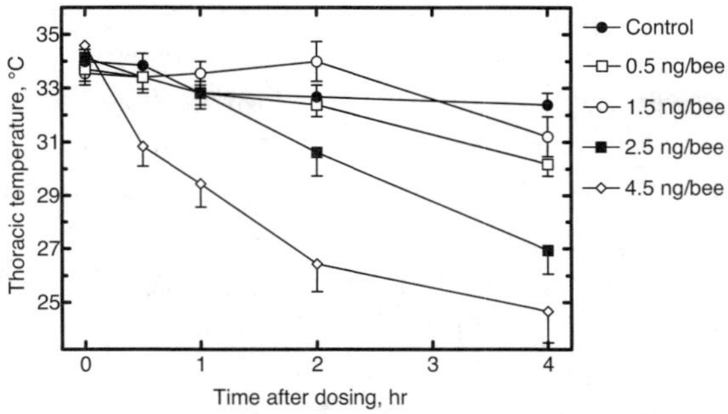


**Figure 8.12 Time-course of thoracic temperature of honeybees sprayed with sublethal doses of deltamethrin while maintained at an ambient temperature of 22°C. (Data from Vandame, R. and Belzunces, L.P. (1998).** ***Neurosci. Lett.*** **251: 57–60.)**

maintained a thoracic temperature that was approximately 10°C above ambient temperature. Deltamethrin, a type II pyrethroid, led to a significant decrease in temperature within 30 min following a dose of 4.5 ng per bee. This dose of deltamethrin is approximately 15 to 20 times lower than that seen in typical field treatments. Exposure to the azole fungicides prochloraz and difenoconazole also lowered thoracic temperature and appeared to interact synergistically when given in a mixture with deltamethrin. The mechanism of action of the hypothermic effect of these pesticides is not clear, but they may interfere with energy availability or directly affect neural mechanisms of thermoregulation. In a survey of investigations on temperature and toxicity in insects, pyrethroids surpassed all other classes of insecticides by exhibiting a negative temperature coefficient (Scott, 1995). Hence, the hypothermic response of bees to the pyrethroid is expected to accentuate toxicity; however, it should be noted that there were no behavioral options to thermoregulation under the laboratory test conditions.

When given the opportunity to move about in a temperature gradient, insects are capable of mounting a thermoregulatory response to some insecticides. Tegowska et al. (2002) determined the effects of two pesticides, indoxycarb, which closes sodium channels, and beta-cylfutryne, a pyrethoid that opens sodium channels, on behavioral thermoregulation in the locust. Locusts dosed with the pyrethroid eventually selected warmer temperatures in the gradient, a behavior that improved their survival when exposed to an insecticide that has a negative temperature coefficient (also see Table 4.2). Indoxycarb led to a marked preference for cooler temperatures, but this behavior did not appear to improve survivability. Behavioral thermoregulation of the cockroach (*Periplaneta americana*) is altered following exposure to an oxadiazine insecticide that blocks sodium channels (Grajpel et al., 2002). The preferred temperature of the cockroach in a temperature gradient decreased from ~25°C to ~20°C following exposure to the insecticide. This behavioral thermoregulatory effect persisted for several hours after exposure. Overall, it appears that some insects are capable of using behavioral thermoregulation to select an appropriate environment to enhance their survival to some insecticides. It would be of future interest to develop a relationship for behavioral temperature selection of agriculturally relevant insects and lethality for a wide selection of insecticides akin to the graphs presented in Figure 8.10. Such research may lead to more effective use of insecticides.

## 8.7 UNICELLULAR ORGANISMS

In a chapter on comparative physiology, this author would be remiss if he did not describe the novel thermotactic responses of the paramecium when subjected to toxic insults. It has been known for over 100 years that these unicellular animals will congregate in a temperature zone of 24 to 29°C when housed in a *micro* temperature gradient (for review, see Malvin et al., 1994). Without a nervous system, these organisms nonetheless exhibit swimming patterns that result in a zone of thermal preference. Moreover, without receptors these organisms are capable of responding to a potentially lethal insult to improve chances of survival. For example, paramecia placed in a temperature gradient and then subjected to hypoxia will move to a cooler area of the gradient. This response affords marked protection from a lack of oxygen (Figure 8.13A). If maintained at their normal preferred temperature of 29°C and subjected to the same hypoxic insult, the paramecia die quickly when compared to those that are maintained at the cooler temperatures (Malvin and Wood, 1992; Malvin, 1998). Again, with no receptors, these organisms nonetheless respond to a potentially lethal insult and improve survival.

Similar to the responses in toads described earlier in this chapter, exposing paramecia to inhibitors of oxidative phosphorylation led to a reduction in selected temperature (Figure 8.13B). Selected temperature decreased by over 10°C in paramecium exposed to 10 mM sodium azide. Since the toxicity to sodium azide in paramecia increases with temperature, the decrease in selected temperature is, like the responses of the multicellular organisms, an apparent adaptive response for survival. Interestingly, the oxidative uncoupler 2,4-dinitrophenol also elicits a decrease in the selected temperature of the paramecium. A similar response to select cooler ambient and body temperatures has also been noted in some species of fish and mammals (see Table 3.9 and Section 8.4). The thermotactic behavior of unicellular organisms such as paramecia could be a useful model to study certain types of toxicants, especially chemicals that target cellular respiration.

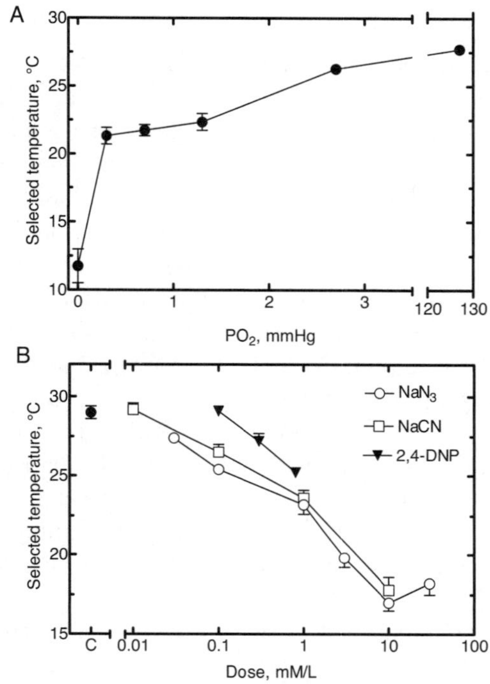


**Figure 8.13 (A) Thermotactic behavior of *Paramecium caudatum* when placed in a micro temperature gradient and subjected to varying levels of hypoxia. (B) Effect of chemical toxicants on temperature preference of paramecia. (Data from Malvin, G.M., Havlen, P., and Baldwin, C. (1994). *Am. J. Physiol.* 267: R349–R352; Malvin, G.M. (1998). *Clin. Exp. Pharmacol. Toxicol.* 25: 165–169.)**

# 8.8 RESPONSES TO WILDLIFE

Free-living birds and mammals are subject to exposure to pesticides and other environmental toxicants. Because they occupy the highest trophic levels of the food web, marine birds and mammals are especially susceptible to the bioaccumulation of PCBs and other organochlorine residues because they feed on fish that have accumulated these toxicants in their tissues (Hop et al., 2002). Aquatic poikilothermic invertebrate and vertebrate species also accumulate organochlorines in their tissues but at a much slower rate than homeothermic species because of their reduced metabolic requirements. There is essentially nothing known about how the tissue accumulation of these toxicants affects the ability of wildlife to thermoregulate. All information on this topic comes from acute or chronic dosing studies of pesticides to wildlife species as will be discussed below.

### 8.8.1 Birds

Nontarget wildlife species, especially birds, are apt to be exposed to a variety of insecticides and herbicides used in agricultural applications (for review, see Grue et al., 1997). Furthermore, there is ample opportunity for wild birds to be exposed to these toxicants during periods of severe heat and cold stress. Additional factors such as food and water availability, reproductive status, and age are also important variables that affect the impact of chemical contaminants on avian thermoregulation. Indeed, in a detailed review of the effects of anticholinesterase (anti-ChE) insecticides on wildlife, Grue et al. (1991) concluded that decreased food intake, hypothermia, and dysregulation in levels of reproductive hormones had the greatest impact on the survival and reproductive fitness of wildlife. The majority of conventional laboratory models of toxicology have been developed to study laboratory rodents and nonhuman primates. A challenge to the ecotoxicologist is to develop appropriate behavioral and physiological tests to determine if environmental pollutants have an impact on the growth, reproduction, and survival of wildlife (Burger and Gochfeld, 2000).

The effects and mechanisms of action of pesticides and other toxicants on bird thermoregulation have not been studied as extensively as in mammals. Like mammals, birds show hypothermic responses to pesticides, but the decrease in core temperature is relatively mild (Rattner and Franson, 1984; Brunet and McDuff, 1997). American kestrels (*Falco sparverius*) dosed with methyl parathion displayed a peak hypothermic response at approximately 2 h after dosing (Figure 8.14A). It is interesting to note that the hypothermic response to methyl parathion is similar at ambient temperatures of 22 and –5°C. In a laboratory rodent, the hypothermic response would be notably exacerbated with a 27°C reduction in ambient temperature (see Chapter 3). The inhibition in plasma cholinesterase (ChE) activity was highly correlated with a reduction in cloacal temperature (Figure 8.14B). As in most rodent studies, plasma ChE must be inhibited by approximately 50% before there is a reduction in body temperature. The hypothermic response to the organophosphate dimethoate in brown-headed cow birds (*Molothrus ater*) housed at an ambient temperature of 25°C peaked at approximately 4 h after oral dosing with 15.8 mg/kg (half the $LD_{50}$). The hypothermic response was slightly aggravated with exposures to an ambient temperature of 5°C, but the recovery was remarkably similar to that of birds exposed to the organophosphate at 25°C. Recovery in brain ChE activity appeared to be exacerbated in cold-exposed birds. The overall effects of inhibition in brain ChE activity and viability and survival of wildlife to anti-ChE pesticides have been thoroughly reviewed (Grue et al., 1991).

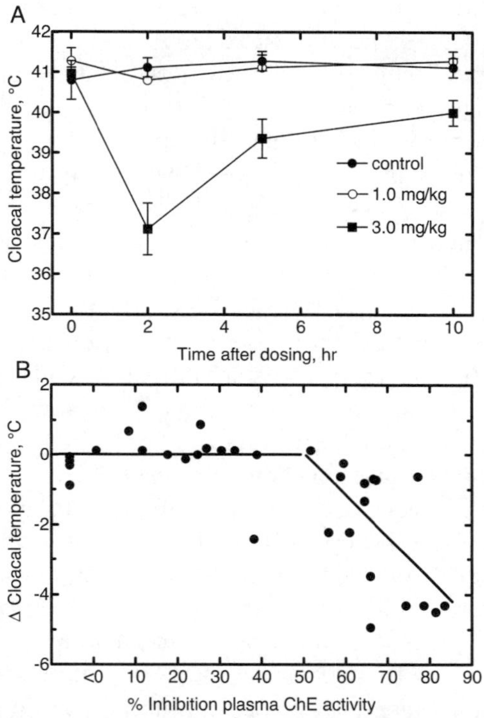


**Figure 8.14 (A) Time-course of cloacal temperature (i.e., core temperature) of American kestrels following exposure to methyl parathion by oral gavage when maintained at an ambient temperature of 22°C. (B) Correlation between inhibition in plasma cholinesterase (ChE) activity and change in cloacal temperature when measured 2 h after exposure to methyl parathion. (Data from Rattner, B.A. and Franson, J.C. (1984). *Can. J. Physiol. Pharmacol.* 62: 787–792.)**

Like the responses in mammals (Chapter 3), inhibition in cholinesterase enzymes by anti-ChE agents leads to an accumulation of acetylcholine and abnormal stimulation of cholinergic pathways. Pigeons injected intraventricularly with acetylcholine or carbachol (a cholinergic agonist) become hyperthermic, but administration of very high doses eventually leads to hypothermia (Chawla et al., 1975; Hillman et al., 1980). Thus, the limited understanding of the role of cholinergic pathways in temperature regulation of birds does not adequately explain their response to the anti-ChE insecticides. There is no evidence that anti-ChE agents evoke a hyperthermic response in birds, but this may be attributed to an inability to detect small elevations in body temperature, a critical issue in rodent studies that is discussed in detail in Chapter 6. Application of radiotelemetry to monitor core temperature in undisturbed birds could lead to a better understanding of the thermoregulatory effects of relatively low levels of anti-ChE insecticides and other toxicants.

The limited studies suggest that the hypothermic response to cholinesterase inhibitors in adult birds is not influenced as much by ambient temperature as one sees in rodents. The insulative quality of feathers in many avian species is likely to be very protective and attenuate effects of environmental cold stress on toxic-induced hypothermia. Juvenile birds with underdeveloped thermoregulatory capacity have been shown to be more sensitive to relatively mild reductions in ambient temperature when exposed to an organophosphate insecticide (Maguire and Williams, 1987). Brain ChE activity of 14-day-old bobwhites (*Colinus virginianus*) was inhibited to a greater degree in birds treated with chlorpyrifos at a temperature of 30°C as compared to 35°C. Thermoregulatory stability is well developed in adult birds, but environmental heat and cold stress can nonetheless influence toxicological responses. For example, Japanese quail (*Coturnix japonica*) were acclimated to ambient temperatures of 4, 26, or 37°C for 10 days and then evaluated for their sensitivity to parathion (Rattner et al., 1987). Chronic heat stress led to a significant 0.7°C increase in body temperature, while cold stress had no effect on body temperature. The parathion $LD_{50}$ was decreased by approximately 50% in heat stressed birds and 30% in cold stressed birds. However, cold exposure was apparently protective at sublethal doses of parathion as based on less inhibition in plasma and brain ChE activity and a marked elevation in liver paroxonase levels, an enzyme involved in metabolic deactivation of parathion. Chronic heat stress led to the greatest inhibition in plasma and brain ChE activity in birds exposed to parathion. These data illustrate how a toxicological effect can be exacerbated in the face of heat or cold stress in a homeotherm even when core temperature is unaffected.

Extrapolating laboratory studies to field conditions is a challenge. Burger and Gochfeld (2000) explored the importance of behavioral thermoregulation in young herring gulls and common terns exposed to lead. The young are generally brooded by the parents, but the presence of predators and other disturbances can force parents away from the nest, leaving the young susceptible to thermal stress. If pollutants have effects on behavioral thermoregulation, then there could be dire consequences for the developing chicks. Tests were performed on birds in their natural habitat, and it was shown that exposure to lead resulted in a deficit in shade-seeking behavior, and this persisted for several weeks after exposure (Burger and Gochfeld, 2000). These shore birds nest on open beaches in full sun, and the ability of young birds to behaviorally thermoregulate is critical for their survival. Future development of tests for thermoregulation in undisturbed wildlife species under natural conditions should lead to a better assessment of the potential hazards of environmental pollutants.

The thermoregulatory system of birds is unique from the standpoint that they have the highest core temperature among all homeothermic species. The internal body temperatures of many avian species is 41 to 42°C, which is 5 to 6°C above that of typical laboratory rodents. In view of their high body temperatures and metabolic rates, one should wonder if they would show greater sensitivity to insecticides and other toxicants. That is, a higher baseline core temperature might lead to greater susceptibility in terms of production of free radical formation, oxidative damage, and related processes; however, this has apparently not been investigated. Birds are often used in toxicological screening and mechanistic studies, and it is important to consider how their unique thermoregulatory characteristics, in particular, their high body temperature, might influence a toxic response.

## 8.8.2 Mammals

The endocrine system represents a critical facet of the thermoeffector systems for controlling heat production and heat loss. Man-made contaminants that either mimic or antagonize the function of endogenous hormones, termed endocrine disrupting chemicals (EDCs), have come under intense scrutiny in recent years. Much of the research has focused on the deleterious effects of EDCs on reproduction in wildlife, domestic species, and humans. It follows that EDCs may also affect thermoregulation.

There has also been considerable interest in studying the EDCs that interfere with thyroid function, including polychlorinated biphenyls (PCBs) and dioxins (see also Section 3.5.1). Many of these compounds are structurally similar to thyroxine ($T_4$) and have been shown to cause marked reductions in blood levels of $T_4$ and triiodothyronine ($T_3$) when administered to wildlife and other species (for review, see Brucker-Davis, 1998). The regulation of normal plasma levels of $T_3$ and $T_4$ is essential for the maintenance of basal metabolic processes and the activation of nonshivering thermogenic mechanisms during periods of acute cold stress. Prolonged cold exposure activates thermoreceptors leading to secretion of thyroid-releasing hormone and thyroid-stimulating hormone and the subsequent elevation in circulating thyroid hormones (Figure 8.15). Many EDCs appear to interact with the thyroid system at the level of control, synthesis, transport, and/or excretion of thyroid hormones, all of which can contribute to deficits in metabolic thermogenesis. Thyroidal EDCs have been shown to alter $T_4$ and $T_3$ levels by affecting the hypothalamo–pituitary axis, the synthesis and secretion of thyroid hormones, and the metabolism and excretion of thyroid hormones (Brucker-Davis, 1998).

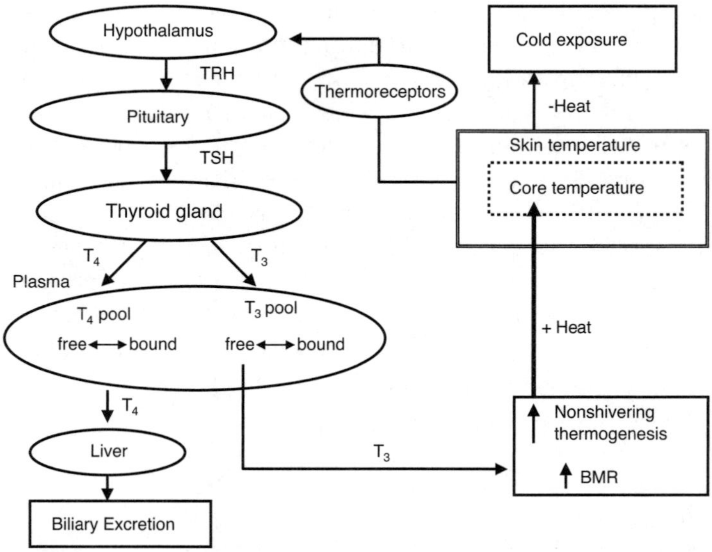


**Figure 8.15 Generalized scheme of how the hypothalamic–pituitary–thyroid axis is called upon during periods of prolonged cold stress to meet the increased demands for heat production. TRH, thyroid-releasing hormone; TSH, thyroid-stimulating hormone; BMR, basal metabolic rate. Modified from several sources (see Tomasi et al., 2001; Crofton, personal communication).**

Most EDC research is performed in laboratory species maintained under relatively nonstressful ambient conditions (i.e., environments where the thyroid axis is not activated above basal levels). However, wildlife should be particularly susceptible to EDCs that affect one or more aspects of the hypothalamic–pituitary–thyroid axis because they rely so heavily on increased thyroid function to maintain an elevated metabolic rate during the cold winter months. There are relatively few studies on thermoregulatory function of wild mammals exposed to EDCs. The cotton rat (*Sigmodon hispidus*) was exposed to vinclozolin, an organochlorine fungicide structurally similar to $T_4$, at a dose that would normally be encountered in a sprayed agricultural field (6.1 mg/kg/day) and a higher dose (100-fold increase) chosen to represent a worse case scenario of bioaccumulation (Tomasi et al., 2001). Metabolic rate was evaluated in adults subjected to chronic dietary exposure beginning at weaning. Neither resting, cold-induced, nor norepinephrine-induced increases in metabolic rate were affected by vinclozolin treatment. This high dose of vinclozolin did lead to a 25% reduction in serum levels of $T_3$ and $T_4$. The decrease in circulating levels of $T_4$ appeared to be compensated by an increase in turnover such that $T_4$ utilization was unaffected. It may be that the cotton rat is a species that is tolerant to this fungicide and is capable of rapidly metabolizing the fungicide. The co-evolution of wild species with natural plant toxins may well lead to adaptations in hepatic P-450 systems that allow them to rapidly metabolize many pesticides (see Chapter 10). It would be of interest to assess how this and other EDCs affect cold acclimation processes while the animals are exposed to the toxicants under a seasonal regime. In addition, antithyroidal compounds, dioxin, and PCBs have been shown to have marked effects on the pre- and post-natal development of thermoregulation in rodents (see Chapter 3). It would be important to determine how chronic exposure to EDCs affects thermoregulatory processes in the offspring of maternally exposed rodents and other wildlife species.

In theory, exposure to thyroid EDCs could lead to a reduction in $T_4$ and $T_3$, and subsequent impairment in the ability to maintain thermogenesis in a cold environment (Tomasi et al., 2001). The thyroid is also critical in the accumulation of fat for winter survival. In addition to heat production, a normal thyroid gland is essential for growth, reproduction, immune function, and many other processes. Hence, any deficiencies in thyroid function brought on by exposure to an EDC could be modulated by the season. For example, someone interested solely in the toxic effects of thyroidal EDCs on reproduction should nonetheless consider how seasonal effects on thermoregulatory requirements will impact the function of the thyroid axis.

# Chapter 9

# Genetic Variability and Molecular Markers

## 9.1 INTRODUCTION

Toxicology as well as all fields of biomedical research has witnessed a revolution in the past decade with the development of genomic, transgenic, molecular, and other technologies to identify the origins of disease. In the study of temperature and toxicological response, genomic and molecular analyses have led to some significant findings, but the overall contribution is meager compared to other fields of study. This chapter endeavors to review the fields of genetics and molecular markers, particularly the expression of stress proteins, and how the work has led to a better understanding of how the thermoregulatory system responds to toxic insults.

## 9.2 GENETIC STRAIN VARIATION

Significant advances in toxicology and pharmacology have been made by comparing the responses between inbred and outbred strains of laboratory rodents when treated with a toxicant or drug. Genetic differences between strains of rodents can allow one to better understand the mechanism of action of a toxicant (for review, see Kacew and Festing, 1996). It is important to determine if the toxicological response of a particular strain of rodent is in any way attributable to intraspecies differences in thermoregulatory control. That is, are genetic differences in the regulated core temperature and/or thermoeffector response to a toxicant responsible for the purported effects of the toxicant? The behavioral and autonomic thermoeffectors that are activated in response to a toxicant will govern the change in body temperature; and it is important to understand if the unique response of a rodent strain to a toxicant is mediated by a particular thermoeffector. Intraspecies differences in sensitivity to a toxicant and the role of thermoregulatory control have been studied using several approaches: comparing the thermoregulatory response of inbred and outbred strains of rodents; selective breeding for a thermoregulatory trait that is responsive to a toxicant and development of strains that are sensitive or resistant to the toxicant; and utilizing genetic variation within a strain as a tool to identify genetic markers involved in the thermoregulatory response to a toxicant.

### 9.2.1 Intraspecies Variation

A considerable data base has been developed on the intraspecies thermoregulatory sensitivity to anti-ChE agents in the rat and mouse. Many studies have focused on the hypothermic response to acute exposure to organophosphate agents. Diisopropyl fluorophosphate (DFP) is an organophosphate that elicits a marked decrease in core temperature that peaks at 4 to 6 h after a subcutaneous injection in the rat (see Chapter 3). In a study using colonic probes, it was shown that the inbred Fischer 344 (F344) strain was relatively insensitive to the acute effects of DFP when compared to the outbred Long-Evans (LE) and Sprague-Dawley (SD) strains (Gordon and MacPhail, 1993). A DFP dose of 1.5 mg/kg led to a 0.3°C decrease in core temperature of the F344 strain but a 0.5 and 1.3°C decrease in the LE and SD strains, respectively, when measured 90 min after injection. Motor activity was reduced in proportion to the reduction in core temperature in all three strains, but strain differences were more apparent in deficiencies in horizontal activity (i.e., movement within the chamber) than with vertical activity (i.e., rearing movement).

In a telemetric study to document the time-course of change in core temperature and heart rate in undisturbed and unrestrained LE, SD, F344, and Wistar (W) rat strains, the LE and SD strains showed the most profound hypothermic response and slowest recovery from DFP compared to the F344 and W strains (Figure 9.1). Interestingly, the SD strain had a marked hypothermic response, but its heart rate was only slightly reduced as compared to the other three strains. That is, the decrease in heart rate that is typical following acute exposure to organophosphates was not necessarily linked to the hypothermic response (Gordon and Watkinson, 1995). Not surprisingly, serum and brain ChE activity was inhibited more

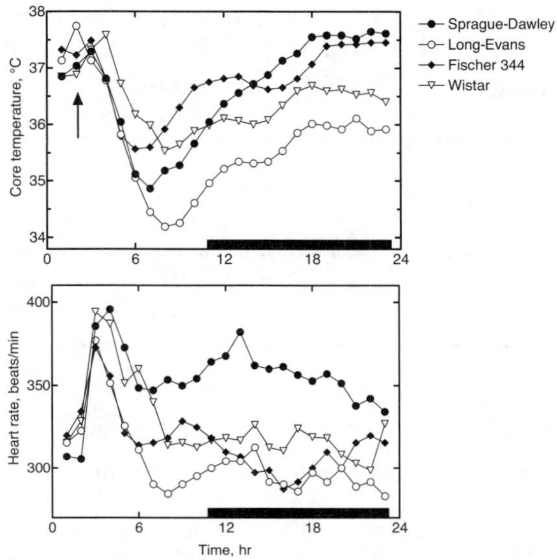


**Figure 9.1 Intraspecies variation in response of core temperature and heart rate to subcutaneous injection of 1.5 mg/kg DFP in four strains of the rat. (Data modified from Gordon, C.J. and Watkinson, W.P. (1995). *J. Toxicol. Environ. Health* 45: 59–73.)**

in the LE and SD strains. Hence, the hypothermic response to DFP within strains of the rat is likely to be linked to the relative inhibition in brain ChE activity (Gordon and Watkinson, 1995).

The hypothermic response to 4.0 mg/kg DFP was affected by genetic strain in the mouse, with the C57 being the most sensitive followed by the DBA and C3H strains (Smolen et al., 1986). A greater hypothermic response to an organophosphate would be expected if a strain had a higher density of muscarinic receptors, yet these strains are equally sensitive to muscarinic and nicotinic stimulation (see Gordon, 1994, for review). Clement (1991) assessed the hypothermic response to the organophosphate sarin in male CD-1 mice and found that they could be classified as either responders or nonresponders. That is, animals showed a minimal or marked hypothermic response to 130 μg/kg of sarin. The occurrence of the hypothermic response was inversely related to serum levels of carboxylesterase (Table 9.1). These enzymes bind to and inactivate circulating organophosphates, thus attenuating their toxicity. Hypothalamic AChE activity of nonresponders was only inhibited by 40% as compared to 74% inhibition in responders, the result being a greater hypothermic response to sarin in the responders (Clement, 1991). Contrarily, serum carboxylesterase levels were highest in the SD strain followed by the LE and F344 strains. This is in direct contrast to the hypothermic

**Table 9.1 Identifying Responders and Nonresponders with a Population of CD-1 Mice Based on Their Hypothermic Response, Hypothalamic AChE Activity, and Circulating Levels of Carboxylesterase Activity Following Subcutaneous Administration of the Organophosphate Sarin (130 μg/kg)[a]**

| *Group* | *AChE Activity, % of Control* | *Core Temperature (°C)* | *Plasma Carboxylesterase Activity[b]* |
|---|---|---|---|
| Responders | 26.3 ± 11.3 | 30.1 ± 0.96 | 81,866 ± 6,776 |
| Nonresponders | 59.6 ± 8.4 | 35.9 ± 0.43 | 92,579 ± 8,049 |

[a] Data given as mean ± S.D

[b] Activity expressed in units of nmol p-nitrophenylacetate hydrolyzed/min/ml.

*Source*: Data from Clement, J.G. (1991). *Biochem. Pharmacol.* 42: 1316–1318.

sensitivity of these rat strains to DFP, as discussed earlier in this chapter. Higher carboxylesterase levels in the rat do correlate with the ability to detoxify CBDP, the active metabolite of tri-o-cresyl phosphate (Clement and Erhardt, 1990).

## 9.2.2 Selective Breeding

Selective breeding has been used to develop strains of rodents with reduced or increased sensitivity to toxicants such as organophosphates and ethanol. Overstreet and colleagues administered DFP to rats of the SD strain and selected rats with the smallest or largest hypothermic response with the goal of developing strains of rats that were sensitive (FSL) or resistant (FRL) to DFP (for review, see Overstreet, 2002). To briefly describe the protocol, mating pairs of rats showing the greatest hypothermic response to DFP produced offspring with a greater hypothermic response compared to litter mates from pairs with an attenuated hypothermic response. Following selective breeding with 15 generations, it was found that heightened sensitivity of the FSL strain to DFP carried over to other cholinergic agents such as oxotremorine and physostigmine. It is interesting to note that selective breeding of the SD strain did not lead to a strain that was more resistant than the original population to DFP. The increase in sensitivity to DFP by selective breeding appears to be a result of an increase in density of muscarinic receptors in the CNS, leading to an augmented heat loss response following exposure to an anti-ChE agent (Overstreet et al., 1988).

Crabbe and others (1987) used hypothermia as a biomarker to develop genetic strains of the mouse that were sensitive or resistant to ethanol.

Strains of mice that showed either an exaggerated (COLD) or attenuated (HOT) hypothermic response to intraperitoneal injection of ethanol were developed with selective breeding. The maximum hypothermic response to ethanol was apparent following two to three generations of genetic selection (Figure 9.2). By the fifth generation, the maximum hypothermic response for one replicate of HOT and COLD mice was –1.8 and –3.5°C, respectively. The pharmacokinetics of ethanol metabolism was similar between the HOT and COLD mice. This would suggest that the genetic selection for thermoregulatory response to ethanol was attributed to differences in the CNS response to ethanol and not to an inability to eliminate ethanol from the circulation. When placed in a temperature gradient and allowed to behaviorally thermoregulate, COLD mice displayed a marked preference for colder temperatures and became more hypothermic than HOT mice following ethanol exposure (O'Connor et al., 1993). The accentuated regulated hypothermic response to ethanol in the COLD mice provides further evidence of genetic selection for altered CNS sensitivity to ethanol.

In addition to differences in pharmacokinetics and CNS sensitivity, intraspecies and interspecies variation in baseline core temperature and control of thermoeffector responses could explain a differential response to a toxicant. However, this is rarely considered in toxicology and pharmacology in spite of differences in core temperature, metabolic rate, capacity for nonshivering thermogenesis, heat and cold tolerance, and

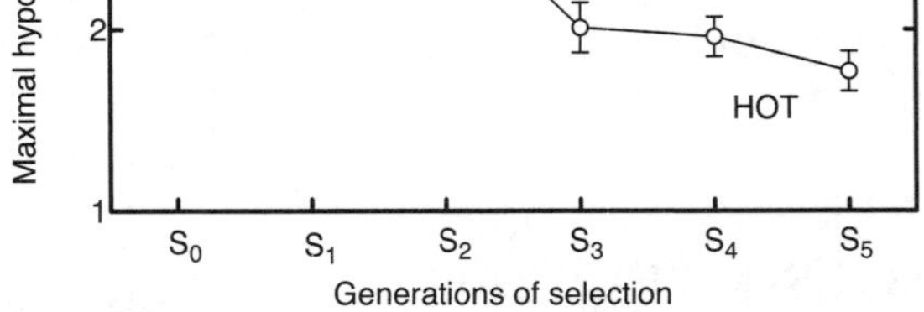


**Figure 9.2 Time-course of hypothermic response to ethanol in mice that are selectively bred for an augmented (COLD) or attenuated (HOT) hypothermic response to ethanol (3.0 g/kg; IP). Differences in thermoregulatory sensitivity to ethanol are seen in the second generation. (Data modified from Crabbe, J.C., Kosobud, A., Tam, B.R., Young, E.R., and Deutsch, C.M. (1987).** ***Alcohol Drug Res.*** **7: 163–174.)**

other differences found within strains of mice, rats, and other rodent species (Gordon 1993; also see Chapter 2). For example, the F344 and SD rat strains have remarkable heat tolerance when compared to other strains such as LE and Wistar (Furuyama, 1982). A toxicant that causes a forced elevation in core temperature would be expected to have a greater impact on a rat strain with a lower resistance to heat stress.

### *9.2.3 Genetic Markers: Quantitative Trait Loci*

The advent of the fields of genomics and proteomics has led to a revolution in understanding the contribution of individual genes to a species' sensitivity and response to toxic agents. However, genomic studies in temperature regulation are meager, and there is little work pertaining to the thermoregulatory response to toxicants. One promising approach is the use of quantitative trait loci (QTL) to evaluate the likelihood of the thermoregulatory as well as other toxicological responses to a toxicant (Crabbe et al., 1996; Crawshaw et al., 2001). QTL is a global approach using a statistical association of a physiological response with genetic markers to determine the genes involved in a physiological response. Crawshaw and colleagues used QTL techniques to assess the genes involved in the thermoregulatory response of the mouse to ethanol. A panel of recombinant inbred mouse strains from C57Bl/6J and DBA/2J progenitor strains were dosed with ethanol while two thermoregulatory responses to ethanol were evaluated. Changes in core temperature were used as a benchmark of ethanol-induced regulated hypothermia, and fluctuations in core temperature mediated by a rapid alteration in ambient temperature were used as an index of thermoregulatory disruption. One major finding using the QTL approach is that ethanol-induced regulated hypothermia and thermoregulatory disruption are represented in different portions of the mouse genome. The QTL approach could also be applied with other toxicants to identify loci of the genome that contribute to the control of a thermoregulatory response.

## 9.3 HEAT SHOCK PROTEINS

The heat shock proteins are ubiquitous and found in various forms in all organisms, from bacteria to humans (for review, see Welch, 1992; Kregel, 2002; Katschinski, 2004). A main function of heat shock proteins is to provide cytoprotection to a variety of insults, but they are also operative in a variety of cellular regulatory pathways (Figure 9.3). A most interesting phenomenon is the cross-tolerance between insults that can be mediated

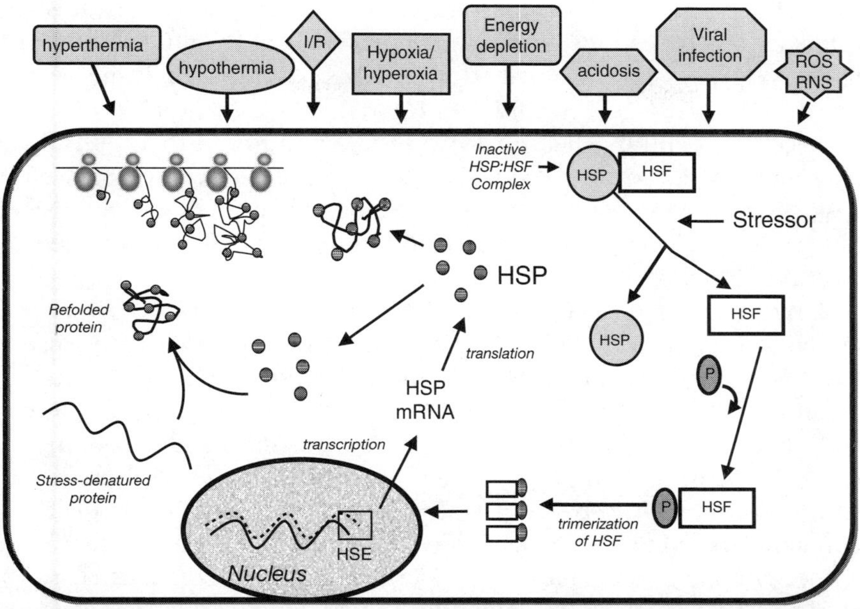


**Figure 9.3 Summary of the physiological roles of the 72-kDa heat shock protein Hsp70 and mechanisms for increased expression in response to a variety of insults. An array of biological and environmental insults are thought to induce the expression of Hsp70, including exposure to xenobiotic agents. HSF, heat shock factors; HSE, heat shock elements; I/R, ischemia/reperfusion; ROS, reactive oxygen species. (Reprinted with permission from Kregel, K.C. (2002). *J. Appl. Physiol.* 92: 2177–2186.)**

by induction of heat shock proteins. The implications of these responses in human health and toxicology are enormous. Moreover, in view of their ubiquitous nature and common mode of expression, the heat shock and related stress proteins have been considered as very promising biomarkers of toxic exposure in invertebrates and vertebrates (for discussion see Welch, 1992; Staempfli et al., 2002).

Heat shock proteins were initially discovered to be expressed in response to heat shock in *Drosophila melanogaster* and were shown to afford thermal tolerance to subsequent heat shock episodes. In addition to hyperthermic insults, expression of heat shock proteins such as Hsp70 is mediated from a variety of stressors including hypoxia, acidosis, energy depletion, reperfusion of ischemic tissues, and infection (Kregel, 2002). Exposing cells *in vitro* to insults such as hypoxia, ethanol, arsenic, and cadmium leads to the induction of heat shock proteins. The induction of heat shock proteins by these toxicants is most likely responsible for the enhanced thermotolerance of the cells when they are exposed to high temperatures in the range of 41 to 45°C (Li, 1983). However, one cannot be certain that the toxicant-induced expression of heat shock proteins is solely responsible for the improved heat tolerance since many regulatory processes are activated in response to the toxic insult. Barnes et al. (2001, 2002) used techniques for regulating expression of Hsp70 induction (i.e., without heat or toxic insult) and found that with higher expression of this cytoprotective protein, thermal tolerance as well as resistance to the genotoxic effects of arsenic increased markedly. Hence, inducing a specific heat shock protein affords protection from chemical and thermal insults.

Many researchers using *in vitro* models have demonstrated a cytoprotective role of heat shock protein induction to a variety of insults. However, there are relatively few *in vivo* studies that document the cytoprotective role of these proteins because it is more difficult to induce controlled elevations in body temperature in mammals (see Salminen et al., 1997). Many of the whole-animal studies have relied primarily on restrained or anesthetized preparations. The usefulness of these approaches is limited because of the stress and other artifacts that affect thermoregulation (see Chapter 7). Merging the fields of thermal physiology and molecular biology would be most beneficial in this endeavor (e.g., Kregel, 2002), especially if radiotelemetry could be used to monitor and control for precise time–temperature challenges to whole animals. Nonetheless, it is clear that a critical elevation in body temperature elicits expression of heat shock proteins and that heat shock may well protect the animal from other insults such as exposure to a toxicant. For example, intraperitoneal administration of amphetamine (15 mg/kg) in the mouse led to a 2.5°C increase in core temperature in less than 1 h after injection (Figure 9.4). The hyperthermic responses were associated with induction of hepatic levels of hsp70i and hsp25 between 6 to 72 h, although one cannot rule out the direct effects of amphetamine on heat shock protein expression. Amphetamine-induced hyperthermia and induction of heat shock proteins afforded protection to the hepatotoxic effects of acute acetaminophen and bromobenze but not carbon tetrachloride or cocaine. Since the hyperthermic response to amphetamine in the mouse is dependent on ambient temperature (Chapter 3), it would be of interest to modulate the thermoregulatory response to amphetamine by raising or lowering ambient temperature and then determine the effects of core temperature on the expression of heat shock protein. That is, the same dose of amphetamine can be given at different ambient temperatures to either raise or lower core temperature; then one can determine if it is in fact the temperature or the amphetamine treatment that induces heat shock proteins and affords tolerance to toxicants and other insults.

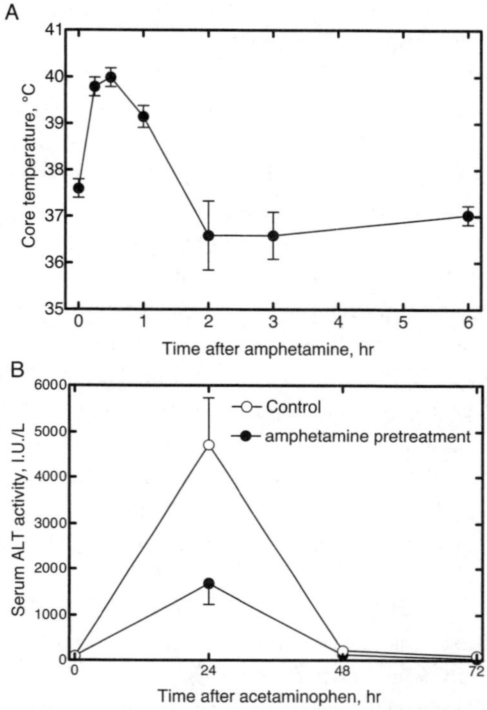


**Figure 9.4 (A) Hyperthermic response of the mouse to amphetamine (15 mg/kg; IP) while maintained at an ambient temperature of 24 to 25°C. (B) Effect of amphetamine hyperthermia pre-treatment on the hepatotoxic effects of acetaminophen (350 mg/kg; IP) as indicated by an increase in serum levels of alanine aminotransferase (ALT). (Data modified from Salminen, W.F., Jr., Voellmy, R., and Roberts, S.M. (1997). *Toxicol. Appl. Pharmacol.* 147: 247–258.)**

## 9.3.1 Endotoxin and Heat Shock

Endotoxin can elicit hypothermic and febrile responses in rodents, depending on the dose and route of administration (see Chapters 5 and 6). The hypothermia resulting from an acute dose of endotoxin is somewhat similar to that following treatment with xenobiotic chemicals in that the rats seek colder ambient temperatures and undergo a regulated hypothermic response (see Chapter 5). Induction of heat shock proteins following endotoxin administration has received some attention, and the responses may be relevant in the study of the physiological responses to xenobiotics. High doses of lipopolysaccharide (LPS) endotoxin administered to rodents are a frequently used model to study the mechanisms of septicemia, a major cause of human death and morbidity. Induction of heat shock protein synthesis in mice and rats appears to afford marked protection to the toxic effects of endotoxemia. Exposing rats to an acute heat stress (47 to 50°C) for approximately 45 min to raise core temperature to 42.6°C resulted in 100% survival of an acute intravenous dose of LPS (20 mg/kg) given 24 h after heat shock as compared to 71% mortality in nonheated animals (Ryan et al., 1992). Mice that were heat shocked by raising their core temperature to between 42.5 and 43°C for 20 min showed a five-fold increase in survival when challenged with a high dose of LPS (20.5 mg/kg; IP). The protective effects of heat shock were maximal at 12 h after hyperthermic treatment and were associated with expression of Hsp72 in various organs, with the largest expression in the gut and lungs and the least amount in the brain and heart (Hotchkiss et al., 1993). Overall, the endotoxin studies may be relevant in understanding the responses to xenobiotics as discussed below. High doses of endotoxin lead to the expression of heat shock proteins while acute heat stress protects animals from the acute effects of endotoxemia. Presumably, the expression of heat shock proteins affords protection from the toxic effects of endotoxins.

Heat shock may also have a role in modulating the febrile response to relatively low doses of endotoxin. As compared to the acute LPS studies described earlier, an LPS dose of only 50 μg/kg is needed to induce a fever in the rat, whereas doses on the order of 1 mg/kg and larger result in hypothermia. Heat shock in the rat (core temperature of ~42°C for ~30 min) leads 24 h later to an exacerbated fever in response to 50 μg/kg LPS (Kluger et al., 1997). This heat shock episode resulted in expression of hepatic Hsp70 and suppression in serum levels of TNFα, a cytokine normally released during fever that naturally acts to lower core temperature (Chapter 6). All together, heat shock and the expression of heat shock proteins appear to exacerbate the febrile response to a low dose of LPS but protect the organisms from large doses of LPS and endotoxemia. One can attempt to draw parallels between the response to xenobiotic chemicals and LPS, but the data base is relatively meager. Moreover, there could be an interplay between exposure to toxicants, fever, and Hsp induction. The development of fever has been shown to activate synthesis of Hsp's in the host cells as well as in pathogen cells, and the induction of Hsp's may be critical in the recovery from infection (Hasday and Singh, 2000).

### *9.3.2* In Vivo *Xenobiotic Studies*

The induction of heat shock protein Hsp72 has been shown to have a possible role in the protection of the CNS from oxygen toxicity (Arieli et al., 2003). Acclimating rats to heat (32°C) for 4 weeks ameliorates the toxic effects of an oxygen-enriched hyperbaric treatment (608 kPa) as evidenced by a doubling in the latency time to the onset of convulsions (Figure 9.5). This degree of heat acclimation is also associated with a marked elevation in whole-brain Hsp72 levels when compared to control rats acclimated to 24°C. Moreover, during the recovery from heat acclimation, rats become more sensitive to oxygen toxicity, and the response is correlated with a recovery in CNS levels of Hsp72. This represents one of the few whole-animal studies showing that relatively mild heat exposure is sufficient for induction of heat shock proteins, a response that appears to provide protection from a toxic insult. That acclimation to a warm environment affords protection to a toxic insult such as hyperoxia would lead one to speculate that protection from other toxic insults could be mediated by a similar heat treatment. If this were the case, it would be especially pertinent in ecotoxicological studies as well as human health effects studies where seasonal acclimatization to warm or cold environments may well lead to shifts in susceptibility to environmental contaminants. At this point, there are considerable data on the role of temperature acclimatization in response to toxicants, but the role of heat shock proteins remains to be assessed (Chapter 7).

There is some evidence that heat stress and exposure to xenobiotics may interact to promote the induction of heat shock proteins. Rats exposed to a combination of severe heat stress (40°C) and carbon monoxide (600 ppm) for 60 min expressed greater levels of hepatic Hsp70 when compared to animals exposed to just the heat stress or carbon monoxide (Tang-Chun et al., 1995). The heat stress treatment was severe, with body

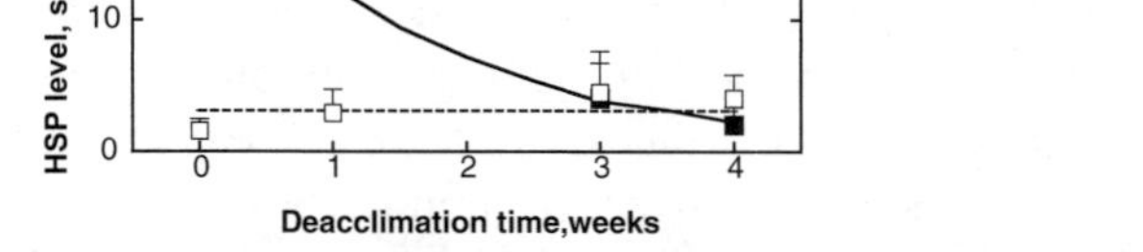


**Figure 9.5 Relationship between oxygen toxicity in the CNS as expressed by the latency to onset of convulsions and whole-brain Hsp72 levels in the rat following 4 weeks of de-acclimation from a warm environment. (Graph modified from Arieli, Y., Eynan, M., Ganez, H., Arieli, R., and Kashi, Y. (2003). *Brain Res.* 962: 15–20.)**

temperature reaching 42°C by 1 h. Nonetheless, this is one of the few *in vivo* studies showing an interaction between heat stress and exposure to a toxicant on the expression of heat shock proteins. These authors also reported that human workers subjected to heat stress and carbon monoxide tended to produce more antibodies to Hsp27 and Hsp70. In another study, subjecting the rat to brief heat stress (core temperature of 41 to 42°C for 15 min) led to marked protection to intrathecal instillation of phospholipase $A_2$ given 18 h after heat shock (Villar et al., 1993). Phospholipase $A_2$ instillation leads to acute lung injury accompanied by inflammation and increased lung lavage cell counts. This response may be similar to that seen when particulate matter (e.g., $PM_{10}$) is instilled intrathecally, an insult that leads to a prolonged reduction in the core temperature of the rat (see Section 3.5.3.3).

What role does the thermoregulatory system play in responding to heat shock, eliciting a stress protein response, and responding to a xenobiotic? *In vivo* expression of heat shock proteins in the rat can be observed within 24 h after dosing with a variety of structurally diverse toxic chemicals, including chlordane, chlorpyrifos, and alachlor (Bagchi et al., 1996). The heat shock proteins (Hsp89α and Hsp89) were expressed in brain and liver the day after treatment with toxicants of doses equal to 25% of the $LD_{50}$. Since these pesticides also elicit oxidative stress and related cellular damage, the expression of heat shock proteins may be an important protective mechanism.

However, these stress protein studies have apparently not considered how the thermoregulatory response to the toxicant interacts with the expression and protective function of the heat shock proteins. That is, the doses of the pesticides used in many of the heat shock protein studies would be expected to elicit a profound hypothermic response that would persist for many hours after exposure. Depending on the toxicant, the hypothermia would be followed by a fever-like increase in core temperature the day after exposure (see Chapters 3 and 6). The thermoregulatory response should lead one to determine how the changes in body temperature elicited by the pesticide may modulate the expression of the heat shock proteins. Clearly, high core temperatures lead to an expression of heat shock proteins, and a hypothermic response following acute exposure to a toxicant would be expected to antagonize the expression of the stress proteins, although it should be noted that hypothermia by itself is an insult that can also induce heat shock protein expression (Kregel, 2002). Hence, the hypothermic and/or hyperthermic responses to a toxic agent may well interact with the expression of heat shock and other stress proteins. Raising or lowering the ambient temperature in animals exposed to xenobiotics to modulate their body temperature would lead to a better

**Table 9.2 Some Genes Other Than Heat Shock Proteins That Have Been Shown to Undergo a Change in Expression with Heat Stress**

| *Functional Class* | *Gene* | *Change in Function* | *Exposure Model* |
|---|---|---|---|
| Acute phase reactant | C-reactive protein | Up | Intact animal |
| Apoptosis inhibitor | *bcl*-2 | Down | Cell culture |
| Cell adhesion | ICAM-1 | Down | Intact animal |
| Cell cycle | p53 | Up | Cell culture |
| Cell differentiation | *mcl*-1 | Up | Cell culture |
| Coagulation | β-Fibrinogen | Down | Intact animal |
| Cytoskeleton | GFAP | Up | Intact animal |
| Heavy metal binding | Metallothionein | Down | Both |
| Redox control | SOD-1 | Up | Cell culture |
| Signal transduction | cNOS | Up | Intact animal |

*Source*: Modified from Sonna, L.A., Fujita, J., Gaffin, S., and Lilly, C.M. (2002). *J. Appl. Physiol.* 92: 1725–1742

understanding of the role of heat shock protein expression in toxicological studies.

Although this section has focused primarily on heat shock proteins, it is important to note that genomic and proteomic technologies have demonstrated that a variety of other genes are expressed by heat and cold stress (Sonna et al., 2002). There are approximately 50 genes not considered to be heat shock proteins that have been shown to change expression with heat stress in either *in vitro* or *in vivo* models (Table 9.2). The expression of a fewer number of genes (<20) is also affected by cold stress (Sonna et al., 2002). Clearly, these genes have major implications for the mechanisms of toxicity as outlined in Table 1.1. An integrative approach to studying the expression or repression of these genes with heat and cold stress will undoubtedly lead to a better understanding of the mechanisms of toxicity.

# Chapter 10

# Natural Toxins and Venoms

## 10.1 INTRODUCTION

Understanding the thermoregulatory effects of natural toxins and venoms is important from several standpoints. Relative to the xenobiotic chemicals such as those summarized in Chapter 3, there is little known about the thermoregulatory effects of toxins. This is somewhat surprising when one considers that humans, domesticated species, and wildlife are subject to exposure to a variety of toxins and venoms under a range of thermally stressful environments. Many toxins have a profound economic impact on key agricultural species and represent a significant human health hazard. Of further interest to the toxicologist is the fact that the biologically effective doses of toxins and venoms are often found to be orders of magnitude lower than that of most xenobiotics.

Exploring the mechanisms of action of toxins and venoms may shed light on the toxicological mechanisms of some xenobiotics. Because animals have co-evolved with plant toxins and animal venoms, most would conclude that the ability to respond to the man-made xenobiotic chemicals is attributable in part to the endogenous responses to toxins and venoms. This is apparent in the development of oxidative pathways (e.g., P-450 systems) in the liver and other tissues that have evolved to detoxify the plethora of plant toxins encountered by herbivorous animals (Brattsten, 1979). The enzymatic pathways that conjugate and excrete toxins and venoms are also operative in the metabolic activation and deactivation of xenobiotics. It follows that other physiological systems, such as temperature regulation, also respond to xenobiotics because of adaptive responses to toxins and venoms. To this end, this chapter endeavors to review the effects of toxins and venoms commonly found in the environment that have major implications for the health and well-being of humans, domestic animals, and wildlife.

## 10.2 FESCUE TOXICOSIS

Fungal toxins found in grain and grass used for grazing can lead to significant thermoregulatory problems in cattle and other grazing species, especially during the summer months (Burke et al., 2001; Al-Tamimi et al., 2003). Infection of tall fescue grass with endophytic fungus can result in a pathological condition known as fescue toxicosis. Environmental temperature is a key facet in the manifestation of fescue toxicosis. During the summer months, the development of the toxicosis is directly related to heat stress conditions, with hyperthermia as the predominant thermoregulatory effect. For example, beef heifers allowed to feed on fescue seed infected with the endophyte *Neotyphodium coenophialium* at a thermoneutral temperature of 19°C display no changes in core temperature or respiratory rate. However, when fed the same diet at an ambient temperature of 31°C (i.e., typical summer conditions), a marked increase in core temperature and respiratory rate was observed while the cattle were allowed to feed on the endophyte-infected diet (Figure 10.1). There did not appear to be much tolerance to the hyperthermic effects of the endophyte since there was a steady elevation in core temperature for at least 18 days of exposure. Feeding cattle the ergopeptine alkaloids (i.e., toxins found in cereal grains infected with *Claviceps purpurea*) at a level of 10 μg/kg/day resulted in greater hyperthermia than in control animals exposed to summer heat stress (Al-Tamimi et al., 2003). The efficacy of these fungal toxins to raise core temperature as well as elicit other pathological effects, such as reduced feed intake and reproductive dysfunction, also occurred at doses of approximately 10 μg/kg. It is remarkable to note that these doses are approximately three orders of magnitude below that needed to elicit thermoregulatory effects for xenobiotic agents in rodents (e.g., Table 3.2).

Time of day is a key facet in the manifestation of thermoregulatory stress in fescue toxicosis (Al-Haidary et al., 2001). Because of their large body mass, mammals such as cattle accumulate large amounts of heat in the daytime during the summer months and dissipate the heat load slowly at night with a resultant peak in core temperature during this period. Cooler ambient temperatures during the night are essential in order for

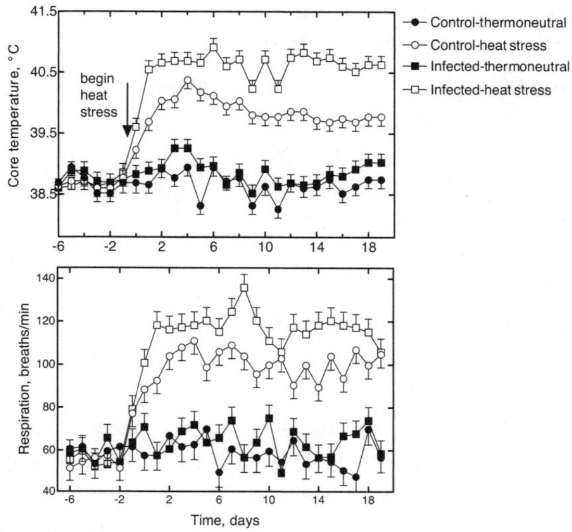


**Figure 10.1 Effects of heat stress on the thermoregulatory response of beef heifers (body weight ~350 kg) while feeding on endophyte-infected feed containing the fungal toxin ergovaline. Cattle were maintained in a thermoneutral (19°C) or heat stress environment (4 h at 31°C during the day, then 4 h at 25°C at night). Estimated dosage of ergovaline is 6.1 to 8.7 μg/kg/day; treatment began on day −27. Core temperature and respiratory rate determined daily at 1,600 h, which represents the higher elevation in temperature during the circadian cycle. (Data modified from Burke, J.M., Spiers, D.E., Kojima, F.N., Perry, G.A., Salfen, B.E., Wood, S.L., Patterson, D.J., Smith, M.F., Luch, M.C., Jackson, W.G., and Piper, E.L. (2001). *Biol. Reprod.* 65: 260–268.)**

cattle to use peripheral vasodilation to effectively dissipate the excess heat load accrued during the daytime. Since the endophyte toxins result in peripheral vasoconstriction in cattle, it is not surprising to find that fescue-toxicosis hyperthermia can be most pronounced in the middle of the night when the cattle that were subjected to daytime heat stress are attempting to lower core temperature at night (Al-Haidary et al., 2001).

The rat has been used as a model to study the thermoregulatory effects of endophyte toxins (Spiers et al., 1995). Rats maintained at a cold ambient temperature of 7 to 9°C and injected intraperitoneally with ergovaline (15 μg/kg), the primary toxin found in endophyte-infected fescue, underwent a decrease in metabolic rate and a reduction in tail skin temperature and became hypothermic. Since rats at such a cold temperature are likely to be maximally vasoconstricted, the additional decrease in tail skin temperature at such a cold ambient temperature is attributed to the decrease in core temperature. Thus, the hypothermic effect of ergovaline during cold exposure is a result of impairment of thermoeffectors for heat production (i.e., shivering and nonshivering thermogenesis). However, when administered the same dose of ergovaline at ambient temperatures near or slightly above the upper critical temperature (31 to 33°C), there was a transient increase in metabolic rate and then a prolonged reduction in tail skin blood flow (Figure 10.2).

Not surprisingly, a peripheral vasoconstrictive response at this warm temperature coupled with a small increase in metabolic rate resulted in a profound hyperthermic response (Spiers et al., 1995). The thermoregulatory effects of the endophyte toxins in the rat maintained at a warm temperature are somewhat similar to those observed in cattle in that the hyperthermia is largely mediated by a restriction in heat loss. Hypothermic responses have not been reported in cattle exposed to these toxins and would likely not be expected because of the large differences in body mass and facility to dissipate heat. That is, small mammals rely more on metabolic rate to thermoregulate in the cold, and any effect of a toxin on metabolism will manifest as hypothermia. However, large mammals rely more on peripheral vasomotor tone to regulate core temperature (Chapter 2).

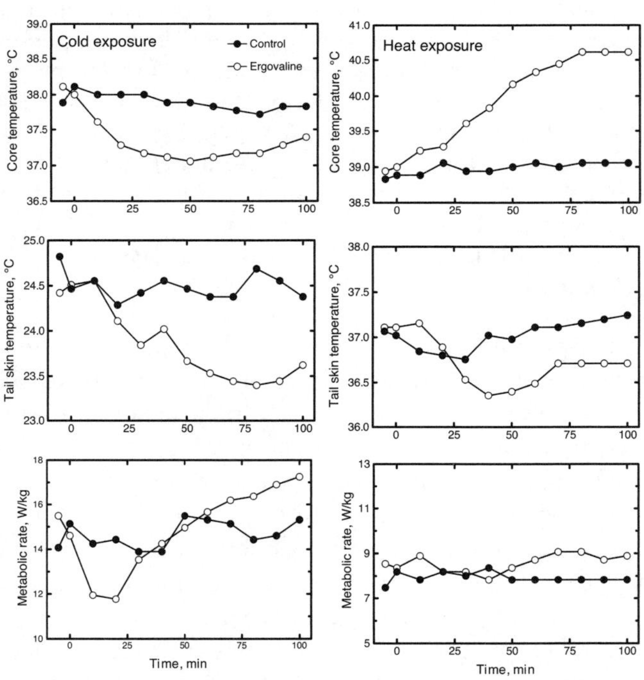


**Figure 10.2 Thermoregulatory response to the fungal toxin ergovaline (15 μg/kg) administered intraperitoneally at 0 min while rats were exposed to ambient cold (6.8 to 9.1°C) or heat stress (31.8 to 32.6°C). (Data modified from Spiers, D.E., Zhang, Q., Eichen, P.A., Rottinghaus, G.E., Garner, G.B., and Ellersieck, M.R. (1995). *J. Anim. Sci.* 73: 1954–1961.)**

The neurochemical mechanisms of the endophyte toxins on temperature regulation are not well known. The hyperthermic response to these fungal toxins appears to involve stimulation of dopaminergic pathways. Administration of a $D_2$ dopaminergic antagonist to cattle alleviates the hyperthermic effects of an endophyte-infected tall fescue diet when housed at a warm ambient temperature of 32°C (Samford-Grigsby et al., 1997). However, the hyperthermic effects of infected diets in sheep were exacerbated by supplementation with a dopaminergic antagonist (Aldrich et al., 1993). It is interesting to note that hyperthermic and peripheral vasoconstrictive effects of these toxins under heat stress conditions are similar to actions of amphetamine-like compounds on thermoregulation (Gordon et al., 1991c). The thermoregulatory effects of the amphetamines appear to be mediated by an increase in CNS levels of serotonin. In the rat maintained in a thermoneutral environment, substituted amphetamines such as methylenedioxymethamphetamine (MDMA) elicited a lethal elevation in core temperature concomitant with peripheral vasoconstriction of the tail. At cold temperatures, the same compounds led to a hypothermic response. To develop a better understanding of the mechanism of action of the fungal toxins, it would be important to determine the activity of neurochemical pathways in the CNS, such as serotonin, in cattle and other animals exposed to the endophyte toxins.

## 10.3 WILDLIFE AND TOXINS

The metabolic rate of mammals is affected by a variety of biotic and abiotic factors, including body size, phylogeny, activity level, climate, and food habitats (McNab, 1986). Natural diet and food availability have marked effects on metabolic rate and, possibly, the regulation of body temperature. In addition to the fungal toxins, herbivorous mammals face possible exposure to many types of secondary compounds in plants that are toxic. These secondary compounds are produced by many plants, presumably as a deterrent against foraging, and will be referred to as toxins in this discussion. Ingestion of these plant toxins is a regular occurrence and is likely to have consequences on basal metabolism and thermoregulation, especially under conditions of heat or cold stress.

Small mammals are metabolic specialists and rely on a high and constant rate of heat production to maintain a normal body temperature. Natural toxins that alter metabolic rate in rodents would be expected to impart significant changes in baseline core temperature or when challenged with heat or cold stress. Some studies have found that natural toxins such as phenolics and tannins alter metabolic rate in wild rodents. For example, a diet containing 6% of the phenol gallic acid fed to the meadow vole (*Microtus pennsylvanicus)* led to a 13.6 to 22.6% increase in basal metabolic rate after 21 days of feeding (Thomas et al., 1988). Tannic acid, a common plant toxin, was found to have no effect on basal metabolic rate but increased the maximum metabolic rate during cold exposure in herbivorous (*Octodon degus*) and omnivorous rodents *(Phyllotis darwini)* when added to their diet (Bozinovic and Novoa, 1997). In addition to the added hepatic metabolism needed to excrete the toxins, the tissue damage caused and ensuing activation of repair mechanisms are thought to be responsible for the increased energy demand associated with ingesting these toxins.

During the winter, the increased demands on heat production combined with a limited food supply are bound to force many mammalian species to ingest plants with toxins that would normally be avoided. Because of the increased demands for metabolic energy to detoxify toxins with hepatic microsomal oxidative pathways and loss of available energy from conjugation (Sorenson, 2005), ingestion of these toxins would be expected to be a detriment to maintaining energy balance. However, it is also intriguing to consider that ingestion of certain toxins may be beneficial to thermoregulation. McLister et al. (2004) found that consumption of the toxins in juniper (*Juniperus monosperma*) differentially affected metabolism and body temperature in two species of the woodrat, *Neotoma albigula* and *N. stephensi* (Figure 10.3). Animals were acclimated to 25 or 18°C and fed a calorically equivalent control or juniper-treated diet for 7 to 10 days and the *minimum cost of thermogenesis* ($C_{min}$) was calculated in animals exposed to an ambient temperature of 15°C with the formula

$$C_{min} = \text{resting metabolic rate}/(\text{core temperature} - \text{ambient temperature}) \quad (10.1)$$

This is also the equation for whole-body thermal conductance, which is a measure of the metabolic costs relative to the temperature gradient between the core and surrounding air (Chapter 2). Juniper consumption adversely affected thermoregulation of warm-acclimated *N. albigula* that were exposed to cold (15°C) but reduced thermoregulatory costs of cold-acclimated animals exposed to the same temperature. Minimum costs for thermogenesis of *N. stephensi* ingesting the juniper toxins was also increased by warm acclimation but unaffected by cold acclimation. The authors noted that *N. albigula* voluntarily consumes more juniper in the winter than summer, suggesting that their behavioral preference for juniper may be an adaptation to reduce the metabolic costs of thermoregulation during winter months (McLister et al., 2004). Consumption of the juniper toxin was associated with a higher body temperature in both species of cold-acclimated wood rats, which may also benefit winter survival. The juniper toxins had no effect on basal metabolic rate in animals maintained

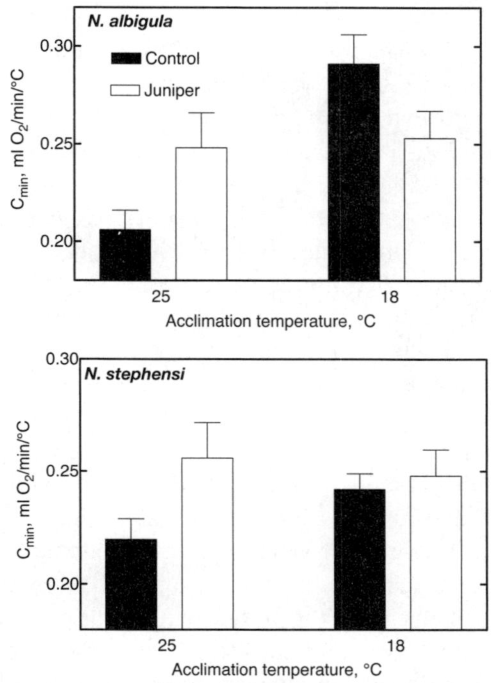


**Figure 10.3 Minimum cost for thermogenesis ($C_{min}$) at an ambient temperature of 15°C in two species of woodrat acclimated to a temperature of 25 or 18°C and fed a control diet or a diet containing juniper toxin. (Data from McLister, J.D., Sorensen, J.S., and Dearing, M.D. (2004). *Physiol. Biochem. Zool.*)**

at thermoneutral temperatures but only when thermoregulation was challenged by exposure to cold stress. The metabolic costs of detoxification of toxins such as that found in juniper may only be apparent when the thermoregulatory system is challenged by heat or cold stress.

## 10.4 ALGAL TOXINS

Unicellular algae in fresh and salt water synthesizes toxins that can be extremely hazardous to the health of humans as well as species that feed directly on these algae and those that are exposed to the toxins via bioaccumulation through the food chain. The toxins produced by marine phytoplankton elicit a variety of pathological effects (Ting and Brown, 2001). The acute and delayed changes in thermoregulation of experimental mammals and humans exposed to murine toxins are profound. Freshwater algal species also produce potent toxins, but their effects on temperature regulation are relatively unknown compared to marine toxins.

Many species of marine toxins represent a significant threat to seafood safety, and their presence must be monitored with vigilance. Assessing the potential toxicity in seafood has required extensive use of the laboratory mouse as a primary test species. Unfortunately, lethality has often been used as a biological end point in the seafood toxicity evaluations. A decrease in body temperature is one of many pathological responses when extracts of tissues suspected of carrying the toxins are administered to laboratory rodents. Understanding the thermoregulatory responses to these toxins should improve the accuracy and sensitivity of using rodents to assess seafood toxicity and may well prevent the use of lethality as an end point. Moreover, characterization of the thermoregulatory responses could lead to a better understanding of the neurotoxic mechanisms of the toxins.

The marine algal toxins that have been studied for effects on thermoregulation, including ciguatoxin, maitoxin, and brevetoxin, are complex cyclic polyether structures. These toxins enhance the activation of voltage-dependent sodium channels, resulting in a variety of pathophysiological sequelae. Moreover, their chemical structure and activity are resistant to heating, which explains why the toxic symptoms persist in victims who have ingested cooked fish. Ciguatera is a toxic syndrome resulting from ingestion of ciguatoxin produced by epibenthic dinoflagellates. These toxins accumulate through the food web and can be concentrated in the tissues of edible tropical fish species. Paradoxical reversal of thermal perceptions is one of the most common symptoms of ciguatera poisoning in humans (Gillespie et al., 1986; Cameron and Capra, 1993). This thermal dysthesia can persist for at least several days following recovery from the acute symptoms of ciguatera, which include vomiting, diarrhea, and severe abdominal pain.

Paradoxical thermal dysthesia is perhaps one of the most unique dysfunctions of the thermoregulatory system that has been documented in humans ingesting a natural toxin. Victims of ciguatera experience painful discomfort when touching cold objects or ingesting cold fluids. Temperatures below 24 to 26°C elicit these painful stimuli, while the perception of warm temperatures appears to be unaffected. Ciguatoxin acts specifically on sodium channels of excitable membranes, and the thermal dysthesia is apparently a result of alteration in function of C-polymodal nociceptors in humans that function at temperatures below ~23°C (Cameron and Capra, 1993). It is interesting to note the lack of information on the effects of ciguatoxins as well as other algal toxins on core temperature in humans. Thermal sensor dysfunction could lead to alterations in core temperature regulation. The recent studies on thermal receptors that have identified novel proteins that confer thermal sensitivity as well as responsiveness of the thermal receptor to chemical stimulation may well provide new research avenues to study the etiology of thermal dysthesia induced by algal toxins (Patapoutian et al. 2003).

The marine algal toxins elicit acute thermoregulatory responses in laboratory rodents that are grossly similar to those of the xenibiotics discussed in Chapter 3, but their relative potency can be six orders of magnitude greater than that of chemical toxicants. Maitoxin is a polyether toxin that is also found in ciguatoxic fish, and the toxin elicits an acute hypothermic response in mice at a dose as low as 338 ng/kg (Gordon et al., 1998). The acute hypothermia elicited by a maitoxin dose of only 338 ng/kg is preceded by a preference for cooler temperatures in a temperature gradient. The hypothermia persists for several hours, even though the exposed mice can select warmer temperatures in the gradient and maintain a normothermic state. Ciguatoxin has also been shown to elicit hypothermia in mice when given orally or intraperitoneally (Lewis et al., 1993).

Brevetoxins are produced by the red tide dinoflagellate *Gymnodinium breve* and elicit acute hypothermic responses in mice and rats. Intravenous infusion of brevetoxin at a dose as low as 12.5 μg/kg leads to a small, transient reduction in core temperature, whereas a larger dose of 25 to 50 μg/kg leads to a prolonged hypothermia lasting several hours (Templeton et al., 1989). The hypothermia is preceded by a precipitous fall in breathing rate, suggesting a reduction in metabolic thermogenesis. When dosed intraperitoneally with brevetoxin at a dose of 180 μg/kg, mice initially prefer floor temperatures that are ~10°C below controls, and their core temperature decreases from 37.5 to 35°C within 30 min (Figure 10.4). When core temperature was monitored with radiotelemetry, a delayed elevation in core temperature was observed that developed on the second day after recovery from the acute hypothermia and persisted for approximately 4 days (Gordon et al., 2001). It is interesting to note that the delayed increase in daytime core temperature was similar to that seen in rats exposed to organophosphate and carbamate insecticides (Chapter 6). Using *c-fos* as a biomarker, it was shown that c-*fos* mRNA accumulation correlates with the reduction in body temperature in mice exposed to ciguatoxin (Peng et al., 1995). In addition, immunohistochemical analysis was used to identify CNS nuclear groups associated with thermoregulation as targets for ciguatoxin.

In view of the thermoregulatory and histochemical analyses, it appears that the thermoregulatory centers are a primary target for these marine algal toxins. The behavioral thermoregulatory studies of the marine algal toxins suggest a regulated hypothermic response following acute exposure. Since the toxicity of many toxicants is ameliorated with reduced temperature (Chapter 4), it is important to understand if the mouse's hypothermic response affects its survival to the toxins. In one study, it was shown that the lethality of a maitoxin-like extract was exacerbated

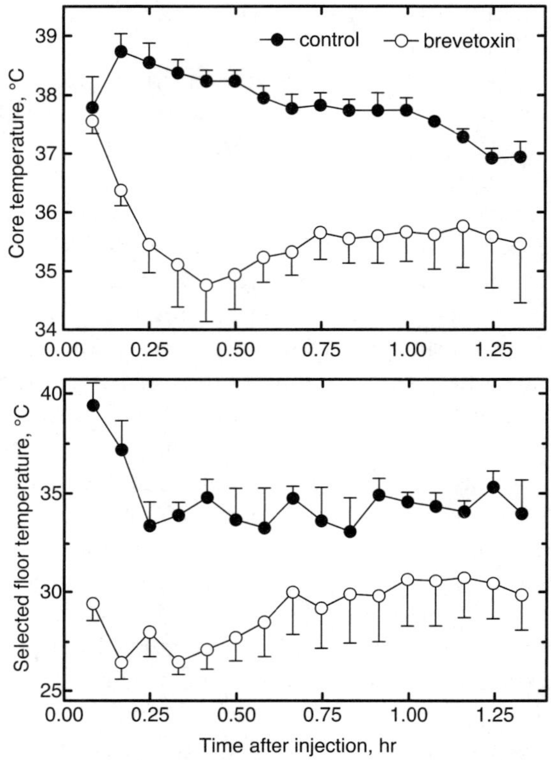


**Figure 10.4 Time-course of core temperature and selected floor temperature of mice placed in a temperature gradient immediately after intraperitoneal administration of the control vehicle or 180 μg/kg brevetoxin. (Data from Gordon, C.J., Kimm-Brinson, K.L., Padnos, B., and Ramsdell, J.S. (2001). *Toxicon* 39: 1367–1374.)**

when the hypothermic response was blocked (Sawyer et al., 1984). This would suggest that the hypothermic response to the toxins may protect small mammals from their acute toxicity.

As many as 50,000 cases of ciguatera are reported annually (Ting and Brown, 2001). This would account for a significant fraction of the total acute poisonings from natural and xenobiotic toxicants. It follows that hypothermia could be considered as a therapy in the treatment in these poisonings. However, little is known about the thermal effects of algal toxins in larger species, including humans. The robust hypothermic response seen in small rodents would most likely be attenuated with an increase in body mass (Chapter 5). In view of the thermal dysthesia that is manifested at reduced temperatures, it is possible that hypothermia may antagonize ciguatera-like symptoms in humans. Nonetheless, understanding the relationships between body temperature and mechanisms of toxicity of the algal toxins could lead to improved methods for treating poisoning victims.

## 10.5 VENOMS

The thermoregulatory effects of venoms from vipers, scorpions, and wasps have occasionally been evaluated in laboratory rodents and other species (Table 10.1). Hypothermia is one of the frequently reported thermoregulatory effects, but there is relatively little known about the mechanism of action. The venom studies generally focus on the dysfunction of systems that are of immediate concern to survival, such as the cardiovascular and respiratory systems. In one of the few systematic studies on the effects of envenomation (i.e., injection of venom) on temperature regulation, the venom from Russel's viper led to a progressive decrease in mouse core temperature for 90 min followed by gradual recovery over the next 120 min (Figure 10.5). Little is known about how ambient temperature affects the thermoregulator responses to these venoms and whether the hypothermic response is forced or regulated. Ambient temperature would undoubtedly modulate the hypothermic response to venoms in rodents as is seen with xenobiotics.

Scorpion venoms have been found to elicit profound fevers in humans. Envenomation from the scorpion *Leiurus quinquestriatus* leads to acute symptoms including hypertension and marked hyperthermia with rectal temperatures often exceeding 41°C (Ismail et al., 1990). For example, a 1-year-old human stung by a scorpion had a core temperature of 41.7°C by 12 h and later lapsed into a coma and died. The hyperthermia induced by scorpion venom can be reduced with acetaminophen, indicating that the elevation in core temperature is mediated by release of prostaglandins in the thermoregulatory centers and could be akin to that of fever (see Figure 6.1). Scorpion venoms also elicit an acute *autonomic storm*, characterized by the release of huge amounts of catecholamines, but a decrease in thyroid hormones (Murthy and Zare, 1998). In view of the marked shifts in core temperature during the early and late stages of exposure to the toxin, there has been an interest in assessing how temperature affects the toxicity of scorpion toxin in experimental animals (Ismail et al., 1990).

The lethal dose of the venom from the scorpion *Androctonus amoreuxi* was determined in rats maintained under hypothermic, normothermic, and hyperthermic conditions (Figure 10.6). The lethal dose was exacerbated in both hypothermic and hyperthermic animals. Increasing the core temperature by 1.9°C led to an 85% reduction in the lethal dose, whereas lowering core temperature by 7.3°C led to a 94% reduction in the lethal

**Table 10.1 Summary of Thermoregulatory Effects of Venoms**

| *Venom Source* | *Dose*[a] | *Species* | *Response* | *Reference* |
|---|---|---|---|---|
| Russel's viper | 50 μg/kg (IP) | Mouse | –2.2°C $T_c$ @ 90 min | Dutta & Chaudhuri, 1991 |
| *Sistrurus malarius barbouri* (snake) | 2 μg/kg (IP) | Mouse | –1.5°C $T_c$ @ 60 min | Assi & Nasser, 1999 |
| *Mesobuthus tamulus* concanesis (scorpion) | 3 mg/kg (SC) | Dog | Decrease in $T_3$ @ 30 min | Murthy & Zare, 1998 |
| *Vespa orientalis* (oriental hornet) | (IP) | Mouse | –8°C $T_c$ @ 180 min | Ishay et al., 1983 |
| *Vespa orientalis* | ~0.3 mg/kg Venom sac extract (IA) | Cat | –0.5°C $T_c$ | Barenholz-Paniry et al., 1990 |
| *Vespa orientalis* | Venom sac extract (IV) | Dog | –0.3–0.7 $T_c$ @ 30 min | Rubin et al., 1993 |
| *Apis mellifera* (honeybee) | 140 μg/kg (IP) | Mouse | –6.7 $T_c$ @ 120 min | Benton et al., 1966 |

[a] Parentheses indicate route of dosing. IP — intraperitioneally; SC — subcutaneous; IA — intra-arterial; IV — intravenous.

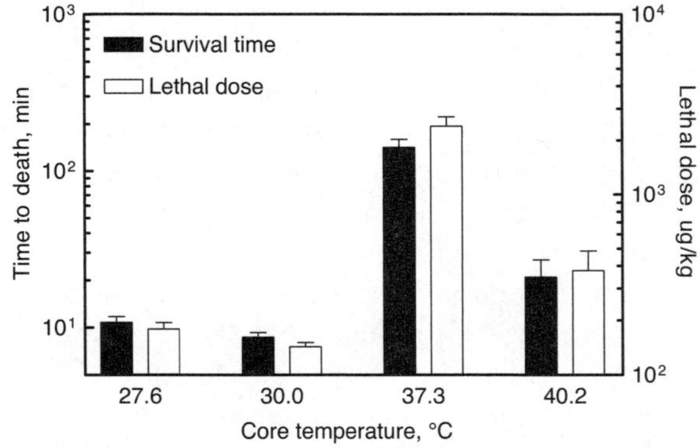


**Figure 10.5 Time-course of core temperature of mice injected intraperitoneally with 50 μg/kg venom from Russel's viper. (Data modified from Dutta, A.S. and Chaudhuri, A.K. (1991). *Ind. J. Exp. Biol.* 29: 937–942.)**

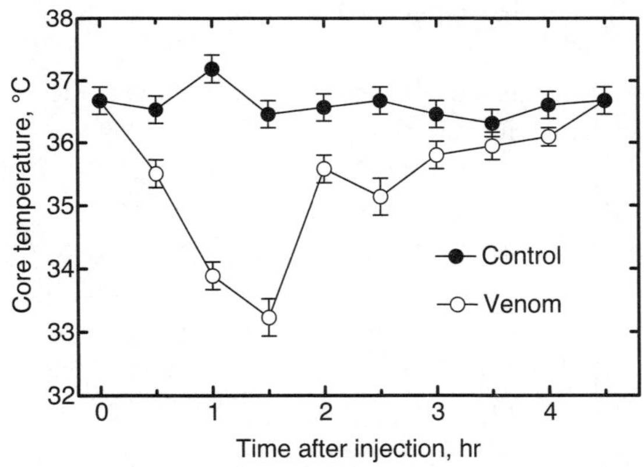


**Figure 10.6 Effect of variations in core temperature on the lethal dose and time to death of rats injected intravenously with scorpion (*Androctonus amoreuxi*) venom. (Data from Ismail, M., Abd-Elsalam, M.A., and Morad, A.M. (1990). *Toxicon* 28: 1265–1284.)**

dose. The core temperature of the hypothermic rats was quite low, and it is possible that the interaction of the scorpion toxin and low body temperature resulted in a hypothermic death. It would be of interest to determine if a moderate level of hypothermia (i.e., ~2°C hypothermia) would be protective as is seen with many xenobiotics (see Chapter 4).

The thermoregulatory and cardiovascular effects of the venom from the oriental hornet (*Vespa orientalis*) have been relatively well studied (Table 10.1). Hornet venoms are a complex mixture of low and high molecular weight substances, including ketones, biogenic amines, kinins, enzymes, and proteins. The sequence of activation and severity of pathophysiological effects of hornet and other venoms can be linked with individual substances in the venom, and the hypothermia is likely attributable to one or more of the constituents in the venom (Rubin et al., 1993). Injection from the extracts of the venom sac of hornets induces a hypothermic response in mice and anesthetized cats and dogs and evokes profound changes in blood pressure and cardiac performance (Ishay et al., 1983; Rubin et al., 1993; Barenholz-Paniry et al., 1990). Ishay et al. (1983) reported that human victims of hornet stings felt cold even at a relatively warm ambient temperature of 28°C. In addition, a drop in core temperature of 0.5 to 1.0°C in humans following a hornet sting was reported. The hypothermic effect of hornet venom in mice has been attributed to a 14–amino acid polypeptide, termed mastoparan, which is also present in the venom of wasp species (Duvdevani et al., 1991). Injection of fractionated wasp venom containing mastoparan led to a reduction in mouse body temperature, but the hypothermic response to mastoparan was more rapid and recovered faster when compared to injection of the unfractionated wasp venom. There are actually several types of mastoparans in wasp and hornet venom that induce mast cell degranulation and have other pharmacological effects. However, there is apparently little known about how these polypeptides are capable of inducing hypothermia in rodents.

Overall, there is little information on the thermoregulatory mechanisms responsible for the change in body temperature following envenomation. The lack of information is surprising despite the fact that it has been known for several decades that the hypothermic response to some venoms may be beneficial for survival. For example, preventing mice from becoming hypothermic following intraperitoneal injection of honeybee and cobra venom led to marked mortality (90 to 95%) when compared to mice that were allowed to develop a moderate hypothermic response for several hours following envenomation (3 to 5% mortality; Benton et al., 1966). The severity of convulsions was noted to be greatest in animals that were unable to become hypothermic. Indeed, Benton et al. (1966) concluded that "it is evident that an initial drop in body temperature is absolutely essential for the survival of the animal."

Unlike the xenobiotics, it is not known if a hypothermic response to envenomation is forced or regulated. Body temperature regulation would not seem to be a priority when dealing with the response to these venoms. The shock, tissue necrosis, cardiovascular disturbances, immune reactions, and impending death would likely place temperature regulation as a low priority for concern. Nonetheless, understanding whether there is a forced or regulated thermoregulatory response could be useful in the treatment of envenomation victims since change in body temperature may influence the course of toxicity to many of these venoms (e.g., see Murthy and Zare, 1998). Measuring key thermoeffectors such as peripheral vasomotor tone, metabolism, and behavioral thermal preference would be enlightening in the understanding of the pathophysiological effects of these venoms.

# REFERENCES

Ahdaya, S.M., Shah, P.V., and Guthrie, F.E. (1976). Thermoregulation in mice treated with parathion, carbaryl, or DDT. *Toxicol. Appl. Pharmacol.* 35: 575–580.

Aldrich, C.G., Rhodes, M.T., Miner, J.L., Kerley, M.S., and Paterson, J.A. (1993). The effects of endophyte-infected tall fescue consumption and use of a dopamine antagonist on intake, digestibility, body temperature, and blood constituents in sheep. *J. Anim. Sci.* 71: 158–163.

Al-Haidary, A., Spiers, D.E., Rottinghaus, G.E., Garner, G.B., and Ellersieck, M.R. (2001). Thermoregulatory ability of beef heifers following intake of endophyte-infected tall fescue during controlled heat challenge. *J. Anim. Sci.* 79: 1780–1788.

Al-Tamimi, H.J., Rottinghaus, G.E., Spiers, D.E., Spain, J., Chatman, D., Eichen, P.A., and Carson, T.L. (2003). Thermoregulatory response of dairy cows fed ergotized barley during summer heat stress. *J. Vet. Diagn. Invest.* 15: 355–360.

Altman, P.L. and Dittmer, D.S. (1966). *Environmental Biology*. Bethesda, MD: Federation of American Society of Experimental Biologists.

Anari, M.R. and Renton, K.W. (1993). Modulatory effect of hyperthermia on hepatic microsomal cytochrome P450 in mice. *Biochem. Pharmacol.* 46: 307–310.

Anderson, J.M. (1971). Assessment of the effects of pollutants on physiology and behavior. *Proc. R. Soc. London B.* 177: 307–320.

Antunes-Madeira, M.C., Videira, R.A., and Madeira, V.M.C. (1994). Effects of parathion on membrane organization and its implications for the mechanisms of toxicity. *Biochem. Biophys. Acta* 1190: 149–154.

Argyropoulos, G. and Harper, M. (2002). Invited review: Uncoupling proteins and thermoregulation. *J. Appl. Physiol.* 92: 2187–2198.

Arieli, Y., Eynan, M., Ganez, H., Arieli, R., and Kashi, Y. (2003). Heat acclimation prolongs the time to central nervous system oxygen toxicity in the rat: Possible involvement of HSP72. *Brain Res.* 962: 15–20.

Aschoff, J. (1981). Thermal conductance in mammals and birds: Its dependence on body size and circadian phase. *Comp. Biochem. Physiol.* 69A: 611–619.

Assi, A.-A. and Nasser, H. (1999). An *in vitro* and *in vivo* study of some biological and biochemical effects of Sistrurus Malrius Barbouri venom. *Toxicology* 137: 81–94.

Astrup, J. (1982). Energy-requiring cell functions in the ischemic brain. *J. Neurosurg.* 56: 482–497.

Avlonitou, E. and Elizondo, R. (1988). Effects of atropine and pyridostigmine in heat-stressed monkeys. *Aviat. Space Environ. Med.* 59: 544–548.

Baer, K.N., Oliver, K., and Pope, C.N. (2002). Influence of temperature and dissolved oxygen on the acute toxicity of profenofos to fathead minnows (*Pimephales promelas*). *Drug Chem. Toxicol.* 25: 231–245.

Baetjer, A.M., Joardar, S.N., and McQuary, W.A. (1960). Effect of environmental temperature and humidity on lead poisoning in animals. *Arch. Environ. Health* 1: 463–477.

Baetjer, A.M. and Horiguchi, S. (1964). Effects of environmental temperature and dehydration on lead poisoning in laboratory animals. *14th Int. Congress Occup. Health Proc.* 62: 795–797.

Baetjer, A.M. and Smith, R. (1956). Effect of environmental temperature on reaction of mice to parathion, an anticholinesterase agent. *Am. J. Physiol.* 186: 39–46.

Bagchi, D., Bhattacharya, G., and Stohs, S.J. (1996). *In vitro* and *in vivo* induction of heat shock (stress) protein (Hsp) gene expression by selected pesticides. *Toxicology* 112: 57–68.

Ballantyne, B. and Marrs, T.C. (1992). Overview of the biological and clinical aspects of organophosphates and carbamates. In *Clinical and Experimental Toxicology of Organophosphates and Carbamates*. Ballantyne, B. and Marrs, T.C., Eds. Oxford: Butterworth-Heinemann, pp. 3–14.

Ballatori, N. and Villalobos, A.R. (2002). Defining the molecular and cellular basis of toxicity using comparative models. *Toxicol. Appl. Pharmacol.* 183: 207–220.

Barenholz-Paniry, V., Ishay, J.S., Freeman, S., and Sohmer, H. (1990). Evoked potential changes in cats following injection of an extract from the venom sac of the oriental hornet (*Vespa orientalis*). *Toxicon* 28: 1317–1324.

Barnes, J.A., Collins, B.W., Dix, D.J., and Allen, J.W. (2002). Effects of heat shock protein 70 (Hsp 70) on arsenite-induced genotoxicity. *Environ. Mol. Mut.* 40: 236–242.

Barnes, J.A., Dix, D.J., Collins, B.W., Luft, C., and Allen, J.W. (2001). Expression of inducible Hsp70 enhances the proliferation of MCF-7 breast cancer cells and protects against the cytotoxic effects of hyperthermia. *Cell Stress Chaperones* 6: 316–325.

Beattie, J.H., Wood, A.M., Trayhurn, P., Jasani, B., Vincent, A., McCormack, G., and West, A.K. (2000). Metallothionein is expressed in adipocytes of brown fat and is induced by catecholamines and zinc. *Am. J. Physiol.* 278: R1082–R1089.

Becker, C.D. and Wolford, M.G. (1980). Thermal resistance of juvenile salmonids sublethally exposed to nickel, determined by the critical thermal maximum. *Environ. Pollut.* (series A) 21: 181–189.

Bejanian, M., Finn, D.A., Syapin, P.J., and Alkana, R.L. (1990). Body temperature and ethanol pharmacokinetics in temperature-challenged mice. *Alcohol* 7: 331–337.

Benignus, V.A. (1994). Behavioral effects of carbon monoxide: Meta analyses and extrapolations. *J. Appl. Physiol.* 76: 1310–1316.

Benton, A.W., Heckman, R.A., and Morse, R.A. (1966). Environmental effects on venom toxicity in rodents. *J. Appl. Physiol.* 21: 1228–1230.

Bhat, R., Turner, S.L., Marks, M.J., and Collins, A.C. (1990). Selective changes in sensitivity to cholinergic agonists and receptor changes elicited by continuous physostigmine infusion. *J. Pharmacol. Exp. Ther.* 255: 187–196.

Blatteis, C.M., Ed. (1998). *Physiology and Pathophysiology of Temperature Regulation.* Singapore: World Scientific.

Bligh, J. (1998). Mammalian homeothermy: an integrative thesis. *J. Thermal Biol.* 23: 143–258.

Bolt, H.M., Roos, P.H., and Thier, R. (2003). The cytochrome P-450 isoenzyme CYP2E1 in the biological processing of industrial chemicals: consequences for occupational and environmental medicine. *Int. Arch. Occup. Environ. Health* 76: 174–185.

Boone, M.D. and Bridges, C.M. (1999). The effect of temperature on the potency of carbaryl for survival of tadpoles of the green frog (*Rana clamitans*). *Environ. Toxicol. Chem.* 18: 1482–1484.

Boulant, J.A. (2000). Role of the preoptic-anterior hypothalamus in thermoregulation and fever. *Clin. Infect. Dis.* Suppl 5: S157–161.

Boulant, J.A., Curras, M.C., and Dean, J.B. (1989). Neurophysiological aspects of thermoregulation. In *Comparative and Environmental Physiology,* Vol. 4. *Animal Adaptation to Cold,* Wang, L.C.H., Ed., pp. 117–160. London: Springer-Verlag.

Boulant, J.A. and Dean, J.B. (1986). Temperature receptors in the central nervous system. *Annu. Rev. Physiol.* 48: 639–654.

Boulant, J.A. and Silva, N.L. (1989). Multisensory hypothalamic neurons may explain interactions among regulatory systems. *News Physiol. Sci.* 4: 245–248.

Bourdeau, P., Haines, J.A., Klein, W., Krishna Murti, C.R. (1989). *Ecotoxicology and climate.* SCOPE 38 IPCS Joint Symposia 9, Chichester, UK, 392 pp.

Bowyer, J.F., Tank, A.W., Newport, G.D., Slikker, W., Jr., Ali, S.F., and Holson, R.R. (1992). The influence of environmental temperature on the transient effects of methamphetamine on dopamine levels and dopamine release in rat striatum. *J. Pharmacol. Exp. Ther.* 260: 817–824.

Boyer, R. and Grue, C.E. (1995). The need for water quality criteria for frogs. *Environ. Health Perspect.* 103: 352–357.

Boyes, W.K. and Moser, V.C. (1988). An alpha 2-adrenergic mode of action of chlordimeform on rat visual function. *Toxicol. Appl. Pharmacol.* 92: 402–418.

Boyes, W.K., Padilla, S., and Dyer, R.S. (1985). Body temperature-dependent and independent actions of chlordimeform on visual evoked potentials and axonal transport in optic system of rat. *Neuropharmacology* 24: 743–749.

Bozinovic, F. and Novoa, F.F. (1997). Metabolic costs of rodents feeding on plant chemical defenses: A comparison between an herbivore and an omnivore. *Comp. Biochem. Physiol.* 117A: 511–514.

Braga, A.L., Zanobetti, A., and Schwartz, J. (2001). The time course of weather-related deaths. *Epidemiology* 12: 662–667.

Branco, L.G. and Malvin, G.M. (1996). Thermoregulatory effects of cyanide and azide in the toad, *Bufo marinus. Am. J. Physiol.* 270: R169–R173.

Brattsten, L.B. (1979). Ecological significance of mixed-function oxidations. *Drug Metab. Rev.* 10: 35–58.

Brenner, I., Shek, P.N., Zamecnik, J., and Shepard, R.J. (1998). Stress hormones and the immunological responses to heat and exercise. *Int. J. Sports Med.* 19: 130–143.

Briese, E. and Hernandez, L. (1996). Ethanol anapyrexia in rats. *Pharmacol. Biochem. Behav.* 54: 399–402.

Broomhall, S. (2002). The effects of endosulfan and variable water temperature on survivorship and subsequent vulnerability to predation in *Litoria citropa* tadpoles. *Aquat. Toxicol.* 61: 243–250.

Brown, J.W., Carroll, R.G., Whitehurst, M.E., and Gordon, C.J. (2001). Thermoregulatory set point decreases following hemorrhage in rats. *FASED J.* 15: A1107.

Brück, K. and Zeisberger, E. (1990). Adaptive changes in thermoregulation and their neuropharmacological basis. In *Thermoregulation: Physiology and Biochemistry*, Schönbaum, E. and Lomax, P., Eds., pp. 255–307. New York: Pergamon Press.

Brucker-Davis, F. (1998). Effects of environmental synthetic chemicals on thyroid function. *Thyroid* 8: 827–856.

Brunet, R. and McDuff, J. (1997). Recovery of brain cholinesterase of brown-headed cowbirds from organophosphorus intoxication: Effect of environmental temperature. *Bull. Environ. Contam. Toxicol.* 59: 285–291.

Buchan, A. and Pulsinelli, W.A. (1990). Hypothermia but not the N-methyl-D-aspartate antagonist, MK-801, attenuates neuronal damage in gerbils subjected to transient global ischemia. *J. Neurosci.* 10: 311–316.

Burgat-Sacaze, V., Braun, J.-P., Benard, P., Eghbali, B., and Rico, A. (1982). Protection against mercuric chloride nephrotoxicity by cold exposure in rats. *Toxicol. Lett.* 10: 151–156.

Burger, J. and Gochfeld, M. (2000). Effects of lead on birds (Laridae): A review of laboratory and field studies. *J. Toxicol. Environ. Health Part B* 3: 59–78.

Burke, D.H. (1978). Cobaltous chloride-induced hypothermia in mice. I. Effect of pretreatment with anticholinergic drugs. *J. Pharmaceut. Sci.* 67: 799–801.

Burke, D.H. and Brooks, J.C. (1979). Cobaltous chloride-induced hypothermia. II. Pretreatment with sympathoplegics, antihistamines, and narcotic antagonists. *J. Pharm. Sci.* 68: 693–696.

Burke, D.H., Brooks, J.C., and Treml, S.B. (1983). Cobaltous chloride-induced hypothermia in mice. III. Effect of pretreatment with 5-hydroxytryptaminergic agents. *J. Pharm. Sci.* 72: 824–826.

Burke, J.M., Spiers, D.E., Kojima, F.N., Perry, G.A., Salfen, B.E., Wood, S.L., Patterson, D.J., Smith, M.F., Luch, M.C., Jackson, W.G., and Piper, E.L. (2001). Interaction of endophyte-infected fescue and heat stress on ovarian function in the beef heifer. *Biol. Reprod.* 65: 260–268.

Burton, A.C. (1975). Variety: The spice of science as well as of life. *Annu. Rev. Physiol.* 38: 1–12.

Burton, A.C. and Edholm, O.G. (1955). *Man in a Cold Environment.* London: Edward Arnold, 273 pp.

Cagen, S.Z., Janoff, A.S., Bus, J.S., and Gibson, J.E. (1976). Effect of paraquat (methyl viologen) on liver function in mice. *J. Pharmacol. Exp. Ther.* 198: 222–228.

Cairns, J., Jr., Heath, A.G., and Parker, B.C. (1975). Temperature influence on chemical toxicity to aquatic organisms. *J. Water Pollut. Control Fed.* 47: 267–280.

Cairns, J., Jr., Heath, A.G., and Parker, B.C. (1975a). The effects of temperature upon the toxicity of chemicals to aquatic organisms. *Hydrobiologia* 47: 135–171.

Calder, W.A. III. (1981). Scaling of physiological processes in homeothermic animals. *Annu. Rev. Physiol.* 43: 301–322.

Cameron, J. and Capra, M.F. (1993). The basis of the paradoxical disturbance of temperature perception in ciguatera poisoning. *J. Toxicol. Clin. Toxicol.* 31: 571–579.

Campen, J.J., Nolan, J.P., Schladweiler, M.C., Kodavanti, U.P., Evansky, P.A., Costa, D.L., and Watkinson, W.P. (2001). Cardiovascular and thermoregulatory effects of inhaled PM-associated transition metals: A potential interaction between nickel and vanadium sulfate. *Toxicol. Sci.* 64: 243–252.

Campen, M.J., Costa, D.L., and Watkinson, W.P. (2000a). Cardiac and thermoregulatory toxicity of residual oil fly ash in cardiopulmonary-compromised rats. *Inhal. Toxicol.* 12 (Suppl. 2): 7–22.

Campen, M.J., Norwood, J., McKee, J.L., Mebane, R., Hatch, G.E., and Watkinson, W.P. (2000). Ozone-induced hypothermia and bradycardia in rats and guinea pigs in nose-only or whole-body inhalation systems. *J. Thermal Biol.* 25: 81–89.

Cancalon, P.F. (1988). Axonal transport in the garfish optic nerve: Comparison with the olfactory system. *J. Neurochem.* 51: 266–276.

Cardell, M., Boris-Moller, F., and Wieloch, T. (1991). Hypothermia prevents the ischemia-induced translocation and inhibition of protein kinase C in the rat striatum. *J. Neurochem.* 57: 1814–1817.

Casarett, L.J. and Doull, J. (1975). Factors influencing toxicology. In *Toxicology: The Basic Sciences of Poisons,* pp. 133–147. New York: Macmillan.

Chaffee, R.R.J. and Roberts, J.C. (1971). Temperature acclimation in birds and mammals. *Annu. Rev. Physiol.* 33: 155–202.

Chagnon, N. and Hlohowskyj, I. (1989). Effects of phenol exposure on the thermal tolerance ability of the central stoneroller minnow. *Bull. Environ. Contam. Toxicol.* 42: 614–619.

Chang, S.-K. and Riviere, J.E. (1991). Percutaneous absorption of parathion *in vitro* in porcine skin: Effects of dose, temperature, humidity, and perfusate composition on absorptive flux. *Fund. Appl. Toxicol.* 17: 494–504.

Chang, S.-K., Brownie, C., and Riviere, J.E. (1994). Percutaneous absorption of topical parathion through porcine skin: *in vitro* studies on the effect of environmental perturbations. *J. Vet. Pharamcol. Ther.* 17: 434–439.

Chawla, N., Johri, M.B., Saxena, P.N., and Singhal, K.C. (1975). Cholinergic mechanisms in central thermoregulation in pigeons. *Br. J. Pharmacol.* 53: 317–322.

Chayoth, R., Kleinman, D., Kaplanski, J., and Sod Moriah, U.A. (1984). Renal clearance of urea, inulin, and p-aminohippurate in heat-acclimated rats. *J. Appl. Physiol.* 57: 731–732.

Chen, C.H. and Engel, S.G. (1990). The susceptibility of membrane vesicles to interaction with ethanol. *Chem. Phys. Lipids* 52: 179–187.

Christensen, C.W. and Fujimoto, J.M. (1984). Cadmium induced hypothermia in mice: Dose dependent tolerance development. *Gen. Pharmacol.* 15: 263–266.

Christoph, R.A. (1989). Organophosphates and carbamates. In *Manual of Toxicologic Emergencies,* 1st ed. Chicago: Year Book Medical Publishers, p. 626.

Clark, W.G. and Lipton, J.M. (1985). Changes in body temperature after administration of amino acids, peptides, dopamine, neuroleptics and related agents: II. Neurosci. Biobehav. Rev. 9: 299–371.

Clark, W.G. and Lipton, J.M. (1985a). Changes in body temperature after administration of acetylcholine, histamine, morphine, prostaglandins and related agents: II. *Neurosci. Biobehav. Rev.* 9: 479–552.

Clement, J.G. (1991). Variability of sarin-induced hypothermia in mice: Investigation into incidence and mechanism. *Biochem. Pharmacol.* 42: 1316–1318.

Clement, J.G. (1991a). Hypothermia: Limited tolerance to repeated soman administration and cross-tolerance to oxotremorine. *Pharm. Biochem. Behav.* 39: 305–312.

Clement, J.G. (1993). Pharmacological nature of soman-induced hypothermia in mice. *Pharmacol. Biochem. Behav.* 44: 689–702.

Clement, J.G. (1993a). Experimentally induced mortality following repeated measurement of rectal temperature in mice. *Lab. Anim. Sci.* 43: 381–382.

Clement, J.G. and Erhardt, N. (1990). Serum carboxylesterase activity in various strains of rats: Sensitivity to inhibition by CBDP. *Arch. Toxicol.* 64: 414–416.

Clemons, G.K. and Wei, D. (1984). Effect of short-term ozone exposure on exogenous thyroxine levels in thyroidectomized and hypophysectomized rats. *Toxicol. Appl. Pharmacol.* 74: 86–90.

Collins, A.C., Smolen, A., Wayman, A.L., and Marks, M.J. (1984). Ethanol and temperature effects on five membrane bound enzymes. *Alcohol* 1: 237–246.

Conn, C.A., Dokladny, K., Menache, M.G., Barr, E.B., Kozak, W., Kozak, A., Wachulec, M., Rudolph, K., Kluger, M.J., and Henderson, R.F. (2002). Effects of sarin on temperature and activity of rats as a model for Gulf War syndrome neuroregulatory functions. *Toxicol. Appl. Pharmacol.* 184: 77–81.

Cook, L.L., Edens, F.W., and Tilson, H.A. (1988). Possible brainstem involvement in the modification of thermoregulatory processes by chlordecone in rats. *Neuropharmacology* 27: 871–879.

Cook, L.L., Edens, F.W., and Tilson, H.A. (1988a). Pharmacological evaluation of central adrenergic involvement in chlordecone-induced hypothermia. *Neuropharmacology* 27: 881–887.

Cook, L.L., Gordon, C.J., Tilson, H.A., and Edens, F.W. (1987). Chlordecone-induced effects on thermoregulatory processes in the rat. *Toxicol. Appl. Pharmacol.* 90: 126–134.

Corbett, D., Evans, S., Thomas, C., Wang, D., and Jonas, R.A. (1990). MK-801 reduced cerebral ischemic injury by inducing hypothermia. *Brain Res.* 514: 300–304.

Cossins, A.R. and Bowler, K. (1987). *Temperature Biology of Animals.* Chapman and Hall: London.

Costa, L.G., Hand, H., Schwab, B.W., and Murphy, S.D. (1981). Tolerance to the carbamate insecticide propoxur. *Toxicology* 21: 267–278.

Costa, L.G. and Murphy, S.D. (1983). Unidirectional cross-tolerance between the carbamate insecticide propoxur and the organophosphate disulfoton in mice. *Fund. Appl. Toxicol.* 3: 483–488.

Costa, L.G., Schwab, B.W., and Murphy, S.D. (1982). Tolerance to anticholinesterase compounds in mammals. *Toxicology* 25: 79–97.

Costa, L.G., Wu, D.S., Olibet, G., and Murphy, S.D. (1989). Formamidine pesticides and alpha 2-adrenoceptors: Studies with amitraz and chlordimeform in rats and development of a radioreceptor binding assay. *Neurotoxicol. Teratol.* 11: 405–411.

Coudray-Lucas, C., Prioux-Guyonneau, M., Sentenac, H., Cohen, Y., and Wepierre, J. (1983). Brain catecholamine metabolism changes and hypothermia in intoxication by anticholinesterase agents. *Acta Pharmacol. Toxicol.* 52: 224–229.

Coudray-Lucas, C., Prioux-Guyonneau, M., Tassel, A., Coq, H.M., Cohen, Y., and Wepierre, J. (1981). Influence of intoxication by anticholinesterase agents on core temperature in rats: Relationship between hypothermia and acetylcholinesterase inhibition in different brain areas. *Acta Pharmacol. Toxicol.* 49: 215–222.

Cowles, R.B. and Bogert, C.M. (1944). A preliminary study of the thermal requirements of desert reptiles. *Bull. Am. Mus. Nat. Hist.* 83: 261–296.

Crabbe, J.C., Kosobud, A., Tam, B.R., Young, E.R., and Deutsch, C.M. (1987). Genetic selection of mouse lines sensitive (COLD) and resistant (HOT) to acute ethanol hypothermia. *Alcohol Drug Res.* 7: 163–174.

Crabbe, J.C., Phillips, T.J., Gallaher, E.J., Crawshaw, L.I., and Mitchell, S.R. (1996). Common genetic determinants of the ataxic and hypothermic effects of ethanol in BXD/Ty recombinant inbred mice: Genetic correlations and quantitative trait loci. *J. Pharmacol. Exp. Ther.* 277: 624–632.

Craig, F.N., Bales, P.D., and Frankel, H.M. (1959). Lethality of sarin in a warm environment. *J. Pharmacol. Exp. Ther.* 127: 35–38.

Craig, F.N., Cummins, E.G., and Sum, V.M. (1977). Environmental temperature and the percutaneous absorption of a cholinesterse inhibitor, VX. *J. Invest. Dermatol.* 68: 357–361.

Craig, T.V., Hunt, W., and Atkinson, R. (1959a). Hypothermia—its use in severe carbon monoxide poisoning. *New Engl. J. Med.* 261: 854–856.

Crawshaw, L.I., Wallace, H.I., Christensen, R., and Crabbe, J.C. (2001). Influence of ethanol on thermoregulation: Mapping of quantitative trait loci. *Physiol. Genomics* 7: 159–169.

Crawshaw, L.I., Wallace, H.L., Crabbe, J.C., Ramos, C., Duerr, J., and O'Connor, C.S. (1997). Ethanol disrupts and decreases the regulated body temperature differentially in C57BL/6J and DBA/2J mice. *Am. J. Physiol.* 273: R696–R702.

Crawshaw, L.I., Wollmuth, L.P., and O'Connor, C.S. (1989). Intracranial ethanol and ambient anoxia elicit selection of cooler water by goldfish. *Am. J. Physiol.* 256: R133–R137.

Cunningham, C.L. and Bischof, L.L. (1987). Stress and ethanol-induced hypothermia. *Physiol. Behav.* 40: 377–382.

Cupp, C.M., Kleiber, G., Reigart, R., Sandfifer, S.H. (1975). Hypothermia in organophosphate poisoning and response to PAM. *J. S. Carolina Med. Assoc.* 71: 166–168.

Curley, A. and Garrettson, L.K. (1969). Acute chlordane poisoning. *Arch. Environ. Health* 18: 211–215.

Curriero, F.C., Heiner, K.S., Samet, J.M., Zeger, S.L., Strug, L., and Patz, J.A. (2002). Temperature and mortality in 11 cities of the eastern United States. *Am. J. Epidemiol.* 155: 80–87.

Curriero, F.C., Samet, J.M., and Zeger, S.L. (2003). Letter to the editor. *Am. J. Epidemiol.* 158: 93–94.

Danel, T., Libersa, C., and Touitou, Y. (2001). The effect of alcohol consumption on the circadian control of human core body temperature is time dependent. *Am. J. Physiol.* 281: R52–R55.

Davidson, I.W.F., Parker, J.C., and Beliles, R.P. (1986). Biological basis for extrapolation across mammalian species. *Reg. Toxicol. Pharm.* 6: 211–237.

Dean, J.C., Salmon, S.E., and Griffith, K.S. (1979). Prevention of doxorubicin-induced hair loss with scalp hypothermia. *New Engl. J. Med.* 301: 1427–1429.

Del Ramo, J., Diaz-Mayans, J., Torreblanca, A., and Núñez, A. (1987). Effect of temperature on the acute toxicity of heavy metals (Cr, Cd, and Hg) to the freshwater crayfish, *Procambarus clarkii* (Girard). *Bull. Environ. Contam. Toxicol.* 38: 736–741.

Dietrich, W.D. (1992). The importance of brain temperature in cerebral injury. *J. Neurotrama* 9: S475–S485.

Domanik, A.M. and Zar, J.H. (1978). The effect of malathion on the temperature selection response of the common shiner, *Notropis cornutus* (Mitchill). *Arch. Environ. Contam. Toxicol.* 7: 193–206.

Dorr, R.T., Alberts, D.S., and Stone, A. (1985). Cold protection and heat enhancement of doxorubicin skin toxicity in the mouse. *Cancer Treat. Rep.* 69: 431–437.

Doull, J. (1972). The effect of physical environmental factors on drug response. *Essays Toxicol.* 3: 37–63.

Dreyer, F. and Schmitt, A. (1983). Transmitter release in tetanus and botulinium A toxin-poisoned mammalian endplates and its dependence on nerve stimulation and temperature. *Pflugers Arch.* 399: 228–234.

Drinkwater, B.L., Raven, P.B., Horvath, S.M., Gliner, G.A., Ruhling, R.O., Bolduan, N.W., and Taguchi, S. (1974). Air pollution, exercise, and heat stress. *Arch. Environ. Health* 28: 177–181.

Dutta, A.S. and Chaudhuri, A.K. (1991). Neuropharmacological studies on the venom of *Vipera russelli*. *Ind. J. Exp. Biol.* 29: 937–942.

Duvdevani, P., Ishay, J.S., and Schejter, A. (1991). Mastoparan induces hypothermia in mice. *Experientia* 47: 460–462.

Dyer, R.S. and Howell, W.E. (1982). Triethyltin: Ambient temperature alters visual system toxicity. *Neurobehav. Toxicol. Teratol.* 4: 267–271.

Epstein, Y., Seidman, D.S., Moran, D., Arnon, R., Arad, M., and Varssano, D. (1990). Heat-exercise performance of pyridostigmine-treated subjects wearing chemical protective clothing. *Aviat. Space Environ. Med.* 61: 310–313.

Everts, J.W. (1997). Ecotoxicology for risk assessment in arid zones: Some key issues. *Arch. Environ. Contam. Toxicol.* 32: 1–10.

Falk, M.H. and Issels, R.D. (2001). Hyperthermia in oncology. *Int. J. Hyperthermia* 17: 1–18.

Fellows, I.W., MacDonald, I.A., and Bennett, T. (1984). The influence of environmental temperature upon thermoregulatory responses to ethanol in man. *Clin. Sci.* 66: 733–739.

Feng, J.D., Price, M., Cohen, J., and Satinoff, E. (1989). Prostaglandin fevers in rats: Regulated change in body temperature or change in regulated body temperature? *Am. J. Physiol.* 257: R695–R699.

Ferrari, L., Salibián, A., and Muiño, C. (1993). Selective protection of temperature against cadmium acute toxicity to *Bufo arenarum* tadpoles. *Bull. Environ. Contam. Toxicol.* 50: 212–218.

Fewell, J.E. and Eliason, H.L. (2002). Fetal exposure to nicotine does not alter the core temperature response of 7- to 8-week-old rats to intracerebroventricular administration of PGE1. *Physiol. Behav.* 77: 269–275.

Fewell, J.E., Eliason, H.L., and Crisanti, K.C. (2001). Prenatal exposure to nicotine attenuates stress-induced hyperthermia in 7- to 8-week-old rats upon exposure to a novel environment. *Physiol. Behav.* 74: 595–601.

Fine, J.M., Gordon, T., Chen, L.C., Kinney, P., Falcone, G., and Beckett, W.S. (1997). Metal fume fever: characterization of clinical and plasma IL-6 responses in controlled human exposures to zinc oxide fume at and below threshold limit values. *J. Occup. Environ. Med.* 39: 722–726.

Fine, J.M., Gordon, T., Chen, L.C., Kinney, P., Falcone, G., Sparer, J., and Beckett, W.S. (2000). Characterization of clinical tolerance to inhaled zinc oxide in naive subjects and sheet metal workers. *J. Occup. Environ. Med.* 42: 1085–1091.

Finn, D.A., Bejanian, M., Jones, B.L., Syapin, P.J., and Alkana, R.L. (1989). Temperature affects ethanol lethality in C57BL/6, 129, LS and SS mice. *Pharmacol. Bioch. Behav.* 34: 375–380.

Folk, G.E., Jr. (1974). *Textbook of Environmental Physiology.* Philadelphia: Lea and Febiger.

Folk, G.E., Jr., Riedesel, M.L., and Thrift, D.L. (1998). *Principles of Integrative Environmental Physiology.* Austin & Winfield, Bethesda, MD.

Forster, R.E., Ferris, B.G., and Day, R. (1946). The relationship between total heat exchange and blood flow in the hand at various ambient temperatures. *Am. J. Physiol.* 146: 600–609.

Francesconi, R., Hubbard, R., and Mager, M. (1983). Malathion administration: Effects on physiological and physical performance in the heat. *Pharm. Biochem. Behav.* 19: 1031–1034.

Francesconi, R., Hubbard, R., and Mager, M. (1984). Effects of pyridostigmine on ability of rats to work in the heat. *J. Appl. Physiol.* 56: 891–895.

Francesconi, R., Hubbard, R., Matthew, C., Leva, N., Young, J., and Pease, V. (1986). Oral pyridostigmine administration in rats: Effects on thermoregulation, clinical chemistry, and performance in the heat. *Pharmacol. Biochem. Behav.* 25: 1071–1075.

Frank, R. and Tankersley, C. (2002). Air pollution and daily mortality: A hypothesis concerning the role of impaired homeostasis. *Environ. Health Perspect.* 110: 61–65.

Fregly, M.J. (1989). Activity of the hypothalamic-pituitary-thyroid axis during exposure to cold. *Pharmacol. Ther.* 41: 85–142.

Freinkel, N., Metzger, B.E., Harris, E., Robinson, S. and Mager, M. (1972). The hypothermia of hypoglycemia with 2-deoxy-D-glucose in normal human subjects and mice. *N. Engl. J. Med.* 287: 841–845.

Fuhrman, F.A. (1946). The effect of body temperature on drug action. *Physiol. Rev.* 26: 247–274.

Fuhrman, G.J. and Fuhrman, F.A. (1961). Effects of temperature on the action of drugs. *Annu. Rev. Pharmacol.* 1: 65–78.

Funckes, A.J., Hayes, G.R., Jr., and Hartwell, W.V. (1963). Urinary excretion of paranitrophenol by volunteers following dermal exposure to parathion at different ambient temperatures. *J. Agric. Food Chem.* 11: 455–457.

Furie, B. and Trubowitz, S. (1976). Insecticides and blood dyscrasias: Chlordane exposure and self-limited refractory megaloblastic anemia. *J. Am. Med. Assoc.* 235: 1720–1722.

Furuyama, F. (1982). Strain difference in thermoregulation of rats surviving extreme heat. *J. Appl. Physiol.* 52: 410–415.

Gallaher, E.J. and Egner, D.A. (1987). Rebound hyperthermia follows ethanol-induced hypothermia in rats. *Psychopharmacology* 91: 34–39.

Gardner, D.R. (1973). The effect of some DDT and methoxchlor analogs on temperature selection and lethality in brook trout fingerlings. *Pest. Biochem. Physiol.* 2: 437–446.

Gautier, H. and Bonora, M. (1994). Ventilatory and metabolic responses to cold and CO-induced hypoxia in awake rats. *Respir. Physiol.* 97: 79–91.

Gearhart, J.M., Jepson, G.W., Clewell, H.J., Andersen, M.E., and Conolly, R.B. (1994). Physiologically based pharmacokinetic model for the inhibition of acetylcholinesterase by organophosphate esters. *Environ. Health Perspect.* 102 (Suppl 11): 51–60.

Gearhart, J.M., Seckel, C., and Vinegar, A. (1993). *In vivo* metabolism of chloroform in B6C3F1 mice determined by the method of gas uptake: The effects of body temperature on tissue partition coefficients and metabolism. *Toxicol. Appl. Pharmacol.* 119: 258–266.

George, F.R. and Collins, A.C. (1985). Ethanol's behavioral effects may be partly due to increases in brain prostaglandin production. *Alcohol Clin. Exp. Res.* 9: 143–146.

Gilbert, M.E., Mack, C.M., and Crofton, K.M. (1989). Pyrethroids and enhanced inhibition in the hippocampus of the rat. *Brain Res.* 477: 314–321.

Gillespie, N.C., Lewis, R.J., Pearn, J.H., Bourke, A.T., Holmes, M.J., Bourke, J.B., and Shields, W.J. (1986). Ciguatera in Australia: Occurrence, clinical features, pathophysiology and management. *Med. J. Aust.* 145: 584–590.

Gliner, J.A., Raven, P.B., Horvath, S.M., Drinkwater, B.L., and Sutton, J.C. (1975). Man's physiologic response to long-term work during thermal and pollutant stress. *J. Appl. Physiol.* 39: 628–632.

Globus, M.Y.-T., Busto, R., Lin, B., Schnippering, H., and Ginsberg, M.D. (1995). Detection of free radical activity during transient global ischemia and recirculation: Effects of intraischemic brain temperature modulation. *J. Neurochem.* 65: 1250–1256.

Gordon, C.J. (1990). Thermal biology of the laboratory rat. *Physiol. Behav.* 47: 963–991.

Gordon, C.J. (1991). Toxic-induced hypothermia and hypometabolism: Do they increase uncertainty in the extrapolation of toxicological data from experimental animals to humans? *Neurosci. Biobehav. Rev.* 15: 95–98.

Gordon, C.J. (1993). *Temperature Regulation in Laboratory Rodents.* New York: Cambridge University Press.

Gordon, C.J. (1993a). Acute and delayed effects of diisopropyl fluorophosphate on body temperature, heart rate, and motor activity in the awake, unrestrained rat. *J. Toxicol. Environ. Health* 39: 247–260.

Gordon, C.J. (1994). Thermoregulation in laboratory mammals and humans exposed to anticholinesterase agents. *Neurotoxicol. Teratol.* 16: 427–453.

Gordon, C.J. (1994a). 24-hour control of body temperature in the rat. II. Diisopropyl fluorophosphate-induced hypothermia and hyperthermia. *Pharmacol. Toxicol. Behav.* 49: 747–754.

Gordon, C.J. (1994b). 24-hour control of body temperature in rats. I. Integration of behavioral and autonomic effectors. *Am. J. Physiol.* 36: R71–R77.

Gordon, C.J. (1996). Homeothermy: Does it impede the response to cellular injury. *J. Thermal Biol.* 21: 29–36.

Gordon, C.J. (1996a). Pharmacological analysis of diisopropyl fluorophosphate: Effects on core temperature, heart rate, and motor activity in the unrestrained rat. *Pharm. Bioch. Behav.* 55: 185–194.

Gordon, C.J. (1997). Behavioral thermoregulatory response to chlorpyrifos in the rat. *Toxicology* 124: 165–171.

Gordon, C.J. (1997a). The role of behavioral thermoregulation as a thermoeffector during prolonged hypoxia in the rat. *J. Thermal Biol.* 22: 315–324.

Gordon, C.J. (2001). The therapeutic potential of regulated hypothermia. *Emerg. Med. J.* 18: 81–89.

Gordon, C.J. (2003). Role of environmental stress in the physiological response to chemical toxicants. *Environ. Res.* 92: 1–7.

Gordon, C.J., Dyer, R.S., Long, M.D., and Fehlner, K.S. (1985). Effect of sulfolane on behavioral and autonomic thermoregulation in the rat. *J. Toxicol. Environ. Health* 16: 461–468.

Gordon, C.J. and Fogelson, L. (1991). Comparative effects of hypoxia on behavioral thermoregulation in rats, hamsters, and mice. *Am. J. Physiol.* 260: R120–R125.

Gordon, C.J. and Fogelson, L. (1993). Relationship between serum cholinesterase activity and the change in body temperature and motor activity in the rat: A dose–response study of the diisopropyl fluorophosphate. *Neurotoxicol. Teratol.* 15: 21–25.

Gordon, T. and Fine, J.M. (1993). Metal fume fever. *Occup. Med.: State of the Art Reviews.* 8: 505–517.

Gordon, C.J. and Heath, J.E. (1986). Integration and central processing in temperature regulation. *Annu. Rev. Physiol.* 48: 595–612.

Gordon, C.J. and Mack, C.M. (2001). Diurnal variation in thermoregulatory response to chlorpyrifos and carbaryl in the rat. *Toxicology* 169: 93–105.

Gordon, C.J. and Mack, C.M. (2003). Influence of gender on thermoregulation and cholinesterase inhibition in the Long-Evans rat exposed to diazinon. *J. Toxicol. Environ. Health A* 66: 291–304.

Gordon, C.J. and MacPhail, R.C. (1993). Strain comparison of DFP neurotoxicity in rats. *J. Toxicol. Environ. Health* 38: 257–271.

Gordon, C.J. and O'Callaghan, J.P. (1995). Trimethyltin-induced neuropathy in the rat: Interaction with thermoregulation. *Neurotoxicology* 16: 319–326.

Gordon, C.J. and Padnos, B. (2002). Dietary exposure to chlorpyrifos alters core temperature in the rat. *Toxicology* 177: 215–226.

Gordon, C.J. and Refinetti, R. (1993). Measurement of behavioral thermoregulation. In *Methods in Neurosciences: Paradigms for the Study of Behavior,* Vol. 14, Conn, P.M., Ed., pp. 266–280. San Diego, CA: Academic Press.

Gordon, C.J and Rowsey, P.J. (1998). Delayed febrile effects of chlorpyrifos: Is there cross-tolerance to bacterial lipopolysaccharide? *Toxicology* 130: 17–28.

Gordon, C.J. and Rowsey, P.J. (1999). Are circulating cytokines interleukin-6 and tumor necrosis factor α involved in chlorpyrifos-induced fever? *Toxicology* 134: 9–17.

Gordon, C.J. and Stead, A.G. (1986). Effect of nickel and cadmium chloride on autonomic and behavioral thermoregulation in mice. *Neurotoxicology* 7: 97–106.

Gordon, C.J. and Stead, A.G. (1986a). Effects of alcohol on behavioral and autonomic thermoregulation in mice. *Alcohol* 3: 339–343.

Gordon, C.J. and Watkinson, W.P. (1988). Behavioral and autonomic thermoregulation in the rat following chlordimeform administration. *Neurotoxicol. Teratol.* 10: 215–219.

Gordon, C.J. and Watkinson, W.P. (1995). Strain differences in the laboratory rat: Impact on the autonomic, behavioral, and biochemical response to cholinesterase inhibition. *J. Toxicol. Environ. Health* 45: 59–73.

Gordon, C.J. and Yang, Y. (1997). Contribution of spontaneous motor activity to the 24 hour control of body temperature in male and female rats. *J. Thermal Biol.* 22: 59–68.

Gordon, C.J. and Yang, Y.-L. (2001). Reduction in open field-induced hyperthermia in the rat exposed to chlorpyrifos, an anticholinesterase pesticide. *J. Thermal Biol.* 26: 313–318.

Gordon, C.J., Fogelson, L., and Highfill, J.W. (1990). Hypothermia and hypometabolism: Sensitive indices of whole-body toxicity following exposure to metallic salts in the mouse. *J. Toxicol. Environ. Health* 29: 185–200.

Gordon, C.J., Fogelson, L., Lee, L., and Highfill, J. (1991b). Acute effects of diisopropyl fluorophosphate (DFP) on autonomic and behavioral thermoregulatory responses in the Long-Evans rat. *Toxicology* 67: 1–14.

Gordon, C.J., Fogelson, L., Mohler, F., Stead, A.G., and Rezvani, A.H. (1988a). Behavioral thermoregulation in the rat following the oral administration of ethanol. *Alcohol. Alcoholism* 23: 383–390.

Gordon, C.J. and Grantham, T.A. (1999). Effect of central and peripheral cholinergic antagonists on chlorpyrifos-induced changes in body temperature in the rat. *Toxicology* 142: 15–28.

Gordon, C.J., Grantham, T.A., and Yang, Y. (1997). Hypothermia and delayed fever in the male and female rat exposed to chlorpyrifos. *Toxicology* 118: 149–158.

Gordon, C.J., Gray, L.E., Jr., Monteiro-Riviere, N.A., and Miller, D.B. (1995). Temperature regulation and metabolism in rats exposed perinatally to dioxin: Permanent change in regulated body temperature? *Toxicol. Appl. Pharmacol.* 133: 172–176.

Gordon, C.J., Kimm-Brinson, K.L., Padnos, B., and Ramsdell, J.S. (2001). Acute and delayed thermoregulatory response of mice exposed to brevetoxin. *Toxicon* 39: 1367–1374.

Gordon, C.J., Lee, K.-L.A., Chen, T.-L.A., Killough, P., and Ali, J.S. (1991a). Dynamics of behavioral thermoregulation in the rat. *Am. J. Physiol.* 261: R705–R711.

Gordon, C.J., Long, M.D., and Dyer, R.S. (1984). Effect of triethyltin on autonomic and behavioral thermoregulation of mice. *Toxicol. Appl. Pharmacol.* 73: 543–550.

Gordon, C.J., Long, M.D., and Dyer, R.S. (1984a). Effect of ambient temperature on the hypometabolic and hypothermic effects of sulfolane in rats. *Arch. Toxicol.* 56: 123–127.

Gordon, C.J., Long, M.D., Fehlner, K.S., and Dyer, R.S. (1986). Sulfolane-induced hypothermia enhances survivability in mice. *Environ. Res.* 40: 92–97.

Gordon, C.J., Long, M.D., and Stead, A.G. (1985a). Thermoregulation in mice following acute chlordimeform administration. *Toxicol. Lett.* 28: 9–15.

Gordon, C.J., McMahon, B., Richelson, E., Padnos, B., and L. Katz (2003). Neurotensin analog NT77 induces regulated hypothermia in the rat. *Life Sci.* 73: 2611–2623.

Gordon, C.J., Mohler, F.S., Watkinson, W.P., and Rezvani, A.H. (1988). Temperature regulation in laboratory mammals following acute toxic insult. *Toxicology* 53: 161–178.

Gordon, C.J., Puckett, E., and Padnos, B. (2002). Rat tail skin temperature monitored non-invasively by radiotelemetry: Characterization by examination of vasomotor response with thermomodulatory agents. *J. Pharmacol. Toxicol. Methods* 47: 107–114.

Gordon, C.J., Watkinson, W.P., O'Callaghan, J.P., and Miller, D.B. (1991c). Effects of 3,4-methylenedioxymethamphetamine on autonomic thermoregulatory responses of the rat. *Pharmacol. Biochem. Behav.* 38: 339–344.

Gordon, C.J., Workman, R.A., and Highfill, J. (1987). Thermoregulation in mice following acute administration of lead acetate. *Neurotoxicol. Teratol.* 9: 191–196.

Gordon, C.J., Yang, Y., and Gray, L.E., Jr. (1996). Autonomic and behavioral thermoregulation in golden hamsters exposed perinatally to dioxin. *Toxicol. Appl. Pharmacol.* 137: 120–125.

Gordon, C.J., Yang, Y., and Ramsdell, J.S. (1998). Behavioral thermoregulatory response to maitotoxin in mice. *Toxicon* 36: 1341–1347.

Gorman, D., Monigatti, J., Glass, B., Gronwall, D., and Beasley, M. (2001). Assessment of pentachlorophenol-exposed timber workers using a test-of-poisoning model. *Int. J. Occup. Environ. Health* 7: 189–194.

Grajpel, B., Kielbasiewicz, E., Kadziela, W., Wojciechowski, M., Piechowicz, B., Tegowska, E., and Stankiewicz, M. (2002). Studies of the effects of a new oxadiazine insecticide on cockroaches. In *Arthropods: Chemical, Physiological and Environmental Aspects,* Konopinska, D., Ed., pp. 261–265. Univ. Wroclaw, Poland.

Gralewicz, S. and Socko, R. (1997). Persisting behavioural and electroencephalographic effects of exposure to chlorphenvinphos, an organophosphorous pesticide, in laboratory animals. *Int. J. Occup. Med. Environ. Health* 10: 375–394.

Green, M.D. and Lomax, P. (1976). Behavioural thermoregulation and neuroamines in fish (*Chromus chromus*). *J. Thermal Biol.* 1: 237–240.

Grue, C.E., Gibert, P.L., and Seeley, M.E. (1997). Neurophysiological and behavioral changes in non-target wildlife exposed to organophosphate and carbamate pesticides: Thermoregulation, food consumption, and reproduction. *Am. Zool.* 37: 369–388.

Grue, C.E., Hart, A.D.M., and Mineau, P. (1991). Biological consequences of depressed brain cholinesterase activity in wildlife. In *Cholinesterase-Inhibiting Insecticides: Their Impact on Wildlife and the Environment,* Mineau, P., Ed., pp. 150–209. Amsterdam: Elsevier.

Guerri, C. and Grisolia, S. (1983). Chronic ethanol treatment affects synaptosomal membrane-bound enzymes. *Pharm. Biochem. Behav.* 18: 45–50.

Gultekin, F., Ozturk, M., and Akdogan, M. (2000). The effect of organophosphate insecticide chlorpyrifos-ethyl on lipid peroxidation and antioxidant enzymes (*in vitro*). *Arch. Toxicol.* 74: 533–538.

Günther, H., Brunner, R., and Kluβmann, F.W. (1983). Spectral analysis of tremorine and cold tremor electromygrams in animal species of different size. *Pflug. Arch.* 399: 180–185.

Günther, F.A., Iwata, Y., Carman, G.E., and Smith, C.A. (1977). The citrus reentry problem: Research on its causes and effects, and approaches to its minimization. *Res. Re.* vol. 67: 1–132.

Gupta, R.C. and Dettbarn, W.-D. (1986). Role of uptake of [C14]Valine into protein in the development of tolerance to diisopropylfluoridate (DFP) toxicity. *Toxicol. Appl. Pharmacol.* 84: 551–560.

Gupta, R.C., Goad, J.T., and Kadel, W.L. (1994). Energy related metabolic alterations in diaphragm muscle resulting from acute methomyl toxicity. *Neurotoxicology* 15: 321–330.

Hahn, G.M. (1979). Potential for therapy of drugs and hyperthermia. *Cancer Res.* 39: 2264–2268.

Hahn, G.M., Ning, S.C., Elizaga, M., Kapp, D.S., and Anderson, R.L. (1989). A comparison of thermal responses of human and rodent cells. *Int. J. Radiat. Biol.* 56: 817–825.

Hahn, G.M., Shiu, E.C., West, B., Goldstein, L., and Li, G.C. (1985). Mechanistic implications of the induction of thermotolerance in Chinese hamster cells by organic solvents. *Cancer Res.* 45: 4138–4143.

Hales, J.R.S. and Iriki, M. (1975). Integrated changes in regional circulatory activity evoked by spinal cord and peripheral thermoreceptor stimulation. *Brain Res.* 87: 267–279.

Halliwell, B. (1992). Reactive oxygen species and the central nervous system. *J. Neurochem.* 59: 1609–1632.

Hammel, H.T. (1968). Regulation of internal body temperature. *Annu. Rev. Physiol.* 30: 641–710.

Haque, N., Rizvi, S.J., and Khan, M.B. (1987). Malathion induced alterations in the lipid profile and the rate of lipid peroxidation in rat brain stem and spinal cord. *Pharmacol. Toxicology* 61: 12–15.

Hart, B.L. (1988). Biological basis of the behavior of sick animals. *Neurosci. Biobehav. Rev.* 12: 123–137.

Hasday, J.D. and Singh, I.S. (2000). Fever and the heat shock responses: Distinct, partially overlapping processes. *Cell Stress Chaperones* 5: 471–480.

Hasday, J.D., Garrison, A., Singh, I.S., Standiford, T., Ellis, G.S., Rao, S., He, J.-R., Rice, P., Frank, M., Goldblum, S.E., and Viscardi, R.M. (2003). Febrile-range hyperthermia augments pulmonary neutrophil recruitment and amplifies pulmonary oxygen toxicity. *Am. J. Pathol.* 162: 2005–2017.

Hassan, R.M., Pesce, A.J., Sheng, P., and Hanenson, I.B. (1981). Correlation of serum pseudocholinesterase and clinical course in two patients poisoned with organophosphate insecticides. *Clin. Toxicol.* 18: 401–406.

Hayssen, V. and Lacy, R.C. (1985). Basal metabolic rates in mammals: Taxonomic differences in the allometry of BMR and body mass. *Comp. Biochem. Physiol.* 81A: 741–754.

Hazel, J.R. (1995). Thermal adaptation in biological membranes: Is homeoviscous adaptation the explanation? *Annu. Rev. Physiol.* 57: 19–42.

Heath, A.G., Cech, J.J., Jr., Brink, L., Moberg, P., and Zinkl, J.G. (1997). Physiological responses of fathead minnow larvae to rice pesticides. *Ecotoxicol. Environ. Saf.* 37: 280–288.

Heath, J.E. and Heath, M.S. (1982). Energetics of locomotion in endothermic insects. *Annu. Rev. Physiol.* 44: 133–143.

Heath, J.E., Williams, B.A., Mills, S.M., and Kluger, M.J. (1972). The responsiveness of the preoptic-anterior hypothalamus to temperature in vertebrates. In *Hibernation and Hypothermia: Perspectives and Challenges,* South, F.E., Hannon, J.P., Willis, J.R., Pengelley, E.T., and Alperts, N.R., Eds., pp. 605–627. Amsterdam: Elsevier.

Heath, S., Bennett, W., Kennedy, J., and Beitinger, T. (1994). Heat and cold tolerance of the fathead minnow, *Pimephales promelas*, exposed to the synthetic pyrethroid cyfluthrin. *Can. J. Fish. Aquat. Sci.* 51: 437–440.

Henderson, G.L. and Woolley, D.E. (1970). Mechanisms of neurotoxic action of 1,1,1-trichloro-2,2bis (p-chlorophenyl) ethane (DDT) in immature and adult rats. *J. Pharmacol. Exp. Ther.* 175: 60–68.

Henderson, R.A., Whitehurst, M.E., Morgan, K.R., and Carroll, R.G. (2000). Reduced oxygen consumption precedes the drop in body core temperature caused by hemorrhage in rats. *Shock* 13: 32–324.

Henderson, R.F., Barr, E.B., Blackwell, W.B., Clark, C.R., Conn, C.A., Kalra, R., March, T.H., Sopori, M.L., Tesfaigzi, Y., Menache, M.G., and Mash, D.C. (2002). Response of rats to low levels of sarin. *Toxicol. Appl. Pharmacol.* 184: 67–76.

Herr, D.W., Boyes, W.K., and Dyer, R.S. (1992). Alterations in rat flash and pattern reversal evoked potentials after acute or repeated administration of carbon disulfide. *Fund. Appl. Toxicol.* 18: 328–342.

Hillman, P.E., Scott, N.R., and van Tienhoven, A. (1980). Effect of 5-hydroxytryptamine and acetylcholine on the energy budget of chickens. *Am. J. Physiol.* 239: R57–R61.

Himms-Hagen, J. (1986). Brown adipose tissue and cold-acclimation. In *Brown Adipose Tissue,* Trayhurn, P. and Nicholls, D.G., Eds., pp. 214–268. London: Edward Arnold.

Himms-Hagen, J. (1990). Brown adipose tissue thermogenesis: Role in thermoregulation, energy regulation and obesity. In *Thermoregulation: Physiology and Biochemistry,* Schonbaum, E. and Lomax, P., Eds., pp. 327–414. New York: Pergamon Press.

Hirshberg, A. and Lerman, Y. (1984). Clinical problems in organophosphate insecticide poisoning: The use of a computerized information system. *Fund. Appl. Toxicol.* 4: S209–S214.

Hohnadel, D.C., Sunderman, F.W., Jr., Nechay, M.W., and McNeely, M.D. (1973). Atomic absorption spectrometry of nickel, copper, zinc, and lead in sweat collected from healthy subjects during sauna bathing. *Clin. Chem.* 19: 1288–1292.

Honma, K. and Hiroshige, T. (1978). Simultaneous determination of circadian rhythms of locomotor activity and body temperature in the rat. *Jpn. J. Physiol.* 28: 159–169.

Hop, H., Borgå, K., Garielsen, G.W., Kleivane, L., and Skaare, J.U. (2002). Food web magnification of persistent organic pollutants in poikilotherms and homeotherms from the Barents sea. *Environ. Sci. Technol.* 36: 2589–2597.

Hopfer, S.M. and Sunderman, F.W., Jr. (1988). Hypothermia and deranged circadian rhythm of core body temperature in nickel chloride-treated rats. *Res. Comm. Chem. Path. Pharm.* 62: 495–505.

Horowitz, M. (2003). Matching the heart to heat-induced circulatory load: Heat-acclimatory responses. *News Physiol. Sci.* 18: 215–221.

Hotchkiss, R., Nunnally, I., Lindquist, S., Taulien, J., Perdrizet, G., and Karl, I. (1993). Hyperthermia protects mice against the lethal effects of endotoxin. *Am. J. Physiol.* 265: R1447–R1457.

Hrbek, J.A., Macakova, J., Komenda, S., Siroka, A., and Rypka, M. (1985). Effects of ethanol (0.3 g, 0.6 g and 0.9 g/kg) on the blood pressure, the body temperature and the heart rate in man. *Aca Univer. Palack. Olum. Fac. Med.* 111: 197–207.

Hrdina, P.D., Singhal, R.L., Peters, D.A.V., and Ling, G.M. (1972). Comparison of the chronic effects of p,p'-DDT and α-chlodane on brain amines. *Eur. J. Pharmacol.* 20: 114–117.

Hrdina, P.D., Peters, D.A.V., and Singhal, R.L. (1974). Role of noradrenaline, 5-hydroxytryptamine and acetylcholine in the hypothermic and convulsive effects of α-chlordane in rats. *Eur. J. Pharmacol.* 26: 306–312.

Hsu, Y.N., Lin, M.T., Hong, J.S., and Tsai, M.C. (1986). Effect of chlordecone exposure on thermoregulation in the rat. *Pharmacology* 32: 292–300.

Hudson, P.M., Chen, P.H., Tilson, H.A., and Hong. J.S. (1985). Effects of *p,p'*-DDT on the rat brain concentrations of biogenic amine and amino acid neurotransmitters and their association with *p,p'*-DDT-induced tremor and hyperthermia. *J. Neurochem.* 45: 1349–1355.

Hugnet, C., Buronfosse, F., Pineau, X., Cadore, J.-L., Lorgue, G., Berny, P.J. (1996). Toxicity and kinetics of amitraz in dogs. *Am. J. Vet. Res.* 57: 1506–1510.

Huttunen, P. and Hirvonen, J. (1977). The effect of ethanol on the ability of guinea-pigs to survive severe cold. *Forensic Sci.* 9: 185–193.

Illievich, U.M., Zornow, M.H., Choi, K.T., Strnat, M.A., and Scheller, M.S. (1994). Effects of hypothermia or anesthetics on hippocampal glutamate and glycine concentrations after repeated transient global cerebral ischemia. *Anesthesiology* 80: 177–186.

Ingram, D.L. and Mount, L.E. (1975). *Man and Animals in Hot Environments.* New York: Springer-Verlag.

Ishay, J.S., Shemesh, E., and Lilling, I. (1983). Hypothermia induced in mice by injection of venom sac extract hor nets (*Vespa orientalis*, vespinae: hymenoptera). *Experentia* 39: 53–55.

Ismail, M., Abd-Elsalam, M.A., and Morad, A.M. (1990). Do changes in body temperature following envenomation by the scorpion *Leiurus quinquestriatus* influence the course of toxicity? *Toxicon* 28: 1265–1284.

IUPS Thermal Commission. (2001). Glossary of terms for thermal physiology. Third edition. Revised by The Commission for Thermal Physiology of the International Union of Physiological Sciences. *Jpn. J. Physiol.* 51: 245–280.

Iwai, N., Watanabe, C., Suzuki, T., Suzuki, K.T., and Tohyama, C. (1988). Metallothionein induction by sodium selenite at two different ambient temperatures in mice. *Arch. Toxicol.* 62: 447–451.

Jaeger, R.J. and Gearhart, J.M. (1982). Respiratory and metabolic response of rats and mice to formalin vapor. *Toxicology* 25: 299–309.

Jang, C.-G., Lee, S.-J., Yang, S.-I., Kim, J.K., Sohn, U.D., and Lee, S.-Y. (2002). Carbon monoxide as a novel central pyrogenic mediator. *Arch. Pharm. Res.* 25: 343–348.

Janssen, R. (1992). Thermal influences on nervous system function. *Neurosci. Biobehav. Rev.* 16: 399–413.

Javaid, M.Y. (1972). Effect of DDT on temperature selection of some salmonids. *Pakistan. J. Sci. Ind. Res.* 15: 171–176.

Kacew, S. and Festing, M.F. (1996). Role of rat strain in the differential sensitivity to pharmaceutical agents and naturally occurring substances. *J. Toxicol. Environ. Health* 47: 1–30.

Kalant, H. and Le, A.D. (1983). Effects of ethanol on thermoregulation. *Pharmac. Ther.* 23: 313–364.

Kanosue, K., Romanovsky, A.A., Hosno, T., Chen, X.-M., and Yoda, T. (1997). "Set-point" revisited. In *Thermal Physiology 1997,* Johannsen, B.N. and Nielsen, R., Eds., pp. 39–43. Copenhagen, Denamark.

Katschinski, D.M. (2004). On heat and cells and proteins. *News in Physiol. Sci.* 19: 11–15.

Katsouyanni, K., Pantazopoulou, A., Touloumi, G., Tselepidaki, I., Moustris, K., Asimakopoulos, D., Poulopoulou, G., and Trichopoulos, D. (1993). Evidence for interaction between air pollution and high temperature in the causation of excess mortality. *Arch. Environ. Health* 48: 235–242.

Katz, L.M., Wang, Y., McMahon, B., and Richelson, E. (2001). Neurotensin analog NT69L induces rapid and prolonged hypothermia after hypoxic ischemia. *Acad. Emerg. Med.* 8: 1115–1121

Kawaguchi, M. and Tsutsumi, S. (1982). Studies on $As_2O_3$-induced rabbit hypothermia and brain monoamines. Report 2: The roles of noradrenaline and dopamine. *Folia Pharmacol. Jpn.* 80: 147–154.

Kenney, W.L. and Munce, T.A. (2003). Invited review: Aging and human temperature regulation. *J. Appl. Physiol.* 95: 2598–2603.

Keplinger, M.L., Lanier, G.E., Deichmann, W.B. (1959). Effects of environmental temperature on the acute toxicity of a number of compounds in rats. *Toxicol. Appl. Pharmacol.* 1: 156–161.

King-Herbert, A.P., Hesterburg, T.W., Thevenaz, P.P., Hamm, T.E., Jr., Moss, W.R., Janszen, D.B., and Everitt, J.I. (1997). Effects of immobilization restraint on Syrian golden hamsters. *Lab. Anim. Sci.* 47: 362–366.

Klaassen, C.D. and Eaton, D.L. (1991). Principles of toxicology. In *Casarett and Doull's Toxicology: The Basic Science of Poisons,* Amdur, M.O., Doull, J., and Klaassen, C.D., Eds., pp. 12–49. New York: Pergamon Press.

Klemmer, H.W., Reichert, E.R., and Yauger, W.L., Jr. (1978). Five cases of intentional ingestion of 25 percent Diazinon with treatment and recovery. *Clin. Toxicol.* 12: 435–444.

Kluger, M.J. (1986). Is fever beneficial? *Yale J. Biol. Med.* 59: 89–95.

Kluger, M.J., Kozak, W., Leon, L.R., Soszynski, D., and Conn, C.A. (1995). Cytokines and fever. *Neuroimmunomodulation* 2: 216–223.

Kluger, M.J., O'Reilly, B., Shope, T.R., Vander, A.J. (1987). Further evidence that stress hyperthermia is a fever. *Physiol Behav.* 39: 763–766.

Kluger, M.J., Rudolph, K., Sosszynski, D., Conn, C.A., Leon, L.R., Kozak, W., Wallen, E.S., and Mosely, P.L. (1997). Effect of heat stress on LPS-induced fever and tumor necrosis factor. *Am. J. Physiol.* 42: R858–R863.

Kobayashi, H., Yuyama, A., Ohkawa, T., and Kajita, T. (1988). Effect of single or chronic injection with a carbamate, Propoxure, on the brain cholinergic system and behavior in mice. *Jpn. J. Pharmacol.* 47: 21–27.

Kolka, M.A. and Stephenson, L.A. (1990). Human temperature regulation during exercise after oral pyridostigmine administration. *Aviat. Space Environ. Med.* 61: 220–224.

Kolka, M.A. and Stephenson, L.A. (1992). Cardiovascular and thermoregulatory responses to repeated anticholinesterase administration. *J. Thermal Biol.* 17: 333–337.

Kozak, W., Archuleta, I., Mayfield, K.P., Kozak, A., Rudolph, K., and Kluger, M.J. (1998). Inhibitors of alternative pathways of arachidonate metabolism differentially affect fever in mice. *Am. J. Physiol.* 275: R1031–R1040.

Kozak, W., Kluger, M.J., Kozak, A., Wachulec, M., and Dokladny, K. (2000). Role of cytochrome P-450 in endogenous antipyresis. *Am. J. Physiol.* 279: R455–R460.

Kozak, W., Kluger, M.J., Tesfaigzi, J., Kozak, A., Mayfield, K.P., Wachulec, M., Dokladny, K. (2000b). Molecular mechanisms of fever and endogenous antipyresis. *Ann. NY Acad. Sci.* 917: 121–134.

Kozak, W., Mayfield, K.P., Kozak, A., and Kluger, M.J. (2000a). Proadifen (SKF-525A), an inhibitor cytochrome P-450, augments LPS-induced fever and exacerbates prostaglandin-$E_2$ levels in the rat. *J. Thermal Biol.* 25: 45–50.

Kozak, W., Poli, V., Soszynski, D., Conn, C.A., and Kluger, M.J. (1997). Sickness behavior in mice deficient in interleukin-6 during turpentine and influenza pneuomonitis. *Am. J. Physiol.* 272: R621–R630.

Kregel, K.C. (2002). Heat shock proteins: Modifying factors in physiological stress responses and acquired thermotolerance. *J. Appl. Physiol.* 92: 2177–2186.

Kunst, A.E., Looman, C.W., and Mackenbach, J.P. (1993). Outdoor air temperature and mortality in the Netherlands: A time-series analysis. *Am. J. Epidem.* 137: 331–341.

Lagerkvist, B., Linderholm, H., and Nordberg, G.F. (1986). Vasospastic tendency and Raynaud's phenomenon in smelter workers exposed to arsenic. *Environ. Res.* 39: 465–474.

Lagerkvist, B., Linderholm, H., and Nordberg, G.F. (1988). Arsenic and Raynaud's phenomenon. *Arch. Occup. Environ. Health* 60: 361–364.

Lallement, G., Foquin, A., Baubichon, D., Burckhart, M.-F., Carpentier, P., and Canini, F. (1998). Heat stress, even extreme, does not induce penetration of pyridostigmine into the brain of guinea pigs. *Neurotoxicology* 19: 759–766.

Leftwich, R.B., Floro, J.F., Neal, R.A., and Wood, A.J. (1982). Dinitrophenol poisoning: A diagnosis to consider in undiagnosed fever. *Southern Med. J.* 75: 182–189.

Leon, L.R. (2002). Cytokine regulation of fever: Studies using gene knockout mice. *J. Appl. Physiol.* June 1992: 2648–2655.

Leon, L.R., Kozak, W., Peschon, J., and Kluger, M.J. (1997). Exacerbated febrile responses to LPS, but not turpentine, in TNF double receptor-knockout mice. *Am. J. Physiol.* 272: R563–R569.

Leow, A.T.C., Towns, K.M., and Leaver, D.D. (1980). The effects of th triethyltin in the rat following systemic and intracerebroentricular injection. *Chem. Biol. Interact.* 31: 233–238.

Lewis, R.J., Wong, A.W., Hoy, W., and Sellin, M. (1993). Ciguatera and mannitol: *In vivo* and *in vitro* assessment in mice. *Toxicon* 31: 1049–1050.

Li, G.C. (1983). Induction of thermotolerance and enhanced heat shock protein synthesis in Chinese hamster fibroblasts by sodium arsenite and by ethanol. *J. Cell. Physiol.* 115: 116–121.

Li, G.C., Hahn, G.M., and Shiu, E.C. (1977). Cytotoxicity of commonly used solvents at elevated temperatures. *J. Cell. Physiol.* 93: 331–334.

Li, X.Y., Gilmour, P.S., Donaldson, K., and MacNee, W. (1997). *In vivo* and *in vitro* proimflammatory effects of particulate air pollution (PM10). *Environ. Health Perspect.* 105 (Suppl 5): 1279–1283.

Lim, D.K., Ito, Y., Stewart, T., Hoskins, B., and Ho, I.K. (1989). Toxicity study of continuous administration of physostigmine salicylate. *Pharmacol. Biochem. Behav.* 31: 627–631.

Lipton, J.M. and Clark, W.G. (1986). Neurotransmitters in temperature control. *Annu. Rev. Physiol.* 48: 613–623.

Livingston, A., Low, J., and Morris, B. (1984). Effects of clonidine and xylazine on body temperature in the rat. *Br. J. Pharmacol.* 81: 189–193.

Lomax, P. and Lee, R.J. (1982). Cold acclimation and resistance to ethanol-induced hypothermia. *Eur. J. Pharmacol.* 84: 87–91.

Lovegrove, B.G., Heldmaier, G., and Ruf, T. (1991). Perspectives of endothermy revisited: The endothermic temperature range. *J. Thermal Biol.* 16: 185–197.

Luheshi, G.N., Stefferl, A., Turnbull, A.V., Dascombe, M.J., Brouwer, S., Hopkins, S.J., and Rothwell, N.J. (1997). Febrile response to tissue inflammation involves both peripheral and brain IL-1 and TNF-$\alpha$ in the rat. *Am. J. Physiol.* 272: R862–R868.

Luna, G.K., Maier, R.V., Pavlin, E.G., Anardi, D., Copass, M.K., and Oreskovich, M.R. (1987). Incidence and effects of hypothermia in seriously injured patients. *J. Trauma* 27: 1014–1018.

Maguire, C.C. and Williams, B.A. (1987). Cold stress and acute organophosphorus exposure: Interaction effects on juvenile northern bobwhite. *Arch. Environm. Contam. Toxicol.* 16: 477–481.

Maickel, R.P., Kinney, D.R., and Nichols, M.B. (1991). Antagonism of physostigmine induced hypothermia and neuroendocrine changes following exposure to different environmental temperatures. *Prog. Neuro-Psychopharmacol. Biol. Psychiatr.* 15: 873–884.

Maickel, R.P., Kinney, D.R., Ryker, D., and Nichols, M.B. (1988). Time course of physostigmine effects on neuroendocrine responding at varying environmental temperatures. *Prog. Neuro-Psychopharmacol. Biol. Psychiatr.* 12: 935–949.

Malberg, J.E. and Seiden, L.S. (1998). Small changes in ambient temperature cause large changes in 3,4-methylenedioxymethamphetamine (MDMA)-induced serotonin neurotoxicity and core body temperature in the rat. *J. Neurosci.* 18: 5086–5094.

Malberg, J.E., Sabol, K.E., and Seiden, L.S. (1996). Co-administration of NMDA with drugs that protect against MDMA neurotoxicity produces different effects on body temperature in the rat. *J. Pharamcol. Exp. Ther.* 278: 258–267.

Malcolm, R.D. and Alkana, R.L. (1983). Temperature dependence of ethanol lethality in mice. *J. Pharm. Pharmacol.* 35: 306–311.

Malvin, G.M. (1998). Thermoregulatory changes by hypoxia: Lessons from the paramecium. *Clin. Exp. Pharmacol. Toxicol.* 25: 165–169.

Malvin, G.M., Havlen, P., and Baldwin, C. (1994). Interactions between cellular respiration and thermoregulation in the paramecium. *Am. J. Physiol.* 267: R349–R352.

Malvin, G.M. and Wood, S.C. (1992). Behavioral hypothermia and survival of hypoxic protozoans paramecium caudatum. *Science* 255: 1423–1425.

Martens, P. (1999). How will climate change affect human health. *Am. Sci.* 87: 534–541.

Martinez, F., Vicente, I., Garcia, F., Penafiel, R., and Cremades, A. (1993). Effects of different factors in lead- and cadmium-induced hypothermia in mice. *Eur. J. Pharmacol.* 248: 199–204.

Martinez-Garcia, F., Martinez-Ruiz, F., Vicente, I., Penafiel, R., and Cremades, A. (1995). Effect of environmental temperature on tissue lead accumulation in mice repeatedly treated with lead acetate. *Eur. J. Pharmacol.* 293: 271–275.

Masaoka, T., Akahori, F., Arai, S., Nomiyama, K., Nomiyama, H., Kobayashi, K., Nomura, Y. and Suzuki, T. (1994). Nine-year chronic toxicity study of cadmium ingestion in monkeys. I. Effects of dietary cadmium on the general health of monkeys. *Vet. Hum. Toxicol.* 36: 189–194.

Matthew, C.B. (1993). Ambient temperature effects on thermoregulation and endurance in anticholinesterase-treated rats. *Life Sci.* 52: 1343–1349.

Maurissen, J.P.J., Shankar, M.R., and Mattsson, J.L. (2000). Chlorpyrifos: Lack of cognitive effects in adult Long-Evans rats. *Neurotoxicol. Teratol.* 22: 237–246.

Mautz, W.J. (2003). Exercising animal models in inhalation toxicology: Interaction with ozone and formaldehyde. *Environ. Res.* 92: 14–26.

Mautz, W.J. and Bufalino, C. (1989). Breathing pattern and metabolic rate responses of rats exposed to ozone. *Respir. Physiol.* 76: 69–78.

McCreery, M.J. and Hunt, W.A. (1978). Physico-chemical correlates of alcohol intoxication. *Neuropharmacology* 17: 451–461.

McDaniel, K.L. and Moser, V.C. (1993). Utility of a neurobehavioral screening battery for differentiating the effects of two pyrethroids, permethrin and cypermethrin. *Neurotoxicol. Teratol.* 15: 71–83.

McDaniel, K.L. and Moser, V.C. (1997). The influence of dosing volume on the toxicity of p,p'-DDT. *Neurotoxicology* 18: 1071–1078.

McDougal, J.N., Marques, P.R., and Burks, T.F. (1983). Restraint alters the thermic response to morphine by postural interferences. *Pharm. Biochem. Behav.* 18: 495–499.

McLister, J.D., Sorensen, J.S., and Dearing, M.D. (2004). The effect of juniper (*Juniperus monosperma*) consumption on the cost of thermoregulation in the woodrats *Neotoma albigula* and *Neotoma stephensi* depends upon acclimation temperature. *Physiol. Biochem. Zool.* 77: 305–312

McMaster, S.B. and Carney, J.M. (1986). Chronic exercise produces tolerance to muscarinic antagonists in rats. *Pharm. Biochem. Behav.* 24: 865–868.

McMaster, S.B. and Finger, A.V. (1989). Effects of exercise on behavioral sensitivity to carbamate cholinesterase inhibitors. *Pharm. Biochem. Behav.* 33: 811–813.

McNab, B.K. (1980). On estimating thermal conductance in endotherms. *Physiol. Zool.* 53: 145–156.

McNab, B.K. (1986). The influence of food habits on the energetics of eutherian mammals. *Ecol. Monogr.* 56: 1–19.

Meeter, E. (1969). The mode of action of cholinesterase inhibitors on the temperature regulation of the rat. *Arch. Int. Pharmacodyn.* 182: 416–419.
Meeter, E. and Wolthuis, O.L. (1968). The effects of cholinesterase inhibitors on the body temperature of the rat. *Eur. J. Pharmacol.* 4: 18–24.
Meeter, E., Wolthuis, O.L., and Van Benthem, R.M. (1971). The anticholinesterase hypothermia in the rat: Its practical application in the study of the central effectiveness of oximes. *Bull. World Health Org.* 44: 251–257.
Meller, D., Fraser, I., and Kryger, M. (1981). Hyperglycemia in anticholinesterase poisoning. *Can. Med. Assoc.* 124: 763–748.
Mercer, J. (2003). Cold: An underrated risk factor for health. *Environ. Res.* 92: 8–13.
Messaad, I.A., Peters, E.J., and Young, L. (2000). Thermal tolerance of red shiner (*Cyprinella lutrensis*) after exposure to atrazine, terbufos, and their mixtures. *Bull. Environ. Contam. Toxicol.* 64: 748–754.
Michelozzi, P., Forastiere, F., Fusco, D., Perucci, C.A., Ostro, B., Ancona, C., and Pallotti, G. (1998). Air pollution and daily mortality in Rome, Italy. *Occup. Environ. Med.* 55: 605–610.
Miller, D.B. and O'Callaghan, J.P. (1994). Environment-, drug- and stress-induced alterations in body temperature affect the neurotoxicity of substituted amphetamine in the C57Bl/6J mouse. *J. Pharmacol. Exp. Ther.* 270: 752–760.
Mills, C.A. (1945). Influence of environmental temperatures on warm-blood animals. *Ann. NY Acad. Sci.* 46: 97–105.
Minard, F.N. and Grant, D.S. (1982). Hypothermia as a mechanism for drug-induced resistance to hypoxia. *Biochem. Pharmacol.* 31: 1197–1203.
Mohler, F.S. and Gordon, C.J. (1988). Thermoregulatory responses of the rabbit to subcutaneous injection of sulfolane. *Arch. Toxicol.* 62: 216–219.
Mohler, F.S. and Gordon, C.J. (1989). Thermoregulatory responses of the rabbit to central neural injections of sulfolane. *Neurotoxicology* 10: 53–62.
Mohler, F.S. and Gordon, C.J. (1991). Hypothermic effects of a homologous series of short-chain alcohols in rats. *J. Toxicol. Environ. Health* 32: 129–139.
Mordenti, J. (1986). Man versus beast: Pharmacokinetic scaling in mammals. *J. Pharmacol. Sci.* 75: 1028–1040.
Morita, H., Yanagisawa, N., Nakajima, T., Shimizu, M., Hirabayshi, H., Okudera, H.O., Hohara, M., Midorikawa, Y., and Mimura, S. (1995). Sarin poisoning in Matsumoto. *Jpn. Lancet* 346: 290–293.
Moser, V.C. (1991). Investigations of amitraz neurotoxicity in rats. IV. Assessment of toxicity syndrome using a functional observation battery. *Fund. Appl. Toxicol.* 17: 7–16.
Moser, V.C. (1995). Comparisons of the acute effects of cholinesterase inhibitors using a neurobehavioral screening battery in rats. *Neurotoxicol. Teratol.* 17: 617–625.
Moser, V.C. and MacPhail, R.C. (1989). Neurobehavioral effects of triadimefon, a triazole fungicide, in male and female rats. *Neurotoxicol. Teratol.* 11: 285–293.

Mullin, L.S. and Krivanek, N.D. (1982). Comparison of unconditioned reflex and conditioned avoidance tests in rats exposed by inhalation to carbon monoxide, 1,1,1-trichloroethane, toluene or ethanol. *Neurotoxicology* 3: 126–130.

Murthy, K.R. and Zare, M.A. (1998). Effect of Indian red scorpion (Mesobuthus tamulus concanesis, Pocock) venom on thyroxine and triiodothyronine in experimental acute myocarditis and its reversal by species specific antivenom. *Ind. J. Exp. Biol.* 36: 16–21.

Myers, R.D. (1980). Hypothalamic control of thermoregulation, neurochemical mechanisms. In *Handbook of the Hypothalamus,* Vol. 3, Part A, Morgane, P.J. and Panskeep, Eds., pp. 83–210. New York: Marcel Dekker.

Myers, R.D. (1981). Alcohol's effect on body temperature: hypothermia, hyperthermia or poikilothermia. *Brain Res. Bull.* 7: 209–220.

Myers, R.D. (1982). The role of ions in thermoregulation and fever. In *Handbook of Experimental Pharmacology,* Vol. 60, Milton, A.S., Ed., pp. 151–186. Berlin: Springer-Verlag.

Nagasaka, T., Hirata, K., Sugano, Y., and Shibata, H. (1979). Heat balance during physical restraint in rats. *Jpn. J. Physiol.* 29: 383–392.

Naitoh, K., Inai, Y., Hirabayashi, T., and Tsuda, T. (2002). Exhalation behavior of four organic substrates and water absorbed by human skin. *Biol. Pharm. Bull.* 25: 867–871.

Namba, T., Nolte, C.T., Jackrel, J., and Grob, G. (1971). Poisoning due to organophosphate insecticides. *Am. J. Med.* 50: 475–492.

Narahashi, T. (2000). Neuroreceptors and ion channels as the basis for drug action: Past, present, and future. *J. Pharmacol. Exp. Therapeut.* 294: 1–26.

Narciso, S.P., Nadziejko, E., Chen, L.C., Gordon, T., and Nadziejko, C. (2003). Adaptation to stress induced by restraining rats and mice in nose-only inhalation holders. *Inhal. Toxicol.* 15: 1133–1143.

National Research Council. (1996). *Guide for the Care and Use of Laboratory Animals.* Washington, DC: National Academy Press.

Nielsen, B. (1971). Thermoregulation during work in carbon monoxide poisoning. *Acta Physiol. Scand.* 82: 98–106.

Noli, M.I., Pavia, M.A. Jr., Mignone, I.R., Brignone, J.A., Hagmuller, K., and Zaninovich, A.A. (1998). Cadmium chloride prevents the rise in rat brown adipose tissue mitochondrial respiration in response to acute cold stress. *Bull. Environ. Contam. Toxicol.* 61: 31–37.

Nomiyama, K., Matsui, K., and Nomiyama, H. (1980). Environmental temperature, a factor modifying the acute toxicity of organic solvents, heavy metals, and agricultural chemicals. *Toxicol. Lett.* 6: 67–70.

Nomiyama, K., Matsui, K., and Nomiyama, H. (1980a). Effects of temperature and other factors on the toxicity of methylmercury in mice. *Toxicol. Appl. Pharmacol.* 56: 392–398.

Norris, E.R. and Elliott, H.W. (1945). Tolerance to arsenic trioxide in the albino rat. *Am. J. Physiol.* 143: 635–639.

O'Callaghan, J.P. and Miller, D.B. (2002). Neurotoxic effects of substituted amphetamines in rats and mice: Challenges to the current dogma. In *Handbook of Neurotoxicology,* Vol. 2, Massaro, E.J., Ed., pp. 269–301. Totowa, NJ: Humana Press.

O'Connor, C.S., Crawshaw, L.I., and Crabbe, J.C. (1993). Genetic selection alters thermoregulatory response to ethanol. *Pharmacol. Biochem. Behav.* 44: 501–508.

O'Connor, C.S., Crawshaw, L.I., Bedichek, R.C., and Crabbe, J.C. (1988). The effect of ethanol on temperature selection in the goldfish, *Carassius auratus. Pharm. Biochem. Behav.* 29: 243–248.

O'Connor, C.S., Crawshaw, L. I., Kosobud, A., Bedichek, R.C., and Crabbe, J.C. (1989). The effect of ethanol on behavioral temperature regulation in mice. *Pharmacol. Biochem. Behav.* 33: 315–319.

Ogilvie, D.M. and Anderson, J.M. (1965). Effect of DDT on temperature selection by young Atlantic salmon, *Salmo salar. J. Fish. Res. Board Can.* 22: 503–512.

Oh, S.H., Deagen, J.T., Whanger, P.D., and Weswig, P.H. (1978). Biological function of metallothionein. V. Its induction in rats by various stresses. *Am. J. Physiol.* 234: E282–E285.

Overstreet, D.H. (2002). Behavioral characteristics of rat lines selected for differential hypothermic responses to cholinergic or serotonergic agonists. *Behav. Genet.* 32: 335–348.

Overstreet, D.H., Kozar, M.D., and Lynch, G.S. (1973). Reduced hypothermic effects of cholinomimetic agents following chronic anticholinesterase treatment. *Neuropharmacology* 12: 1017–1032.

Overstreet, D.H., Russel, R.W., Crocker, A.D., Gillin, J., and Janowsky, D.S. (1988). Genetic and pharmacological models of cholinergic supersensitivity and affective disorders. *Experentia* 44: 465–472.

Paladino, F.V. and Spotila, J.R. (1978). The effect of arsenic on the thermal tolerance of newly hatched muskellunge fry (*Esox masquinongy*). *J. Thermal Biol.* 3: 223–227.

Parsons, M.T., Owens, C.A., and Spellacy, W.N. (1987). Thermic effects of tocolytic agents: Decreased temperature with magnesium sulfate. *Obstet. Gynecol.* 69: 88–90.

Patapoutian, A., Peier, A.M., Story, G.M., and Viswanath, V. (2003). ThermoTRP channels and beyond: Mechanisms of temperature sensation. *Nat. Rev.* 4: 529–539.

Pauluhn, J. and Schmuck, G. (2003). Critical analysis of potential body temperature confounders on neurochemical endpoints caused by direct dosing and maternal separation in neonatal mice: A study of bioallethrin and deltamethrin interactions with temperature on brain muscarinic receptors. *J. Appl. Toxicol.* 23: 9–18.

Peng, Y.G., Taylor, T.B., Finch, R.E., Moeller, P.D., and Ramsdell, J.S. (1995). Neuroexcitatory actions of thermoregulation. *Neuroreport* 26: 305–309.

Peris, J. and Cunningham, C.L. (1986). Handling-induced enhancement of alcohol's acute physiological effects. *Life Sci.* 38: 273–279.

Pertwee, R.G. and Tavendale, R. (1979). Effects of delta9-tetrahyrocannabinol, 2,4-dinitrophenol and pentolinium tartrate on behavioural thermoregulation in mice. *Br. J. Pharamcol.* 66: 39–50.

Peters, D.A.V., Hrdina, P.D., Singhal, R.L., and Ling, G.M. (1972). The role of brain serotonin in DDT-induced hyperpyrexia. *J. Neurochem.* 19: 1131–1136.

Peterson, R.H. (1973). Temperature selection of Atlantic salmon (*Salmo salar*) and brook trout (*Salvelinus fontinalis*) as influenced by various chlorinated hyrdocarbons. *J. Fish. Res. Board Can.* 30: 1091–1097.

Peterson, R.H. (1976). Temperature selection of juvenile atlantic salmon (*Salmo salar*) as influenced by various toxic substances. *J. Fish. Res. Board Can.* 33: 1722–1730.

Phillips, P.K. and Heath, J.E. (1995). Dependency of surface temperature regulation on body size in terrestrial mammals. *J. Thermal Biol.* 20: 281–289.

Pittet, J.-F., Pastor, C.M., and Morel, D.R. (2000). Spontaneous high systemic oxygen delivery increases survival rate in awake sheep during sustained endotoxemia. *Crit. Care Med.* 28: 496–503.

Pondal, M.G., Paier, B., Noli, M.I., Tudino, M.B., Stripeikis, J.D., Hagmuller, K., and Zaninovich, A.A. (1995). Changes induced by zinc on thyroxine deiodination by rat liver *in vivo* and *in vitro*. *APPTLA* 45: 35–41.

Pope, C.N. and Chakraborti, T.K. (1992). Dose-related inhibition of brain and plasma cholinesterase in neonatal and adult rats following sublethal organophosphate exposures. *Toxicology* 73: 35–43.

Postlethwait, E.M., Langford, S.D., and Bidani, A. (1994). Determinants of inhaled ozone absorption in isolated rat lungs. *Toxicol. Appl. Pharmacol.* 125: 77–89.

Potter, C.L., Sipes, I.G., and Russell, D.H. (1983). Hypothroxinemia and hypothermia in rats in response to 2,3,7,8-tetrachlorodibenzo-*p*-dioxin administration. *Toxicol. Appl. Pharmacol.* 69: 89–95.

Prasad, A.S., Schulert, A.R., Sandstead, H.H., Miale, A., Jr., and Farid, Z. (1963). Zinc, iron, and nitrogen content of sweat in normal and deficient subjects. *J. Lab. Clin. Med.* 62: 84–89.

Precht, H., Christophersen, J., Hensel, H., and Larcher, W. (1973). *Temperature and Life*. New York: Springer-Verlag.

Prosser, C.L. (1973). *Comparative Animal Physiology,* 3rd ed. Philadelphia: W.B. Saunders.

Prosser, C.L. (1986). *Adaptational Biology: Molecules to Organisms*. New York: Wiley & Sons.

Quan, N. and Blatteis, C.M. (1989). Intrapreoptically microdialyzed and microinjected norepinephrine evokes different thermal responses. *Am. J. Physiol.* 257: R816–R821.

Rainham, D.G. and Smoyer-Tomic, K.E. (2003). The role of air pollution in the relationship between a heat stress index and human mortality in Toronto. *Environ. Res.* 93: 9–19.

Rattner, B.A. and Franson, J.C. (1984). Methyl parathion and fenvalerate toxicity in American kestrels: Acute physiological responses and effects of cold. *Can. J. Physiol. Pharmacol.* 62: 787–792.

Rattner, B.A., Becker, J.M., and Nakatsugwa, T. (1987). Enhancement of parathion toxicity to quail by heat and cold exposure. *Pest. Biochem. Physiol.* 27: 330–339.

Rausch, R.N., Crawshaw, L.I., and Wallace, H.L. (2000). Effects of hypoxia, anoxia, and endogenous ethanol on thermoregulation in goldfish, *Carassius auratus. Am. J. Physiol.* 278: R545–R555.

Ray, D.E. (1980). Selective inhibition of thermogenesis by tributyl S,S,S,-phosphorotrithioate (DEF). *Br. J. Pharmacol.* 69: 257–264.

Ray, D.E., Roubicek, C.B., and Hamidi, M. (1968). Organ and gland weights of rats chronically exposed to 22° and 35°C. *Growth* 32 :1–12.

Rebagliati, I., Raices, M., Ricci, C., Weisstaub, A., Hagmuller, K., and Zaninovich, A.A. (2001). Effects of zinc on brown fat thermal response to cold in normal and triiodothyronine-treated hypothyroid rats. *Bull. Environ. Contam. Toxicol.* 67: 641–648.

Rebert, C.S., Matteucci, M.J., and Pryor, G.T. (1989). Acute electrophysiologic effects of inhaled toluene on adult male Long-Evans rats. *Pharm. Biochem. Behav.* 33: 157–165.

Reed, W.G. and Krentz, M.J. (1991). Accidental hypothermia. In *Fever: Basic Mechanisms and Management.* Mackowiak, P., Ed., pp. 289–296. New York: Raven Press.

Refinetti, R. and Menaker, M. (1992). The circadian rhythm of body temperature. *Physiol. Behav.* 51: 613–667.

Reynolds, W.W. and Casterlin, M.E. (1982). The pyrogenic responses of nonmammalian vertebrates. In *Pyretics and Antipyretics,* Milton, A.S., Ed., Ch. 9. Berlin: Springer-Verlag.

Rezvani, A.H., Mack, C.M., Crovi, S.I., and Myers, R.D. (1986). Central $Ca^{++}$ channel blockade reverses ethanol-induced poikilothermia in the rat. *Alcohol* 3: 273–279.

Rezvani, A.H., Mack, C.M., deLacy, P.A., and Janowsky, D.S. (1990). Verapamil effects on physiological and behavioral responses to ethanol in the rat. *Alcohol Alcoholism* 25: 51–58.

Richter, E.D., Gordon, M., Halamish, M., and Gribetz, B. (1981). Death and injury in aerial spraying: Pre-crash, crash, and post-crash prevention strategies. *Aviat. Space Environ. Med.* 52: 53–56.

Roberts, D.E., Sawka, M.N., Young, A.J., and Freund, B.J. (1994). Pyridostigmine bromide does not alter thermoregulation during exercise in cold air. *Can. J. Physiol. Pharmacol.* 72: 788–793.

Romanovsky, A.A., Shido, O., Sakurada, S., Sugimoto, N., and Nagasaka, T. (1996). Endotoxin shock: Thermoregulatory mechanisms. *Am. J. Physiol.* 270: R693–R703.

Romanovsky, A.A., Shido, O., Sakurada, S., Sugimoto, N., and Nagasaka, T. (1997). Endotoxin shock-associated hypothermia: How and why does it occur? *Ann. NY Acad. Sci.* 813: 733–737.

Romm, E. and Collins, A.C. (1987). Body temperature influences on ethanol elimination rate. *Alcohol* 4: 189–198.

Rowsey, P.J. and Gordon, C.J. (1997). Tolerance to the hypothermic and hyperthermic effects of chlorpyrifos. *Toxicology* 121: 215–221.

Rowsey, P.J. and Gordon, C.J. (2000). A peripheral mechanism of fever: Differential sensitivity to the antipyretic action of methyl scopolamine. *Autonom. Neurosci.: Basic Clin.* 85: 148–155.

Rowsey, P.J., Metzger, B.L., and Gordon, C.J. (2001). Effects of exercise conditioning on thermoregulatory response to anticholinesterase insecticide toxicity. *Biol. Res. Nursing* 2: 267–276.

Rowsey, P.J., Metzger, B.L., Carlson, J., and Gordon, C.J. (2003). Effects of exercise conditioning on thermoregulatory responses to repeated administration of chlorpyrifos. *Environ. Res.* 92: 27–34.

Rowsey, P.J., Yang, Y.-L., and Gordon, C.J. (2002). Peripheral cholinergic pathway modulates hyperthermia induced by stress in rats exposed to open-field stress. *J. Appl. Physiol.* 92: 789–794.

Rozman, K., Pereira, D., and Iatropoulos, M.J. (1986). Histopathology of interscapular brown adipose tissue, thyroid, and pancreas in 2,3,7,8-tetrachlorodibenzo-p-dioxin (TCDD)-treated rats. *Toxicol. Appl. Pharmacol.* 82: 355–359.

Rubin, E. and Rottenberg, H. (1983). Ethanol and biological membranes: Injury and adaptation. *Pharm. Biochem. Behav.* 18: 7–13.

Rubin, Y., Duvdevani, P., and Ishay, J.S. (1993). Cardiovascular haemodynamics of oriental venom sac extract. *Pharmacol. Toxicol.* 72: 268–272.

Rupniak, N.M., Tye, S.J., Brazell, C., Heald, A., Iversen, S.D., and Pagella, P.G. (1992). Reversal of cognitive impairment by heptyl physostigmine, a long-lasting cholinesterase inhibitor, in primates. *J. Neurol. Sci.* 107: 246–249.

Russell, R.W., Booth, R.A., Lauretz, S.D., Smith, C.A., Jenden, D.H. (1986). Behavioral, neurochemical and physiological effects of repeated exposures to subsymptomatic levels of the anticholinesterase, soman. *Neurobehav. Toxicol. Teratol.* 8: 675–685.

Ryan, A.J., Flanagan, S.W., Moseley, P.L., and Gisolfi, C.V. (1992). Acute heat stress protects rats against endotoxin shock. *J. Appl. Physiol.* 72: 1517–1522.

Ryhanen, R., Honkakoski, P., Harri, M., Ylitalo, P., and Hanninen, O. (1988). Effect of the cold environment on organophosphate toxicity and inhibition of cholinesterase activity. *Gen. Pharmac.* 19: 741–745.

Saadeh, A.M., Al-Ali, M.K., Farsakh, N.A., and Ghani, M.A. (1996). Clinical and sociodemographic features of acute carbamate and organophosphate poisoning: A study of 70 adult patients in North Jordan. *Clin. Toxicol.* 34: 45–51.

Saiki, C. and Mortola, J.P. (1997). Effect of 2,4-dinitrophenol on the hypometabolic response to hypoxia of conscious adult rats. *J. Appl. Physiol.* 83: 537–542.

Salgado, V.L., Herman, M.D., and Narahashi, T. (1989). Interactions of the pyrethroids fenvalerate with nerve sodium channels: Temperature dependence and mechanism of depolarization. *Neurotoxicology* 10: 1–14.

Salgado, V.L., Irving, S.N., and Miller, T.A. (1983). The importance of nerve terminal depolarization in pyrethroid poisoning of insects. *Pest. Biochem. Physiol.* 20: 169–182.

Salminen, W.F., Jr., Voellmy, R., and Roberts, S.M. (1997). Protection against hepatotoxicity by a single dose of amphetamine: The potential role of heat shock protein induction. *Toxicol. Appl. Pharmacol.* 147: 247–258.

Samford-Grigsby, M.D., Larson, B.T., Forcherio, J.C., Lucas, D.M., Paterson, J.A., and Kerley, M.S. (1997). Injection of a dopamine antagonist into Holstein steers to relieve symptoms of fescure toxicosis. *J. Anim. Sci.* 75: 1026–1031.

Sarkar, S.N., Balasubramanian, S.V. and Sikdar, S.K. (1993). Effect of fenvalerate, a pyrethoid insecticide on membrane fluidity. *Biochem. Biophys. Acta* 1147: 137–142.

Satinoff, E. (1974). Neural integration of thermoregulatory responses. In *Limbic and Autonomic Nervous Systems Research,* DiCara, I.V., Ed., pp. 41–82. New York: Plenum.

Sawyer, P.R., Jollow, D.J., Scheuer, P.J., York, R., McMillan, J.P., Withers, N.W., Fudengerg, H.H., and Higerd, T.B. (1984). Effect of ciguatera-associated toxins on body temperature in mice. In *Seafood Toxins,* Ragelis, E., Ed., pp. 321–329. ACS Symposium Series 262.

Saxena, P.N., Raza, S.S., Attri, A., Agarwai, R., Gupta, S., and Saksena, M. (1991). Central hyperthermic effect of arsenic in rabbits. *Ind. J. Med. Res.* 94: 241–245.

Schmidt-Nielsen, K. (1964). *Desert Animals: Physiological Problems of Heat and Water.* New York: Oxford University Press.

Schmidt-Nielsen, K. (1975). *Animal Physiology: Adaptation and Environment.* London: Cambridge University Press.

Schmidt-Nielsen, K. (1985). *Scaling: Why is Animal Size so Important.* London: Cambridge University Press.

Scimeca, J.A., Little, P.J., and Martin, B.R. (1985). Relationship between the pharmacological effects and the biodisposition of [$^3$H]diisopropylfluorophosphate in mice after inhalation. *Toxicol. Appl. Pharmacol.* 79: 502–510.

Scott, J.G. (1995). Effects of temperature on insecticide toxicity. *Rev. Pestic. Toxicol.* 3: 111–125.

Seo, B.-W. and Meserve, L.A. (1995). Effects of maternal ingestion of aroclor 1254 (PCB) on the development pattern of oxygen consumption and body temperature in neonatal rats. *Bull. Environ. Contam. Toxicol.* 55: 22–28.

Sessler, D.I. (1994). Thermoregulation and heat balance: General anesthesia. In *Thermal Balance in Health and Disease,* Zeisberger, E., Schonbaum, E., and Lomax, P., Eds., pp. 251–265. Basel: Birkhauser.

Sharma, H.S. and Dey, P.K. (1987). Influence of long-term acute heat exposure on regional blood-brain barrier permeability, cerebral blood flow and 5-HT level in conscious normotensive young rats. *Brain Res.* 424: 153–162.

Shibasaki, M., Wilson, T.E., Cui, J., and Crandall, C.G. (2002). Acetylcholine released from cholinergic nerves contributes to cutaneous vasodilation during heat stress. *J. Appl. Physiol.* 93: 1947–1951.

Shoubridge, E.A. and Hochachka, P.W. (1980). Ethanol: Novel end product of vertebrate anaerobic metabolism. *Science* 209: 308–309.

Simon, E. (1999). Thermoregulation as a switchboard of autonomic nervous and endocrine control. *Jpn. J. Physiol.* 49: 297–323.

Sinclair, J.D., and Taira, T. (1988). Hangover hyperthermia in rats: Relation to tolerance and external stimuli. *Psychopharmacology* 94: 161–166.

Singer, R., Harker, C.T., Vander, A.J., Kluger, M.J. (1986). Hyperthermia induced by open-field stress is blocked by salicylate. *Physiol. Behav.* 36: 1179–82.

Slade, R., Watkinson, W.P., and Hatch, G.E. (1997). Mouse strain differences in ozone dosimetry and body temperature changes. *Am. J. Physiol.* 272: L73–L77.

Slikker, W., III, Desao, V.G., Duhart, H., Feuers, R., and Iman, S.Z. (2001). Hypothermia enhances bcl-2 expression and protects against oxidative stress-induced cell death in Chinese hamster ovary cells. *Free Rad. Biol. Med.* 31: 405–411.

Smith, M.J. and Heath, A.G. (1979). Acute toxicity of copper, chromate, zinc, and cyanide to freshwater fish: Effect of different temperatures. *Bull. Environm. Contam. Toxicol.* 22: 113–119.

Smolen, T.N., Smolen, A., and Collins, A.C. (1986). Dissociation of decreased numbers of muscarinic receptors from tolerance to DFP. 25: 1293–1301.

Sogorb, A., Andreu-Moliner, E.S., Almar, M.M., del Ramo, J., and Núñez, A. (1988). Temperature-toxicity relationships of fluvalinate (synthetic pyrethroid) on Procambarus clarkii (Girard) under laboratory conditions. *Bull. Environ. Contam. Toxicol.* 40: 13–17.

Song, J.H. and Narahashi, T. (1996). Modulation of sodium channels of rat cerebellar Purkinje neurons by the pyrethroid tetramethrin. *J. Pharm. Exp. Ther.* 277: 445–453.

Sonna, L.A., Fujita, J., Gaffin, S., and Lilly, C.M. (2002). Invited review: Effects of heat and cold stress on mammalian gene expression. *J. Appl. Physiol.* 92: 1725–1742.

Sorenson, J.S., McLister, J.D., and Dearing, M.D. (2005). Plant secondary metabolites compromise energy budgets of a specialist and generalist mammalian herbivore. *Ecology*. 86: in press.

Spencer, P.S., McCauley, L.A., Lapidus, J.A., Lasarev, M., Joos, S.K., and Storzbach, D. (2001). Self-reported exposures and their association with unexplained illness in a population-based case-control study of gulf war veterans. *J. Occup. Environ. Med.* 43: 1041–1056.

Spiers, D.E., Zhang, Q., Eichen, P.A., Rottinghaus, G.E., Garner, G.B., and Ellersieck, M.R. (1995). Temperature-dependent responses of rats to ergovaline derived from endophyte-infected tall fescue. *J. Anim. Sci.* 73: 1954–1961.

Staempfli, C., Blooten, K., and Tarradellas, J. (2002). Hsp70 instability and induction by a pesticide in Folsomia candida. *Biomarkers* 7: 68–79.

Steiner, A.A. and Branco, L.G. (2000). Central CO-heme oxygenase pathway raises body temperature by a prostaglandin-independent way. *J. Appl. Physiol.* 88: 1607–1613.

Steiner, A.A., Colombaria, E., and Branco, L.G. (1999). Carbon monoxide as a novel mediator of the febrile response in the central nervous system. *Am. J. Physiol.* 277: R499–R507.

Stemler, F.W., Corcoran, K.D., Parrish, J.H., Hurt, H.H., Tezak-Reid, T., Kaminskis, A., and Jaeger, J.J. (1990). Effects of physostigmine on the cardiopulmonary system of conscious pigs. *Fund. Appl. Toxicol.* 14: 96–103.

Stitt, J.T. (1979). Fever versus hyperthermia. *Fed. Proc.* 38: 39–43.

Stohs, S.J. and Bagchi, D. (1995). Oxidative mechanisms in the toxicity of metal ions. *Free Rad. Biol. Med.* 18: 321–336.

Stolwijk, J.A.J. and Hardy, J.D. (1974). Regulation and control in physiology. In *Medical Physiology,* Vol. 2, Mountcastle, V.B., Ed. pp. 1343–1358. St. Louis: C.V. Mosby.

Sunderman, F.W., Jr., Dingle, B., Hopfer, S.M., and Swift, T. (1988). Acute nickel toxicity in electroplating workers who accidently ingested a solution of nickel sulfate and nickel chloride. *Am. J. Indust. Med.* 14: 257–266.

Suzuki, T., Kohda, K., and Kawazoe, Y. (1987). Temperature-dependence of cytotoxicity of several genotoxicants in Chinese hamster V79 cells: Bleomycin, paraquat, and some N-alkyl-N-nitrosureas. *Biochem. Biophys. Res. Comm.* 146: 67–72.

Swanson, K.L. and Woolley, D.E. (1982). Comparison of the neurotoxic effects of chlordecone and dieldrin in the rat. *Neurotoxicology* 3: 81–102.

Takayoshi, S., Kohda, K., and Kawazoe, Y. (1987). Temperature-dependence of cytotoxicity of several genotoxicants in Chinese hamster V79 cells: Bleomycin, paraquat, and some N-alkyl-N-nitrosureas. *Biochem. Biophys. Res. Comm.* 146: 67–71.

Takehiro, T.N., Shida, K., and Lin, Y.C. (1979). Effects of 2,4-dinitrophenol on the body temperature cardiopulmonary functions in unanesthetized rats, *Rattus rattus. J. Thermal Biol.* 4: 297–301.

Tamborini, P., Sigg, H., and Zbinden, G. (1990). Acute toxicity testing in the nonlethal dose range: A new approach. *Regul. Toxicol. Pharmacol.* 12: 69–87.

Tang-Chun, W.U., Han-Zhen, H.E., Tanguay, R.M., Yang, W.U., Dai-Gen, X.U., William, C.R., Shen, W.U., Feng, J., and Zhang, G. (1995). The combined effects of high temperature and carbon monoxide on heat stress response. *J. Tongji Med. Univ.* 15: 178–183.

Tattersall, G.J. and Milsom, W.K. (2003). Transient peripheral warming accompanies the hypoxic metabolic response in the golden-mantled ground squirrel. *J. Exp. Biol.* 206: 33–42.

Taylor, C.R. (1977). Exercise and environmental heat loads: Different mechanisms for solving different problems? *Int. Rev. Physiol. Environ. Physiol. II* 15: 119–146.

Tegowska, E., Brajpel, B., Stankiewicz, M., Piechowicz, B., Wojciechowski, M., Kadziela, W., Olszak, R. (2002). Thermoregulatory behavior of ectotherms and pesticides. In *Arthropods: Chemical, Physiological and Environmental Aspects,* D. Konopinska, Ed., pp. 255–260. Univ. Wroclaw, Poland.

Templeton, C.B., Poli, M.A., and LeClaire, R.D. (1989). Cardiorespiratory effects of brevetoxin (PbTx-2) in conscious, tethered rats. *Toxicon* 27: 1043–1049.

Tesfaigzi, Y., Kluger, M., and Kozak, W. (2001). Clinical and cellular effects of cytochrome P-450 modulators. *Respir. Physiol.* 128: 79–87.

Thomas, D.W., Samson, C., and Bergeron, J.-M. (1988). Metabolic costs associated with the ingestion of plant phenolics by *Microtus pennsylvanicus. J. Mamm.* 69: 512–515.

Thorne, P.S., Yeske, C.P., and Karol, M.H. (1987). Monitoring guinea pig core temperature by telemetry during inhalation exposures. *Fund. Appl. Toxicol.* 9: 398–408.

Thornhill, J.A., Cooper, K.E., and Veale, W.L. (1979). The effect of repeated periods of restraint and continued exposure to various ambient temperatures on the core temperature responses of drug-naive rats. *Can. J. Physiol. Pharmacol.* 57: 1011–1017.

Ting, J.Y. and Brown, A.F. (2001). Ciguatera poisoning: A global issue with common management problems. *Eur. J. Emerg. Med.* 8: 295–300.

Tomasi, T.E., Ashcraft, J., and Britzke, E. (2001). Effects of fungicides on thyroid function, metabolism, and thermoregulation in cotton rats. *Environ. Toxicol. Chem.* 20: 1709–1715.

Turco, R.P. (2002). *Earth Under Seige: From Air Pollution to Global Change,* 2nd ed. London: Oxford University Press.

Uhari, M., Pakarinen, A., Hietala, J., Nurmi, T., and Kouvalainen, K. (1983). Serum iron, zinc, ferritin, and ceruloplasmin after intense heat exposure. *Eur. J. Appl. Physiol.* 51: 331–335.

Vandame, R. and Belzunces, L.P. (1998). Joint actions of deltamethrin and azole fungicides on honeybee thermoregulation. *Neurosci. Lett.* 251: 57–60.

van den Bercken, J. and Akkermans, L.M. (1971). Negative temperature coefficient of the action of DDT in a sense organ. *Eur. J. Pharmacol.* 16: 241–244.

Van Miert, A.S. (1989). Extrapolation of pharmacological and toxicological data based on metabolic weight. *Arch. Exp. Veterinarmed.* 43: 481–488.

Villar, J., Edelson, J.D., Post, M., Mullen, B.M., and Slutsky, A.S. (1993). Induction of heat stress proteins is associated with decreased mortality in an animal model of acute lung injury. *Am. Rev. Respir. Dis.* 147: 177–181.

Vijverberg, H.P.M. and van den Bercken, J. (1990). Neurotoxicological effects and the mode of action of pyrethroid insecticides. *Crit. Rev. Toxicol.* 21: 105–126.

Viswanathan, P.N. and Krihna Murti, C.R. (1989). Effect of temperature and humidity on ecotoxicology of chemicals. In *Ecotoxicology and Climate,* Bourdeau, P., Haines, J.A., Klein, W., and Krishna Murti, C.R., Eds., pp. 139–154. New York: John Wiley & Sons.

Vocci, F. and Farber, T. (1988). Extrapolation of animal toxicity data to man. *Regul. Toxicol. Pharmacol.* 8: 389–398.

Walker, S.M., Ackland, T.R., and Dawson, B. (2001). The combined effect of heat and carbon monoxide on the performance of motorsport athletes. *Comp. Biochem. Physiol. Part A* 128: 709–718.

Walls, C.B. and Glass, W.I. (1998). Health effects of occupational pentachlorophenol exposure in timber sawmill employees: A preliminary study. *N.Z. Med. J.* 111: 362–364.

Wang, L.C.H. and Lee, T.F. (1989). Temperature regulation. In *Psychoendocrinology,* Brush, R. and Levine, S., Eds., pp. 437–539. New York: Academic Press.

Wang, Z., Matsushita, T., Aoyagi, K., Sugaya, T., Matsushima, S., and Wakatsuki, T. (1991). Statistical analyses of clinical cases of skin lesions from agricultural chemical in Japanese farmers. *J. Jpn. Assoc. Rural Med.* 40: 909–915.

Watanabe, C. and Suzuki, T. (1986). Sodium selenite-induced hypothermia in mice: Indirect evidence for a neural effect. *Toxicol. Appl. Pharmacol.* 86: 372–379.

Watanabe, C., Suzuki, T., and Matsuo, N. (1990a). Toxicity modification of sodium selenite by a brief exposure to heat or cold in mice. *Toxicology* 64: 245–253.

Watanabe, C., Weiss, B., Cox, C., and Ziriax, J. (1990). Modification by nickel of instrumental thermoregulatory behavior in rats. *Fund. Appl. Toxicol.* 14: 578–588.

Watenpaugh, D.E. and Beitinger, T.L. (1985). Temperature tolerance of nitrite-exposed channel catfish. *Trans. Am. Fish. Soc.* 114: 274–278.

Watkinson, W.P. and Gordon, C.J. (1993). Caveats regarding the use of the laboratory rat as a model for acute toxicological studies: Modulation of the toxic response via physiological and behavioral mechanisms. *Toxicology* 81: 15–31.

Watkinson, W.P., Aileru, A.A., Dowd, S.M., Doerfler, D.L., Tepper, J.S., and Costa, D.L. (1993). Acute effects of ozone on heart rate and body temperature in the unanesthetized, unrestrained rat maintained at different ambient temperatures. *Inhal. Toxicol.* 5: 129–147.

Watkinson, W.P., Highfill, J.W., and Gordon, C.J. (1989). Modulating effect of body temperature on the toxic response produced by the pesticide chlordimeform in rats. *J. Toxicol. Environ. Health* 28: 483–500.

Watkinson, W.P., Highfill, J.W., Slade, R., and Hatch, G.E. (1996). Ozone toxicity in the mouse: Comparison and modeling of responses in susceptible and resistant strains. *J. Appl. Physiol.* 80: 2134–2142.

Watkinson, W.P., Wiester, M.J., and Highfill, J.W. (1995). Ozone toxicity in the rat. I. Effect of changes in ambient temperature on extrapulmonary physiological parameters. *J. Appl. Physiol.* 78: 1108–1120.

Weber, L.W.D., Haart, T.W., and Rozman, K. (1987). Effect of 2,3,7,8-*p*-dioxin (TCDD) on thermogenesis in brown adipose tissue of rats. *Toxicol. Lett.* 39: 241–248.

Weihe, W.H. (1973). The effect of temperature on the action of drugs. *Annu. Rev. Pharmacol.* 13: 409–425.

Weiss, H.S., Pitt, J.F., Kerr, K.M., Hakaim, A.G., Weisbrode, S.E., and Daniel, F.B. (1981). Sensitivity of mice to benzo(a)pyrene skin cancer following acclimation to cool and warm temperature. *Proc. Soc. Exp. Biol. Med.* 167: 122–128.

Welch, W.J. (1992). Mammalian stress response: Cell physiology, structure/function of stress proteins, and implications for medicine and disease. *Physiol. Rev.* 72: 1063–1081.

Wenger, B., Quigly, M.D., and Kolka, M.A. (1993). Seven-day pyridostigmine administration and thermoregulation during rest and exercise in dry heat. *Aviat. Space Environ. Med.* 64: 905–911.

Wester, R.C., Quan, D., and Maibach, H.I. (1996). In vitro percutaneous absorption of model compounds glyphosate and malathion from cotton fabric into and through human skin. *Food Chem. Toxicol.* 34: 731–735.

Wheeler, T.G. (1989). Soman toxicity during and after exposure to different environmental temperatures. *J. Toxicol. Environ. Health* 26: 349–360.

Whittow, G.C. (1970). *Comparative Physiology of Thermoregulation,* Vol. 1, *Invertebrates and Nonmammalian Vertebrates.* New York: Academic Press.

Wiester, M.J., Watkinson, W.P., Costa, D.L., Crissman, K.M., Richards, J.H., Winsett, D.W., and Highfill, J.W. (1996). Ozone toxicity in the rat. III. Effect of changes in ambient temperature on pulmonary parameters. *J. Appl. Physiol.* 81: 1691–1700.

Wilkinson, S.P., McHugh, P., Horsely, S., Tubbs, H., Lewis, M., Thould, A., Winterton, M., Parson, V., and Willianms, R. (1975). Arsine toxicity aboard the Asiafreighter. *Br. Med. J.* 3: 559–563.

Wood, S., Rom, W.N., White, G.L., and Logan, D.C. (1983). Pentachlorophenol poisoning. *J. Occupat. Med.* 25: 527–530.

Wood, S.C. (1991). Interactions between hypoxia and hypothermia. *Annu. Rev. Physiol.* 53: 71–85.

Wood, S.C. and Malvin, G.M. (1991). Physiological significance of behavioral hypothermia in hypoxic toads (Bufo marinus). *J. Exp. Biol.* 159: 203–215.

Woolley, D.E. (1973). Studies on 1,1,1-trichloro-2,2bis (p-chlorophenyl) ethane (DDT)-induced hyperthermia: Effects of cold exposure and aminopyramine injections. *J. Pharmacol. Exp. Ther.* 184: 261–268.

Wright, B.E. and Katovich, M.J. (1996). Effect of restraint on drug-induced changes in skin and core temperature in biotelemetered rats. *Pharm. Biochem. Behav.* 55: 219–225.

Wright, G.L. and Lessler, M.A. (1979). Interaction of lead and high ambient temperatures in the rat. *Environ. Res.* 18: 377–384.

Wu, I.-W., Lin, J.-L., and Cheng, E.-T. (2001). Acute poisoning with the neonicotinoid insecticide imidacloprid in N-methyl pyrrolidone. *Clin. Toxicol.* 39: 617–621.

Yamaguchi, S., Shimojo, N., Sano, K., Kano, K., Hirota, Y., and Saisho, A. (1984). Effects of environmental temperature on the toxicity of methylmercury in rats. *Bull. Environ. Contam. Toxicol.* 32: 543–549.

Yamamoto, H., Fujii, K., and Hayakawa, T. (1995). Inhibitory effect of cold stress on lung tumours induced by 7,12-dimethylbenz[a]anthracene in mice. *J. Cancer Res. Clin. Oncol.* 121: 393–396.

Yamauchi, C., Fujita, S., Obara, T., and Ueda, T. (1981). Effects of room temperature on reproduction, body and organ weights, food and water intake, and hematology in rats. *Lab. Anim. Sci.* 31: 251–258.

Yamauchi, C., Fujita, S., Obara, T., and Ueda, T. (1983). Effects of room temperature on reproduction, body and organ weights, food and water intakes, and hematology in mice. *Exp. Anim.* 32: 1–11.

Yang, Y. and Gordon, C.J. (1996). Ambient temperature limits and stability of temperature regulation in telemetered male and female rats. *J. Therm. Biol.* 21: 353–363.

Yang, Y. and Gordon, C.J. (1997). Regulated hypothermia in the hypothyroid rat induced by administration of propylthiouracil. *Am. J. Physiol.* 272: R1390–R1395.

Yeary, R.A., Eaton, J., Gilmore, E., North, B., and Singell, J. (1993). A multiyear study of blood cholinesterase activity in urban pesticide applicators. *J. Toxicol. Environ. Health* 39: 11–25.

Ylach, K.D., Boles, J.W., and Stiles, B.G. (2000). Telemetric evaluation of body temperature and physical activity as predictors of mortality in a murine model of Staphylococcal enterotoxic shock. *Comp. Med.* 50: 160–166.

Zeevalk, G.D. and Nicklas, W.J. (1993). Hypothermia, metabolic stress and NMDA-mediated excitotoxicity. *J. Neurochem.* 61: 1445–1453.

Zimmerberg, B., Ballard, G.A., and Riley, E.P. (1987a). The development of thermoregulation after prenatal exposure to alcohol in rats. *Psychopharmacology* 91: 479–484.

Zimmerberg, B., Beckstead, J.W., and Riley, E.P. (1987). Prenatal alcohol exposure and thermotaxic behavior in neonatal rats. *Neurotoxicol. Teratol.* 9: 281–286.

Zimmerberg, B., Tomlinson, T.M., Glaser, J., and Beckstead, J.W. (1993). Effects of prenatal alcohol exposure on the development pattern of temperature preference in a thermocline. *Alcohol* 10: 403–408.

Zvi, B.Z. and Kaplanski, J. (1980). Effects of chronic heat exposure on drug metabolism in the rat. *J. Pharm. Pharmacol.* 32: 368–369.

# Index

## A

## B



## C

## D



## E

## I–J

## S

## T

## U



## V



## W